Economics in Commercial Fisheries

The Author

K.P. Biswas, M.Sc., Ph.D, D.F.Sc (Bom), EF (West Germany), Former Joint Director Fisheries (L-I), Principal, Fisheries Training Institute, Government of Orissa, part time Lecturer in Fisheries College, Orissa University of Agriculture and Technology, Manager Fisheries, Orissa Maritime and Chilka Area Development Corporation, Bhubaneswar, Director of Fisheries, Andaman and Nicobar Island, Visiting Professor, University of Fisheries and Animal Sciences, Govt. of West Bengal and at present as Visiting Professor of Marine Science Department. Calcutta University has been associated with fish and fishery science and its development for more than 48 years. He has specialized in electrofisheries, sea fishing, fish breeding, industrial fisheries and has published 140 research papers, review papers and articles on fish and fisheries, besides publishing eight books in the field of fisheries.

The author has worked as a member of the Fact Finding Committee of Chilka Lake Fisheries, appointed by the Hon'ble High Court of Orissa and also as a member of the Expert Committee of Government of Orissa on the environmental impact of trawl fishery on the coastal ecology and estuaries with special reference to migration and nesting of Olive Redley turtles in Orissa coast.

He is now actively engaged as a Fellow Member and Vice President of Zoological Society, Kolkata, Fellow Member of Indian Association of Biological Sciences, Life Member of Indian Science Congress Association, Indian Science News Association, Bangiya Bignan Parisad, Centre For Applied Science and Technology and published about fifty four articles and research papers on fisheries, aquatic environment and marine ecology between 1998 to 2005.

Economics in Commercial Fisheries

by

K.P. Biswas

2021

DAYA PUBLISHING HOUSE

A Division of

Astral International Pvt. Ltd.

New Delhi - 110 002

First Published, 2006
Reprinted, 2021

ISBN 978-81-7035-430-7 (HB)

Publisher's note:
Every possible effort has been made to ensure that the information contained in this book is accurate at the time of going to press, and the publisher and author cannot accept responsibility for any errors or omissions, however caused. No responsibility for loss or damage occasioned to any person acting, or refraining from action, as a result of the material in this publication can be accepted by the editor, the publisher or the author. The Publisher is not associated with any product or vendor mentioned in the book. The contents of this work are intended to further general scientific research, understanding and discussion only. Readers should consult with a specialist where appropriate.
Every effort has been made to trace the owners of copyright material used in this book, if any. The author and the publisher will be grateful for any omission brought to their notice for acknowledgement in the future editions of the book.

Published by : **Daya Publishing House®**
A Division of
Astral International Pvt. Ltd.
– ISO 9001:2015 Certified Company –
4736/23, Ansari Road, Darya Ganj
New Delhi-110 002
Ph. 011-43549197, 23278134
E-mail: info@astralint.com
Website: www.astralint.com

Printed at : **Replika Press Pvt. Ltd.**

Preface

Among the prerequisites for the development of a production activity, business profitability takes a prominent place. Fishery enterprises, even the smallest and simplest ones are no exception to the rule. Fishermen are certainly convinced of this, like all producers, they work to earn a living and, if possible, to raise their income. But very often they lack the means of measuring what causes them to make or lose money and of working out for themselves a method for making their calculations. To improve fishermen's income, the minimum basic concepts of economics-accounting and management methods are necessary.

The exploitation of fisheries resources and their management were once considered only a biological aspect of fisheries. This has now evolved into "bio-economics" though being jointly considered with the fisheries of economics.

The concept of "bio-economics" as used in any project includes the investigation of both the "bio-economic" as well as the "socio-economic" aspects of the subject. Such a combined investigation is now becoming known as a "bio-socio-economic" study.

It is becoming more and more evident that an understanding of the socio-economic factors in small-scale fisheries, as well as the participation of small-scale fisher folk in these processes, are necessary for the effective development and management of small-scale fisheries. This envisages not only the investigators understanding the ideas, views and needs of the fisher folk, but also the fisher folk understanding the concepts of resources and their management.

By linking the socio-economic component of the fisheries with the bio-economic assessment in a study, a better understanding could be achieved of the communities, the exact way the components of the communities are affected and what assistance or support they need. It would also be possible for the investigation to identify potential alternatives or additional income-generating opportunities for the community and, at the same time, find out what the appropriate media should be to enlighten the fisher folk on the fishing resources and their management.

K.P. Biswas

— *Acknowledgement to* –

Prof. Y.A. Zamadar
Dr. P.P. Biswas

— *Dedicated to* –

Mrs. Manju Biswas

Contents

Chapter 1

Economics

Considering the sort of activities with which economics are involved, it will be of interest to consider some definitions of the subject. According to Encyclopedia, economics is a social science devoted in studying the production, distribution and consumption of wealth. It consists of the disciplines of micro-economics, the study of individual producer, consumer or market and macro-economics, the study of whole economics or systems, in particular areas, such as, taxation and public spending. A.C. Pigon defined economics as a means of studying how to increase total production in order to raise the standard of living. Alfred Marshall looked on economics as "it examines that part of individual and social action connected with the attainment of the material requisites of well being" so that the subject becomes "on the one side, a study of wealth and the other, the study of man". Economics has developed various system of analysis regarding production and consumption, how people could get their income and how they spend it. Such economic analysis can indicate the way to improve an economic situation and point out the strong and weaker factors of the production system.

A Branch of Science

The fundamental facts of economics is observations from every day life. It is sometimes said to be a positive science as it considers things as they are and not as they ought to be. In order to formulate a set of economic principles, certain basic assumptions are made like other branches of science. These are (a) men in business always aims to make maximum amount of profit, (b) people will always buy commodities of the same quality in the cheapest market and (c) an individual will always select the commodity which yields the greatest amount of satisfaction. Starting from a given hypothesis some accepted facts of everyday experience, logical deductions are made, theories being deduced by logical reasoning. From these deductions economic laws are formulated as it happens like other scientific laws.

Economics and Social Sciences

It is difficult to dissociate Economics from Sociology. The social sciences concern themselves with the study of different aspects of human behaviour. The early philosophers like Plato and Aristotle recognized no clear demarcation between ethics, politics and economics, to them all the above subjects were simply the different aspects of one great comprehensive subject–philosophy.

Economic issues were touched upon by the Greek philosophers only when incidental to ethical or political problems. For example, it was on the ethical ground Plato objected to the payment of interest on borrowed money. Medieval philosophers also regarded economics as subsidiary to theology.

With the advent of the Age of Mercantilism, the subject of economics was surfacing up as great expansion of trade took place. When the state intervened to regulate trade and industry to promote the prosperity of the nation, then the economic matters became political questions.

Increasing knowledge, however, made the study of man and his environment too vast for a single subject and in course of time the social sciences had to split up into different sub-sciences. Economics, now a days has so much expanded that it is necessary to specialize small branches of the subject. But still many economic issues are often inextricably entangled with social or political implications.

Economics in Fisheries

Fish and fisheries play an important role in the well being of national economy and in particular, crucial role in the well being of coastal state's economy.

But "the lacunae of valid statistical and economic data have not only made planning unrealistic in these areas but also rendered appraisal of the performance of fishery development projects quite difficult." (Krishna Kumar, 1980).

The study has revealed that the productivity, profitability and objective contributions to the national economy of fishery units of artisanal type are on an average, as good as, or better than, those of mechanized fishing and high-tech culture fisheries–though the same can not be said of the earnings of individual fisherman.

There come the need for cost earning studies, the information on costs and earnings, in order to assess the techno–economic and socio–economic performance of different fishery/fishing systems, whether in the artisanal or in the high–tech fisheries. The information gained through costs and earnings studies are relevant to fisheries management they may reveal those fisheries, where the need for introducing fishery management is most urgent. They may also provide useful indications of the likely future need for management measures arising from the introduction of improved technology.

In case of small scale or artisanal fisheries, their real importance and potential often cannot be appreciated, because of the lack of detailed information on the operations, economics of individual fishery units and fishermen and of the area and fleet as a whole. Their contribution to the economy as well as the efficiency of their culture methods and fishing techniques cannot be assessed without hard facts and economic analysis. Here the most of fishermen/fishery operators do not co-operate, fishermen due to their ignorance and the operators because of the fear of publicity and taxation likes to keep the actual cost and earnings data as secret.

Economics of fisheries, therefore, concentrate on the issues like profitability in fish production, marketing system of fishes, study of demand and utilization pattern, most effective use of the fishery technology and implementation of fishery policies. It also study the crucial issues like management of finance, insurance, social problems of fishermen and impact of organisational set up in fishery.

The multiple objectives of fisheries economy include enhancement of food supply, creation of employment opportunities, raising national income, regional development, foreign exchange earnings and promotion of recreational activities.

Fishery economics encompasses the concept of fisheries management. The term "Fisheries management" is used to denote the both aspects of management *i.e.*, a comprehensive plan of development of a fishery and application of restrictive (regulatory) measures.

Economics and Policies in Commercial Fisheries

For centuries the Canadian Grand Banks, off the coast of Newfoundland, were prime fishing grounds. The region's abundance of Atlantic Cod (*Gadus morhua*) supported entire fishing communities. At its height, in 1968, the industry employed 40,000 people and landed more than 800,000 tonnes of fish.

But the factory trawlers that subsequently moved onto the banks exacted dreadful toll. Stocks collapsed in 1992. Canada's Department of Fisheries and Ocean belatedly closed the fishery. Thousands of fishermen and workers in the fish processing industry lost their jobs, others redirected their efforts to catch crabs and shrimp.

The Grand Banks disaster shows just how badly fisheries policies can go wrong and unfortunately the same mistakes are being made across the world's oceans. In many cases scientists are warning that fish populations are being over exploited. But all too often, their advice of setting lower catch quotas, reducing the size of fishing fleets and using less harmful fishing gear is ignored. It seems that short-term economic interests steamroller scientific arguments.

Given such failures, some conservation biologists are now arguing that fisheries scientists must abandon their focus on individual stocks and adopt a whole–ecosystem perspective. But whatever methods are used to determine the advice given to policy makers, scientists need to find ways to involve fishermen in their work, to break down the us–them interaction that helps to foster the current gulf between science and policy. It is utterly important to get fishermen's legitimate interest involved.

In the case of Grand Banks, fisheries scientists knew that stocks were declining but were some what reassured by the relatively healthy catches still being landed. They were however neglecting to consider the fact that fishermen were spending more time at sea, with improved equipments, fishing selectively in warmer waters where the remaining fish were congregating. Across most of the banks, there was barely an adult cod to be found.

Analyzing 90 fish stocks world wide using the largest available data set, fishery scientist found many of the stocks had experienced massive declines due to over fishing. With the possible exception of fast–maturing species, such as. Atlantic herring (*Clupea harengus)* these stocks had shown little sign of recovery as much as 15 years after their collapse. The life history of the species matter. Small, early maturing mid–water species like herring might recover faster than late maturing bottom–living species such as cod.

FAO's catch statistics from 1950 to 1994 reveals that the world's fishing fleets have been steadily fishing down the food web towards lower trophic levels which is likely to make ecosystems inherently more vulnerable to damage by reducing the complexity of marine food webs.

Make or Break for Europe's Fisheries

"The desperate race for fish has to stop" declared the European Union's Commissioner for Agriculture and Fisheries in May, 2002, launching a far reaching proposal to reform the European Unions (EU) Common Fisheries Policy (CFP).

Directly or indirectly Europe's fishing industry employs more than 15 million people. But the commissioner wanted to reduce EU's total fishing effort by up to 60 per cent from January 2003. He aims to cut the current fleet of almost 1,00,000 fishing vessels by 10 per cent, while forcing the remainder to reduce their activity. Subsidies worth US $450 million for 2003–06 currently earmarked for renewal and modernization of vessels, would be redirected to pensions and retraining for fishermen.

The commissioner intends to set up inspection scheme to tackle illegal fishing and the misreporting of catches. Sick of the annual "political horse trading" over national catch quotas, the commissioner also wants to introduce management plans lasting several year's in which advice won't be twisted by competing national agendas. And to bring policy–making closer to fishermen, he aims to create regional advisory councils in which stakeholders can submit their own ideas about fisheries management.

It is a bold plan, and according to fisheries scientists, is necessary if Europe's fisheries are to escape destruction. Since 1991 more than 70,000 fishermen have been driven out of a job by dwindling stocks and diminishing catches–adult cod are only half as abundant in Europe's fishing grounds as they were in the 1970s.

Time and again, efforts to reduce EU's over exploitation of its fisheries have floundered mired by opposition from countries with large fishing industries, such as, Spain, Portugal and France. And with commissioner's plan still being discussed by EU's member states it could yet be blocked or watered down considerably.

If Europe can't get its fisheries in order its problems seem certain to be exported with European fishermen increasingly looking further a field for their catches, the EU is already paying compensation to some West African Countries for fishing rights' in their coastal waters.

The public, not industry, is the owner of the resources. But so far, all forms of fisheries management have been industry–friendly to a misplaced degree.

One positive signs include the growing number of nations that are adopting aspects of the FAO's Code of Conduct for Responsible Fisheries, a 1995 document that suggests legal technical and economic arrangements for national authorities seeking to put their fisheries on a more sustainable footing. Iceland and New Zealand, for instance, have reduced their fleet sizes and are strictly enforcing total allowable catches (TACs). The same two countries have also led the way in dividing their TACs into individual transferable catch quotas (ITQs) for each vessel–New Zealand introduced ITQs in 1986; Iceland, after a successful experiment with its herring fishery, applied ITQs across the board in 1990. Without ITQs, fishermen race against one another to land as many fish as possible early in the season. Total catches, as a result, frequently exceed the TAC and illegal fishing is common. With the security of ITQs, however, individual boats no longer compete as fiercely, so fishermen can spread their harvest over the season and sell any unused portions of their ITQ. The net out come is that total catches tend to be lower.

These policies, according to analyses by the organisation for Economic Co-operation and Development have actually increased the profitability of the two countries fishing sectors. Some stocks that are fished by their vessels are now showing signs of recovery.

It is believed that involving the public in fisheries conservation debate will be crucial. Consumer's preference for products labelled as eco–friendly, for instance, could be a powerful factor in changing attitudes within the fishing industry, particularly if the public is informed of the consequences for consumers choice if fisheries are not put on a sustainable track. If the fishing fleets refuse to withdraw from large parts of the oceans, future generations won't have the option of dining on cod or other familiar favourites.

Fisheries Sector in the National Economy

In India, the fisheries sector contributes only a very minor share in its national economy. In Japan, fisheries contribute a very considerable percentage of its national economy. In other western countries, like, Norway, Iceland, Sweden, Peru and England a substantial part of the national income is derived through fisheries. The slow development in the contribution of fisheries to national economy in India is attributed to several causes. This was an industry, which was handled by the poorest section of the nation and the fishing profession itself was considered something as quite backward; and naturally modern developments in fishery had very little impact on the people who are engaged in the trade. It was only after the second world war, the people started realising the importance of fisheries in India, and accordingly, Govt. of India and the various State Govts. in their plan schemes started giving considerable importance in the development of fisheries. The need become more because of the necessity to give protein rich food to the nation, which is traditionally deficient in protein rich food.

According to the National Income Statistics, Indian fisheries contributed a total net value of Rs. 70.04 crores as compared to Rs. 29.72 crores from eggs and poultry, Rs. 90.70 crores from meat and meat products, Rs. 167.93 crores from forestry and Rs. 759.60 crores from milk and milk products at current prices during the year 1958-59.

The contribution of fisheries to national income has gone upto Rs. 196 crores in 1969-70. The particulars of the contribution of fisheries sector to the national income in the past few years are furnished below.

Year	*National Income*	*Share of Fisheries Sector*	*Percentage*
1967-68	286,330	1540	0.54
1968-69	289,360	1730	0.60
1969-70	314,320	1960	0.62

Indian fisheries contribute about 0.6 per cent of the national income and 1.3 per cent of the income from Agriculture as a whole. The vast resources, which are capable of yielding 10 million tones of fish per year (Mitra, 1965), as against only an average annual production of 1.34 million tonnes per year, will go a long way in increasing the national income. Since most of the additional production shall be derived from the marine fisheries and that too from the deeper and distant grounds, fishing costs are likely to increase. Under such circumstances, even if the value of the additional net product is taken at 6½ times of the income from the fisheries sector, *i.e.*, Rs. 70 crores, the prospective net national income from fisheries can be roughly estimated at about Rs. 450 crores at 1959-60 price level.

But the contribution of fishery sector to the net domestic product has shown an eight–fold increase from Rs. 8.06 billion in 1980-81 to Rs. 67.5 billion in 1993-94 at current prices when compared to only the four fold increase in agriculture during the same period. The share of fisheries in GDP from agriculture has almost doubled from 1.97 per cent in 1980-81 to 3.89 per cent in 1993-94.

Indian Fisheries: As Foreign Exchange Earner

In recent years, particularly after the devaluation of the rupee in June, 1966, fisheries has become an export oriented industry, whose earnings of foreign exchange have jumped from Rs. 4.13 crores in 1961 to Rs. 2503.6 crores in 1993-94. Exports of marine products from India have made a significant progress in 1993-94. The export rose to 243960 tonnes valued to Rs. 2503.62 crores. Frozen shrimps, which constitute 80 to 88 per cent of the total export, are mainly responsible for this significant rise. In contrast, dried as well as canned items have suffered badly in the export front. Export growth of Indian marine products for past few years are as below.

Year	*Qty (tonnes)*	*Value (Rs. in crores)*
1950-51	19651	2.46
1955-56	24000	3.9
1961-62	15732	3.92
1965-66	15295	7.06
1970-71	35883	35.07

Contd...

Year	Qty (tonnes)	Value (Rs. in crores)
1975-76	54500	124.5
1980-81	75591	234.84
1981-82	70105	286.00
1982-83	78175	361.36
1983-84	92691	373.00
1984-85	86187	384.29
1985-86	83651	397.99
1986-87	85843	460.67
1987-88	97179	531.20
1988-89	99777	597.85
1989-90	110843	634.76
1990-91	135849	897.55
1991-92	171820	1375.89
1992-93	208602	1767.43
1993-94	243960	2503.62
1994-95		3575.27

Bulk of the export was due to the frozen shrimp and USA, Japan and Australia continued to be the major buyers. The major item of export being frozen shrimp, frozen lobster tail, frozen frog legs, canned prawn, dried prawn, dried fish and shark fins and fish maws.

The major items of marine products shipped through different ports of India are as follow.

Name of Port	Items of Export
Cochin	Frozen prawn, frozen frog legs, frozen lobster tails, canned prawns and dried prawns.
Bombay	Frozen prawn, frog legs, lobsters tails, dried fish, shark skin and fish maws.
Chennai	Frozen prawn, shark skin, fish maws.
Tuticorin	Dried fish.
Mangalore	Frozen prawn.
Kolkata	Frozen prawn, frog legs.
Ratnagiri	Frozen prawn.
Vishakhapatnam	Frozen prawn.
Calicut	Fish oil.

Public Revenue from Fishery Resources

Public revenue from fisheries has already been thought of as early as the time of East India Company in our country, when a head-tax was levied. In Thailand under the Fishery Revenue Act, the revenue is collected. Some of the revenue potentialities from Indian fisheries are indicated below:

1. 20 per cent income of gram panchayts in Orissa state come from fish culture.
2. Reservoir fisheries and sport fisheries offer a very good prospect as a source of public revenue.
3. 85 per cent of the total public revenue from fisheries in U.P. come from reservoir.
4. Rehand reservoir alone has contributed 60 per cent of the revenue from reservoir fisheries.
5. The Chilka lake fisheries in Orissa state add a revenue to the tune of 6.75 lakhs annually from its lease alone.

Employment Potential in Fisheries Sector

In absence of dependable statistical data regarding fishermen population, variations in the natural resources, fisheries developmental stages and fish processing and marketing techniques in different countries, it is difficult to predict regarding job potentialities in fishing industry. However a reasonable estimate indicate that the industry provides gainful employment of full time and part time fishermen in the country in addition to create employment opportunities in ancillary industries like manufacture of boats, nets etc.

Allied Industries

With the development of the fishing industry to economic scale, important allied industries like, boat building, marine engines, fishing equipments, floats, propeller, winches, wire ropes, ecosounders, sonars, nylon manufacturing, net making, freezing, canning, fish meal, fish oil, cold storage, ice plants come up progressively. With the increase in technical know how and exploitation in further and deeper grounds, these industries expand. This not only provide employments in the private sector, but also utilises the other resources of the country for the industry. An idea of the importance of ancillary industry in the national economy can be had from the mere fact that there are 18 ice plant and cold storages, 28 boat building yards, 21 freezing and canning plants under the fold of fishing industry in the Kerala state alone.

Demand Projections

Fewer attempts have so far been made for projecting the demand of fish in India. The reason seems to be the lack of availability of adequate data at the national level. In the absence of data, it was not possible to give quantitative estimates of demand of fish at the end of Five Year Plan periods. For the first time, in National Sample Survey Report No 200, per capita consumption of fish at national level in quantity as well as value terms, both for rural and urban population, under different per capita expenditure classes has been published. In view of the availability of such data, Dayal (1973) has worked out the total demand of fish for consumption in India to be 17 lakh tonnes in the year 1973-74, which will raise to 21 lakh tonnes in the year 1978-79. It will further go upto 23 lakh tonnes in the year 1980-81. However, if we provide 9.8 per cent for wastage, the demand at the production point will be 19, 23 and 26 lakh tonnes in the year 1973-74, 1978-79 and 1980-81 respectively.

Population and Total Demand of Fish

	1965-66	*1968-69*	*1973-74*	*1978-79*	*1980-81*	*2001-2002*
Population of India (million)	484.66	517.73	576.76	632.66	653.14	over 1000 million
Total demand of fish for human consumption (lakh tonnes)	11.57	14.06	17.01	21.37	23.32	45.82

Entry of Corporate Houses in Fisheries

Over the past decade foreign exchange requirements, favorable state policies and liberal investment incentives have attracted substantial investment by International and/or national corporations who are emerging as dominant sets of actors in the production, financing and marketing of shrimp.

With India's entry into feed and seed ancillary industries, there has been an increase in the number of joint ventures with foreign companies. 10 out of 23 joint ventures in 1993-94 were with South-East Asian Companies-in a bid to bring in foreign exchange and new technologies, C P Foods has entered into research and development to increase the quality and quantity of shrimp with Indian Companies.

Indian Business Houses in Aquaculture

Indian business houses, involved in aquaculture, are able to secure land and finance, which are not available to smaller farms through the support of government agencies. These houses, in addition to develop prawn farms of their own have contract farms where support, feed, seed, is provided to the farmers and they in return must sell their crop back to these houses (BPL, ADAK, CP, Aquastar, etc). All prices of both input and of the produce are decided by them.

Rank Aqua

India's largest aquaculture company. The company, which first began in Nellore with 32 acres in Andhra Pradesh, has steadily grown and has now built a feed mill and a hatchery Rank Aqua with the assistance of Hanaqua of Taiwan are now focussing on shrimp processing and feed under the name of Smoken Marine Foods. Rank Aqua entered into the feed industry as it made economic sense to produce feed for its own farms and also due to high profit margins and low risk, the company is guaranteed at least a one crore rupees sale since large shrimp farmers are usually required to store an inventory of at least 300 tonnes of feed.

Indian Tobacco Company (ITC)

To diversify with in aquaculture, Minota Aquatech, the ITC group's aquaculture company has developed its own Research and Development Centre to look at prawn health and nutrition and decided to open the centre and charge user fees for services. The services the centre provides includes site selection, quality of soil and water and feed quality. The company hopes to utilise the centre for the development of prawn feed, which will induce prawns to spawn.

Sea Gold

Sea Gold has established itself as a completely export-oriented semi-intensive operation that will include a hatchery and its own processing plant. The company is being promoted by King Fisheries, are of the leading producers and exporters of prawn.

Role of Fisheries in Indian Economy

India has been one of the top seven fish producing countries of the world during last decade. At present India's fish production touches 3.8 million tonnes.Out of the present production about 62 per cent comes from the marine sector and 38 per cent from the inland sector, while that of the world fish production about 88 per cent and 12 per cent comes from sea and inland waters respectively. Apart from fish production the country have earned foreign exchange worth Rs. 3270 crores during 1994-95. The added fish production has helped the much needed protein rich food to the people. It has been estimated that an average Indian needs 56 gram of fish in his balanced diet. To meet the requirement the country requires over 15 million tonnes of fish each year. Considering the vast potential, the aquaculture sector can boost up their fish production to meet the protein requirement of the country.

The fisheries sector occupies a prominent place in the country's economy. Fish culture is becoming more and more alluring due to its low capital investment, short gestation period and generation of high profit. Its importance from social and economic view point goes to augmentation of nutritional level, employment generation, earning foreign exchange. It is a suitable proposition for rural development and to improve economic condition of the rural people. Above all, development of aquaculture (Blue revolution) is one of the means to achieve the much needed social goal of "Food for All" and "Work for All". Aquaculture is an economic means for low cost protein food production, when capture fisheries is threatened by over fishing and lack of management of fish stocks in natural waters. There is a vast potential of water resources in India, accounting about 9.4 million hectares comprising of ponds, swamps, reservoirs, brackish waters, inland and coastal soils, which can be suitably utilized for production of fin and shell fishes through aquaculture technology.

Inland Fisheries Scenario

Inland fisheries in India is mainly carp based, and carps alone contributed over 1.285 million tones (89.3 per cent of the total

Deep Sea Fishing Trawler–Catching Fish from Sea

Beach Landing Fishing Craft of Artisanal Fisherman, Orissa

production of 1.439 million tonnes during 1993). The country possesses vast fresh water resources, comprising of 2.25 million ha of ponds and tanks, 1.3 million ha of beels and derelict waters, 2.09 million ha of lakes and reservoirs. Additionally 0.12 million km of

Salmon Catching by Shore Seine from Sea

irrigation canal and 2.3 million ha of paddy fields, a part of which is available to fish culture. Out of 7.53 lakh ha. of ponds and tanks, only 1.5 lakh ha are presently under aquaculture.

With a catchment area of 3.12 million sq km, the 14 large, 4 medium and innumerable minor rivers have highly diverse fish fauna. The total length of the rivers and canals has been estimated at 1.7 lakh km. The various estuarine systems cover an area of over 2.7 million ha have been identified as an important source of wild fish and shell fish seed.

There is a good scope for increasing fish production from reservoirs, oxbow lakes, canals and tanks. The unconventional culture systems like integrated rice-cum-fish farming, fish-cum-live stock farming etc. hold tremendous promise in bridging the gap between the demand and supply of fish. Such farming approach is well suited for the unemployed rural youth in Indian context. A good number of village women can be absorbed in this sector. It has been estimated that the per capita land availability which was 0.48 ha, in 1950 has decreased to 0.20 ha in 1981 and likely to further decrease to 0.14 ha by 2000 AD due to rapid growth of population. Maximum utilization of available land and water is possible through integrated aquaculture. Food production per unit area and benefit–

cost ratio could be enhanced in this way. The practice of cultivating carps, murrels and catfishes etc. seems to be easier in comparison to agriculture and horticulture. Women can choose this activity for their self employment and can earn lucrative income there from. During the last three decades the fish culture that was once restricted to the eastern states of the country (West Bengal, Orissa and Assam) has become a major economic activity in the states like Andhra Pradesh, Punjab, Maharashtra, Haryana and Gujrat. The brackish water tiger prawn are highly priced due to good demand in foreign markets. At present, India is among the four important countries in the world which export large quantities of prawn in the world markets. The gestation period is also quite less in brackish water prawn farming. Being profitable, as well as foreign exchange earner, day by day more and more common people and industrial houses have taken up prawn farming. Rich fresh water resources of the country provide good scope to develop culture of giant fresh water prawns. The giant fresh water prawn or "scampi" has a very good demand within and outside the country due to its large size and premium price. Generally after a culture period of six months, scampi grows to a weight of 80-120 gm. with proper management and feeding, thus making it commercially attractive.

Pearls have also a good demand in export market. Good success has been achieved in pearl production from fresh water mussels (*Lamellideus* spp.). The technology is very suitable for rural people. Culture of fishes and prawns in enclosures is a lucrative practice. Different species of catfishes, prawns, murrels are cultured profitably and such culture may provide gainful employment.

Lacustrine fisheries resources of the country are very rich which comprise of upland lakes, oxbow lakes and the reservoirs. The fish yield from the Indian reservoirs is frustratingly low. It is estimated that the country's 113 major and minor rivers along with their tributaries have a combined length of 29000 km. It has been estimated in some selected stretches of the river Ganges, Brahmaputra, Godavari and Krishna the fish yield varies from 0.64 to 1.6 tonnes/km. Riverine fisheries still contribute significantly to the total inland fish production and provide a means of livelihood to thousands of poor fishermen. Number of fishermen per km. of river stretches varies from 3.2 in Narmada, to 7.8 in the Ganges, the average being 6.5.

For development of inland fisheries, several programmes are being implemented by the government since Sixth-Five Year Plan. These are development of aquaculture through Fish Farmers Development Agencies (FFDAS) and National Programme of Fish Seed Development (NPFSD). So far 365 FFDAS have already been established. The FFDAS arrange required inputs like fish feed, credit and fish seed for the fish farmers.

The fish producing potential of freshwater, specially, rivers, lakes and reservoirs of the country is under utilized as the natural spawning of the most valuable fresh water species is being increasingly hindered by human intervention.

Marine Fisheries Scenario

India with a coastline of 7517 km. continental shelf of 0.451 million sq. km. and Exclusive Economic Zone (EEZ) of 2.02 million sq. km. has a rich marine fishery potential. Marine fishery scientists estimated the potential yield of Indian EEZ is at a level of 4.5 million tonnes. India is now harvesting a little over one third of estimated potential of EEZ. Even this quantity comes largely from coastal waters. The present contribution from the off–shore fishery sector is less than one per cent of the total marine fish landings. However, the export of marine products has grown at a rapid pace in the last three years and the trend is likely to continue in future. Marine products export have gone up from Rs. 13.75 billion in 1991–92 to around Rs. 3750 crores in 1994-95, there by showing a healthy growth of 30 per cent. The potential is however, still very great if right promotional policies for the deep sea fisheries development are adopted.

The marine fishery resources of India in the four main regions, North West, South West, South East and North East are:

1. Major pelagic resources of oil sardines, mackerel, seer fishes, tunas, anchovies and ribbon fishes.
2. Demersal fishery resources of deep sea prawns, cephalopods, perches, scianids, cat fishes, polynemids, flat fishes, pomfrets, eels, sharks, rays.
3. Mid water fishery resources, such as, Bombay duck, Silver bellies, Horse mackerels.
4. Crustacean fishery consisting of prawns, shrimps, lobsters and crabs.

5. Molluscs such as, chank, oysters, mussels, calms, squids and cuttle fishes.
6. Sea weed resources.

Indian west coast gives higher fish production because of notable upwellings under the influence of strong south west monsoon and the effect of Somali currents. Generally fishing activities on this coast are confined to a narrow coastal belt of 10-15 nautical miles within the depth of 70 meters. Remote sensing technique is now utilized for locating potential areas of fish accumulation for the benefit of the fishermen.

The government in recent years has taken several steps to increase fish production and boost export of marine products. The measures include induction of offshore pelagic boats to fish for the under exploited pelagic species, introduction of several schemes to bring down the cost of fishing operation in the artisanal and small mechanized sector, reimbursement of central excise duty on high speed diesel used by small mechanized boats, motorization of traditional crafts and improvement of fishing gear to make fishing operations viable. The Marine Products Export Development Authority (MPEDA) is also taking steps to boost export earnings from marine products.

According to Jacob et. al. (1982), there are about 2500 marine fishing villages and 1500 fish landing centres along the coastal areas of India. The marine fishermen population in the country is about 21.2 lakhs of whom 4.9 lakhs are active fishermen taking part in marine fish production.

Marine industry could grow at a faster pace and exports from the sector could have been tripled if the industry is given proper encouragement and incentives. The Federation of Indian Chambers of Commerce and Industry (FICICI), The Confederation of Indian Food Trade and Industry (CIFTI) and the Sea Food Export Association have urged the government to set up a separate Ministry for marine industry.

Inspite of huge marine resources, India contributes only 3.3 per cent of the total world catches of marine resources. However, mariculture technology have been developed in respect of several commercially important marine organisms. Special mention to be made of breeding and production of number of species of prawns, edible oysters, pearl oyster, mussels, calms and some finfishes. Hatchery techniques and farming practices have been developed for most of these species. Additionally research has been focussed on sea ranching of marine animals with dual purpose of increasing production from natural stock and also to prevent the declining trends in some of the resources.

The National Marine Living Resources Data Centre (NMLRDC) has been established at Central Marine Fisheries Research Institute (CMFRI) to develop a data base on marine fisheries resources collected by the institute and other agencies. Detailed data on length, weight, biological characteristics, catch per unit effort, species composition of catches, total catch gear wise and species wise have been computerised. Besides the government of India established various institutions which are directly or indirectly connected with marine fisheries investigations. These are Fishery Survey of India, CMFRI, CIFT, Central Institute of Fisheries Nautical Engineering Training, Central Institute of Coastal Engineering for fishery (CICEF) and CIFE. These institutes have significantly contributed for the development of marine fishing in their own way.

Investment Plan

The investment plan of the entire fisheries sector would reflect total investment credit need to the extent of Rs. 8436 crores and bank finance amounting Rs. 6041 cores.

Sector	*Total investment Credit for the Sector*	*Credit Flow Through Non Banking Financial Institutions*	*Credit flow Through Public Sector Banks*
Fresh Water fisheries	4934	987	3947
Brackish water fisheries	1906	379	1528
Marine fisheries	1596	1032	566
	8436	**2398**	**6041**

Economic Role of Inland Fisheries

Inland fisheries play an important role in the economic and community life of rural areas. In the general field of the economy, inland fisheries industry play in certain countries not a leading role, but at least that of one of the principal characters. These industries also play a supporting role in connection with a number of other industries and of communal activities. These fishery activities can therefore be recommended, not only on their direct economic advantages but also on the indirect advantages to the economy and their importance in other directions.

The importance of these industries as a distinct sector of the economy may be considered from various points of view of which probably the most important are the:

1. Production of food and other commodities and contribution to the national income;
2. Opportunity offered for employment;
3. Market provided for producer equipment and supplies; and
4. Opportunity created for recreation and sport.

Considering the use of fish catches in many western countries, where the bulk of the catch is from marine sources is processed only about 50 per cent of the round fresh weight is considered edible and the rest being either wasted or used for oil or meal.

In contrast, most of the freshwater fish is produced in areas where food habit patterns encourage the human consumption of the whole fish in the form of either fresh or cured products. It thus appears that the contribution of the fresh water fisheries to the total amount of aquatic animal food actually used by humans is quite high, and is likely to be approximately 40 per cent of the total on the edible weight basis of the quantities consumed.

Statistics on man power and equipment engaged in inland fisheries are even more scanty than the statistics on production. The degree of industrialization of operations is very much smaller for inland than for marine fisheries and considerable quantities of fish are taken from inland waters by subsistence operations; not only does the catch from such operations fail to find a place in the official records of catch, but the operations themselves and the labour force and the equipment also pass without notice. It is to be accepted that the number of people who devote a proportion of their energies to fishing operations in inland waters represents a very considerable total of man–days per year.

The principal objective of inland fisheries activity is to produce food and at times, such food is produced so that it may be sold and thus bring money income to its producer. The production of food and earning of money will be more significant when brought from the neglected waters, they might have on a particular project, or on the economy of a community. The general effect which is common to practically all inland fishery projects, will be signified by the term "balancing the economy" and it is to be understood that this means balancing the energy economy of the rural area as well as balancing its money economy.

In the task of balancing the energy economy, the inland waters play an important role in accepting and effectively using materials from the land, some of which are wastes from the processes on the land, and some are unavoidable losses from the land which then are turned to advantage in the water, and, on the other hand, these waters play an important role in contributing materials to the land. Finally fishery activities can, chiefly as an extension of this balance of the energy economy, contribute significantly to programmes for land reclamation and improvement.

Because the emphasis in land use has been on agricultural uses, for crops or for stock, there has been a tendency to regard water masses as lost surface unless they have formed part of irrigation systems and thus have been a source of water for agricultural purposes, the potential value of such bodies of water has been neglected. With only a few exceptions, almost every body of water is potentially a producer of foodstuffs; in each body of water there is a process of organic production corresponding to the best growth of plants on dry land, and if this process is directed properly through appropriate links in a food chain, such water masses may produce more protein per unit area than the adjacent stretches of dry land. So the land surface which is denied to agriculture by water which overlies it is not lost to food production but, on the contrary, may produce more food than any type of agricultural operation which might be contemplated for it, and accordingly it should be granted its proper place in the rural economy.

The point at which to begin in a programme of development of inland fisheries is with existing swamps and marshlands. Most often swamp lands are entirely waste, valuable only for a certain amount of wild life which is found there. But it is unnecessary that such lands should remain unproductive. Modern practice of such situations is to develop a systematic project for drawing the area into a selected part (sump) which lying lowest in the area, can accept the water from the rest and become a permanent body of water of useful depth. The drained area is available then for agricultural purposes and the new body of water may be used for fish culture.

The brackish water marshland area, on the sea coast and along the bank of estuaries, covered with mangrove and associated kinds of plants and accompanied by a typical fauna which includes percomorph fish (especially mullets), and siluroid fish, crabs and shrimps, mussels and oysters. They have some value for the catch of fish etc. and for the mangrove wood, the nipa palm and so forth. Such situations, are, however of considerable value for brackish water fish cultivation and proved to be extremely productive, and of high economic importance for exportable shrimp culture.

Mechanization in Indian Fishing Industry

Most of the fishermen in developing countries, including India use fairly primitive boats powered by sail or oar, as their ancestors have done for hundreds of years. The range of a fishing boat is thus determined by the wind force or sheer human muscular power. Along many coasts, there are good fishing grounds just beyond the effective range of sail boats or canoes. In many cases these rich grounds can be reached if the fishermen exert a special effort or if the wind is particularly steady. Once on these grounds a skillful man may catch as much as in few hours as less ambitious inshore fishermen might catch in a whole day. But price for these few hour of good fishing is not only toil, but danger. What if storm strikes when his crude boat is far off shore? What if night falls and land marks disappear before the lights of the village are safely in sight? But even if all does go well, the fishermen frequently enters harbour, after hours with sail and paddle under the tropical sun, only to find that much of his prize catch is a putrid, unmarketable mess.

Mechanization is the obvious answer to this problem. The long term solution is a fleet of well designed, diesel powered fishing boats, coupled with an extensive training programme. Progress on these lines is of necessarily slow, because of the costs involved for developing nations.

A short term compromise, which has had considerable success in many parts of the world is motorization of the local fishing crafts.

The Food Agricultural Organisation (FAO) who has experiments on a small scale with out board motorization since 1951, has operated out board motorization schemes under Expanded Programme of Technical Assistance and under its Freedom from hunger campaign. Notable success has been achieved, particularly in Sri Lanka. The catch comparison between 20 motorized and non–motorized craft in Sri Lanka, 1962 were,

	Motorized catamaran		*Non-motorized catamaran*	
	Catch in lb per day's Fishing	*Value in Rupees*	*Catch in lb per day's Fishing*	*Value in Rupees*
Average per month	1765	1410	446	353
Average fishing days per month	22		20	

In India the motorization of the indigenous craft as the first step in the introduction of fully motorized fishing is best suited to "Lodhias" and "Machuas" in Gujrat coast, Satpati and Versova boats in Bombay, Tuticorin boats in Gulf of Mannar, Andhra "Nava" in Telugu coast, Batchari in Bengal coast. Even the motorization with out board engines have been done in Verabal of Gujrat coast.

Till 1968-69, a negligible portion (6737) boats were motorized out of 93676 fishing crafts available in different states of our country. The reason for the slow progress in mechanisation programme that the capital investment for motorization is seldom available to fishermen having small boats. There are several reasons for this. First the government of developing nations are usually short of capital. Second, fishing in these countries are usually undeveloped and fish production is uncertain and unsteady. Third unlike the farmer and his land, the fishermen have no "mortageable interest" in the sea and its produce.

Also the big joint-stock companies, like Ross Group or Associated Fisheries of England, the co-operative joint management, Productive-associations etc. of the type found in Japan are yet to come on commercial scale in India.

Companies Invested in Fishing in India

In recent years, some interest by the private industrialist, have been noticed to invest in Indian fishing industry to catch exportable prawn. A few of them are mentioned below.

Union Carbide

Initial investment 3/4 million pounds. They procured two American trawlers of 76 feet OAL and one trawler of Indian built. Plans to have 50-60 trawlers with 10 million pounds and two to three shore plants.

Delhi Cloth Mills

A project to invest 3 million pounds for 30 trawlers and shore facilities.

Sriram Refrigeration Industries

A similar project plan of DCM to collaborate with American investors.

Britania Biscuits

Project to purchase three foreign trawlers one Indian and shore establishment.

Tata Oil Mills

Two 72 feet Mexican trawlers with on board freezing facilities and one 57 feet Indian trawler with shore plant.

Birla Group

A shore plant at Tamilnadu under the name of Eastern Sea Foods Private Ltd and the other at Mangalore under Mulbery Aquatic Products Ltd. Two Indian and six foreign trawlers.

Chapter 2

Primary Production

Oceans having an area of 361059 × 10^3 km^2 account for 70 per cent of the total surface of the earth. A gentle slope extends out from most shorelines to 100-200 m. This section is called continental shelf, and it is a topography common to almost all the sea areas in the world. The continental shelf is thought to have been a part of the land at the time of glacial age. Out from the continental shelf, the sea bottom slopes down rather steeply. This section is called continental slope. Reaching a depth of about 3000 m. the sea bottom resumes a gentle, slope until if reaches a depth of about 6000 m which is called 'deep sea floor'. Sometimes, there are trenches which are 6000-10000 m. deep at points on the deep sea floor.

Biologists and oceanographers call the waters above continental shelves "coastal waters" differentiating them from "high sea waters". Although the coastal waters constitute in terms of area, only 9.9 per cent of all the oceans, they are very rich in biota, with the productivity of the living resources being estimated at approximately two times that of the high sea waters.

The coastal waters are constantly supplied inflowing land waters with nutrient salts, facilitating propagation of phytoplanktons and zooplanktons on which fishes and other marine animals feed.

Coastal waters often have upwellings or vortexes which cause vertical mixing or agitation of sea waters, providing biological environments which facilitate the propagation and growth of planktons. If may be said that the original forms of life on the earth were microscopic water plants called phytoplanktons.

Phytoplanktons drift in enormous quantities in the surface waters penetrated by sunrays and form carbohydrates and other organic materials through photosynthesis, using the solar energy and nutrient salts in the water.

The most important and most numerous phytoplanktons are diatoms. They are primary producers in the marine ecosystem and therefore often called "marine pasture".

All forms of life have undergone a long evolutional process to reach their present forms and places in the ecosystem. Each species has its own habitat suitable to its life cycle. Organisms cannot live in isolation. Except for plants which are capable of forming nutritive substances for themselves, all animals live on other organisms, while they are in turn, preyed on by some other animals.

There is also competition between different animals for common preys. These food relations are most important among living organisms. Animals and plants inhabiting given areas have certain "food chain" relations, which find each animal and plant holding a certain ecological position in the habitat.

However, food chain relations are never uniform. In each individual habitat, there is a unique ecosystem consisting of certain food chains. Even the same habitat, the patterns of food chains also vary with the seasons, as animals feed on different organisms during different stages of their life history *e.g.*, fry, yearlings and adults.

The patterns of food chains reflect the structure of reproduction of the given area. Therefore, it is essential to understand the ecosystem in which a given fishery resource exist, their increase is to be ensured.

In natural ecosystem, in the oceans of the world 10 billion tons (in-dry weight) of phytoplankton are produced in a year.

Phytoplanktons together with zooplanktons are valuable foods for juveniles and larvae of fishes. There are many fishes feeding on the phytoplankton even in their adult stage. It is well known that spotlined sardine feeds on diatoms and flagellates, sweetfish feeds on diatoms and blue green algae and milkfish feeds on blue green algae.

Also many snails live on seaweeds. Abalone feeds on brown algae and wreath shell feeds on attached diatoms and red algae.

Seaweed forests in the water are used as spawning grounds by fishes and other aquatic animals. The young of this animals are protected from natural enemies by growing up within the seaweed forest. Yellowtail has a unique trait that the juveniles grow as they migrate together with the drifting weeds. Driven by world population growth and increased global demand for fishery products, fishing pressure has been rapidly increasing over the past years. Marine captures totalled 86.0 million M.T. in 2000, a level close to the historic peak of 86.4 million M.T. in 1997 following a decline to 79.2 million MT in 1998. Capture of inland fishery resources in turn increased from 7.5 million MT in 1997 to 8.8 million MT in 2000.

Aquaculture's contribution to global fisheries landings continues to grow, increasing from 5.3 per cent in 1970 to 32.2 per cent of total fisheries landings by weight in 2000. Total aquaculture production in 2000 was reported as 45.71 million MT by weight.

Over 75 per cent of the world marine fisheries catch, over 80 million MT per year is sold on international markets in contrast to other food commodities such as rice.

Marine fishing process became industrialized in the early nineteenth century when English fishermen started operating steam trawlers, soon rendered more effective by power winches and, after the First World War, diesel engines. The aftermath of the Second World War added another dividend to the industrialization of fishing, freezer trawlers, radar and acoustic fish finders.

Fisheries in the early 1950s were at the on set of a period of extremely rapid growth, both in the Northern Hemisphere and along the coast of the developing countries. Every where that industrial scale fishing, mainly trawling, but also purse seining and long lining was introduced. It competed with small-scale or artisanal fisheries. This is especially true for tropical shallow waters (10-100m.), where artisanal fisheries targeting food fish for local consumption, and trawlers targeting shrimps for export and discarding the associated by-catch, compete for the same resource.

Throughout the 1950s and 1960s, the huge increase of global fishing effort led to an increase in catches, so rapid that their trend exceeded human population growth, encouraging an entire generation of managers and politicians to believe that launching more boats would automatically lead to higher catches.

Despite the collapses, which started with Peruvian anchoveta, in 1971-72 and declining trend of catch in the late 1980s and early 1990s the global expansion of fishing effort continued and trade in fish products intensified to the extent that they have now become some of the most globalized commodities, whose price increased much faster than the cost of living index. In 1996, FAO published a chronicle of global fisheries showing that a rapidly increasing fraction of world catches originate from stocks that are depleted or collapsed.

But misreporting of catch data by countries with large fisheries, combined with the large and widely fluctuating catch of the species, like Peruvian anchoveta can cause globally spurious trends. Such trends influence unwise investment decisions by firms in the fishing sector and by banks and prevent the effective management of international fisheries.

Trends in Marine Fish Catch

Oceanic Regime

Out of a total ocean surface of about 360 million square kilometer the neritic environment over the continental shelves covers almost 32 million square kilometer and hence the oceanic region accounts for over 91 per cent of the world oceans.

Although mean productivity per unit area is much lower in the oceans than on land, their very large surface means that the oceans still account for at least a third of the annual global carbon fixation. For this reason, oceanic communities contribute significantly to global process.

Oceanic phytoplankton is responsible for the primary production of the oceans and constitute the basis of the food chain in the high seas. Primary production is restricted to the so-called euphotic layer or upper part of the photic zone, where sufficient sunlight penetrates to allow photosynthesis. The depth of the euphotic zone depends on the amount of suspension and detritus present in the water and can vary from 40-50 m. in turbid waters to over 100 m. where the waters are particularly clear. Production is also limited by the availability of inorganic nutrients.

Below the photic layer there is the aphotic zone, where no light arrives and primary production is absent. Organisms living in this zone which are not performing vertical migration to upper waters are exclusively carnivores, suspension or detritus feeders.

Great quantities of nutrients are continually lost to the aphotic zone and are not available for photosynthesis, however large scale ocean circulation linked to Earth's rotation climatic cycles (seasons) and topography of the ocean basin allow periodical or semi-permanent mixing of superficial and nutrient rich deep waters. This phenomenon (upwelling) is the cause of extremely high productivity of some fishing areas.

Oceanic resources include species that are distributed beyond the continental shelf (living in the epipelagic, mesopelagic and bathypelagic zones) although they may spend part of their life cycles in the coastal areas. Exploitable species living in the these zones are fishes, crustaceans cephalopods and marine mammals. Fishes have the greatest importance both in number of species and in terms of fishery revenue. Fish species have been classified on the basis of zone, in which they are usually caught by commercial fisheries.

The two groups of species (epipelagic and deep water) are often different in terms of importance, technology, history and value. The valuable and still developing fisheries for tuna and tuna-like species constitute the bulk of fisheries targeting epipelagic species, although other epipelagic resources, such as cephalopods (short-lived and with rapid turn over) might sustain expanding ocean fisheries, being able to respond promptly to favourable environmental changes.

On the other hand, most oceanic deep-water resources are very dispersed and difficult to harvest and several fisheries on these resources have been discontinued because they were not economically viable. Catches of some deep-water species (blue whiting, *Micromesistius poutassou* which constitute almost half of the deep water catches in the last 20 years) are mostly destined for reduction into fish meal because of the rapid deterioration of their flesh, the presence of parasite, and the low market value for the fresh or processed product. Deep-water species are in general characterized by slow growth rates and late age at first maturity (*e.g.* 25 years for orange roughy, *Hoplostethus atlanticus)* which led to weak biological compensation of fishing mortality.

Oceanic resources are usually exploited by long-range fleets operating in areas where target species concentrate for feeding or reproduction. The more rapid increase of world fishery fleet sizes as compared to catches and the contemporaneous depletions of some coastal resources have contributed to the increase of fishing effort in oceanic areas. Given the complex interrelations between economic and political factors and the scarce knowledge of oceanic stocks, the issue of oceanic resources management is increasingly coming to international attention in the light of growing world human population and limited food fish supplies. Furthermore, considering the oceanic species live in a virtually boundless seas and national jurisdictions, their management has necessitates international co-operation. For these reasons, issues concerning highly valuable oceanic stocks, such as, tuna species are of paramount importance and complexity.

Marine Ecosystem Based Capture Fisheries

The Johannesburg Plan of Implementation of the World Summit on Sustainable Development noting the Reykjavik Declaration on Responsible Fisheries in the Marine Ecosystem, set the goal of encouraging the application of the ecosystem approach to responsible fisheries by 2010. This is an internationally agreed starting point for a new approach to fisheries management and fishery related studies utilizing a multinational interdisciplinary approach, which integrates information concerning productivity, ecology, fisheries, socio-economic aspects and governance.

Initially 49 Large Marine Ecosystems were identified and then an additional 50th was proposed on the basis of consideration of distinct bathymetry, hydrography, productivity and trophically dependent populations.

Recently the definition of Large Marine Ecosystems has been refined as "Large Marine Ecosystems are regions of ocean space encompassing coastal areas from river basins and estuaries to the seaward boundaries of continental shelves and the outer margins of the major current systems."

The first cluster comprised only one Large Marine Ecosystem East Bering Sea in Pacific Ocean of Northern Hemisphere with subarctic climate. Catches of Gadiformes (Cods, Hakes, Haddocks) are predominant in this ecosystem. Other groups of some importance are flat fishes, salmons and crustaceans. This ecosystem is characterized by extreme environment at high latitude, in which temperature, currents and seasonal oscillations influence the productivity. According to Sea WiFS global primary productivity estimates this ecosystem has been classified as a moderately high productive ecosystem. The Second cluster consisting of gulf of Alaska, in Pacific Ocean of Northern Hemisphere having sub arctic climate, adjacent to Eastern Bering Sea is also monotypic. The gulf of Alaska is a highly productive ecosystem. It also presents significant upwelling phenomenon linked to the presence of the counterclockwise gyre of the Alaska current.

List of 50 Large Marine Ecosystems

No	*Name*	*No*	*Name*
1.	Eastern Bering Sea	26.	Black Sea
2.	Gulf of Alaska	27.	Canary Current
3.	California Current	28.	Guinea Current
4.	Gulf of California	29.	Benguela Current
5.	Gulf of Mexico	30.	Agulhas Current
6.	Southeast US Continental Shelf	31.	Somali Coastal Current
7.	Northeast US Continental Shelf	32.	Arabian Sea
8.	Scotian Shelf	33.	Red Sea
9.	Newfoundland Shelf	34.	Bay of Bengal
10.	West Greenland Shelf	35.	South China Sea
11.	Insular Pacific-Hawaiian	36.	Sulu-Celebes Sea
12.	Caribbean Sea	37.	Indonesian Sea
13.	Humbolt Current	38.	Northern Australian Shelf
14.	Patagonian Shelf	39.	Great Barrier Reef
15.	Brazil Current	40.	New Zealand Shelf
16.	Northeast Brazil Shelf	41.	East China Sea
17.	East Greenland Shelf	42.	Yellow Sea
18.	Iceland Shelf	43.	Kuroshio Current
19.	Barents Sea	44.	Sea of Japan
20.	Norwegian Shelf	45.	Oyashio Current
21.	North Sea	46.	Sea of Okhotsk
22.	Baltic Sea	47.	West Bering Sea
23.	Celtic–Biscay Shelf	48.	Faroe Plateau
24.	Iberian Coastal	49.	Antartic
25.	Mediterranean Sea	50.	Pacific Central American Coastal

The catch composition of this ecosystem differs from all other ecosystems in being characterized by a strong prevalence of the freshwater and diadromous group, linked to rich salmon fisheries. Recent researches have hypothesized changes in the future production of salmons as a consequence of long term shifts in the plankton biomass in the last decades. However recent catch trends are rather stable.

Third cluster of Large Marine Ecosystems, consisting of California current in Pacific Ocean, Northeast US continental Shelf, Scotian Shelf and Newfoundland Shelf, all in Atlantic Ocean of Northern Hemisphere bearing temperate climate in first three and subarctic in the last one are historically most productive ecosystems of the northern hemisphere, three in the Northwest Atlantic and one in the North east Pacific are all classified as moderately high productive ecosystems, with the exception of the North east US continental Shelf, which is considered as highly productive and is structurally more complex than other three, with marked temperature and climate changes, river run off, estuarine exchanges, tides and complex circulation regimes. The California current ecosystem is a transition ecosystem between subtropical and sub arctic water masses with an upwelling coastal phenomenon that determines strong inter annual oscillations of the productivity, of the ecosystem and consequently, of the catch levels of different species groups.

The catch composition of this cluster is quite diverse as six species groupings (cods, hakes, haddocks, herrings, sardines, anchovies marine crustaceans marine molluscs, flounders, halibuts, soles and miscellaneous demeral fishes) contribute, on average among the four marine ecosystems at least 10 per cent of the total shelf catches. However the trend charts show the marked decreases of cods, hakes and haddocks in the Atlantic ecosystems in the early 1990s upto the cod collapse in 1993-94, while in the same years Gadiformes catches (mainly of Merluccius products) in the California current increased and have remained high since then. An increase of crustacean catches in the three Atlantic ecosystems can be noted in recent years although it is not clear if this is due to ecological or to economical reasons.

The cluster 4, the second biggest, with eight large marine ecosystems (Gulf of California in Pacific ocean in northern hemisphere with temperate climate, Gulf of Mexico in Atlantic ocean in northern hemisphere with tropical climate, Humboldt current in Pacific ocean in southern hemisphere with mixed climate, Baltic Sea in Atlantic Ocean in northern hemisphere with temperate climate, Black Sea in northern hemisphere with temperate climate, Canary current in Atlantic ocean in northern hemisphere, with temperate climate, Guinea Current in Atlantic Ocean with tropical climate and Pacific Central American Coastal in Pacific Ocean in northern hemisphere with tropical climate) which although in cluster can be subdivided into two main sub-groups enclosed and semi-enclosed seas (Gulf of California, Baltic Sea, and Black sea), which are strongly influenced by human induced eutrophication, river-runoff and by a lack of rapid exchange with the adjacent oceans and upwelling ecosystems, (two in the Pacific ocean; Humboldt current and Pacific central. American coastal, and two in the Atlantic Ocean, Canary current and Guinea Current) that show important upwelling and other seasonal nutrient enrichments. The Gulf of Mexico, although partially isolated from the Atlantic Ocean and water enters into it from the Yucatan Channel and exits from the Straits of Florida creating the loop current which is associated to nutrients flow and upwelling phenomenon can not be considered as a semi-enclosed sea. Further, this large scale and complex marine ecosystems is affected by such levels of enriching river runoff (especially from Mississippi) that large hypoxic areas have been detected in the gulf in recent years.

All these ecosystems are characterized by predominant catches of small pelagic clupeoids (herrings, sardines and anchovies) that represent over half of the total identified shelf catches in all the marine ecosystems. Catch trends show that ups and downs do not occur only in ecosystems driven by upwelling regimes but that also enclosed and semi-enclosed ecosystems have a high variability in catches. The ecosystems in this cluster, five in numbers, (Southeast U.S. continental Shelf in Atlantic Ocean of Northern hemisphere with temperate climate, West Greenland Shelf in Atlantic Ocean of Northern hemisphere with sub-arctic climate, Agulhas current in Indian ocean of Southern hemisphere, with mixed climate, Northern Australian Shelf in Pacific Ocean of Southern hemisphere with tropical climate and Great Barrier Reef in Pacific Ocean of Southern hemisphere with tropical climate) are distinguished by a very high percentage of crustacean catches. The second species group in terms of catches is clupeoids (herrings, sardines and anchovies) in the

South east U.S. continental Shelf, flat fishes (Flounders, halibuts and soles) in the West Greenland Shelf and non-oceanic tunas in Agulhas current and molluscs in the Northern Australian Shelf and Great Barrier Reef. Catch trends in recent years are very diverse.

These ecosystems are characterized by a rather wide range of productivity levels, from low (West Greenland Shelf) and moderate (Southeast U.S. continental Shelf and Agulhas Current) to moderately–high and high productivity (Great Barrier Reef and Northern Australian Shelf respectively). Geographically, with the exception of the West Greenland Shelf and partially of the Northern Australian Shelf, these marine ecosystems all lay along the eastern margins of the continents. Nutrient enrichment and mixing are due to different factors; offshore upwelling regime, although not as intense as in the higher latitude regions, in the Southeast US continental Shelf; tidal effects in the Great Barrier Reef, changes in sea and air temperature in the West Greenland Shelf; current–associated in the Agulhas Current and tidal mixing, monsoons and tropical cyclones in the Northern Australian Shelf.

This is the cluster with the highest number of Large Marine Ecosystems. These nine ecosystems (Insular Pacific-Hawaiian in Pacific Ocean of Northern hemisphere with tropical climate, Northeast Brazil Shelf in Atlantic Ocean with tropical climate, North Sea in Atlantic Ocean of Northern hemisphere with temperate climate, Somali Coastal current in Indian Ocean of Northern hemisphere with tropical climate, Arabian Sea in Indian Ocean of Northern hemisphere with tropical climate, Red Sea in Indian Ocean of Northern hemisphere with tropical climate, Bay of Bengal in Indian Ocean of Northern hemisphere with tropical climate, South China Sea in Pacific Ocean of Northern hemisphere with tropical climate and Sulu-Celebes Sea in Pacific Ocean of Northern hemisphere with tropical climate) are probably less characterized than others and for this reason they have been grouped together by the clustering routine. Geographically this cluster groups all tropical ecosystems, with the sole exception of North Sea and it includes four out five of the Indian Oceans marine ecosystems. The general greater marine bio diversity of tropical regions is so reflected in catch composition. The main distinguishing feature is the high catch percentages for miscellaneous coastal fishes and miscellaneous pelagic fishes. The catches of herrings, sardines and anchovies and of marine crustaceans in the nine ecosystems exceed 10 per cent on average. Most of these ecosystems are characterized by fishing activities mainly concentrated, for different reasons on the coastal areas and this explain the high percentages of miscellaneous coastal fish catches. Catch trends in 1990-99 period are quite diverse and it is difficult to identify a common pattern.

Primary production ranges from low (Insular Pacific-Hawaiian and Sulu-Celebes Sea) to high (North Sea, Northeast Brazilian Shelf and Arabian Sea) with the remaining ecosystems classified as moderately or moderately-high (South China Sea) productive.

The region of Pacific-Hawaiian (including shelf areas of several other Pacific Islands) is dominated by the equatorial currents system. Fishery production in the Insular Pacific-Hawaiian and Sulu-Celebes Sea ecosystems is mostly concentrated in the coastal waters as the islands are usually surrounded by very narrow shelf areas.

The Northeast Brazil Shelf is characterized by high levels of nutrients in the inner part of the shelf. The North Sea includes one of the most diverse coastal regions of the world, with a great variety of habitats. Three of the Indian Ocean ecosystems (Somali Coastal Current, Arabian Sea and Bay of Bengal) are influenced by monsoons. In the Somali Coastal Current and in the Arabian Sea, the south west monsoon from May to October cause seasonal upwelling that are on the other hand lacking in the Bay of Bengal. In the Arabian Sea about 65 per cent of fish landings derived from artisanal fisheries and this would explain the prevalence of coastal species catches but it may also be influenced by the presence of low-oxygen water, which restricts productivity at depth of 200 m and more. The elongated and narrow shape, semi-enclosed character and circulation pattern of the Red Sea protect the coast from storms and provide habitats for a large number of marine coastal species. Different sub-systems within the ecosystem have been identified in the South China Sea. In the cluster consisting of four large marine ecosystems (Caribbean Sea in Atlantic Ocean of Northern hemisphere with tropical climate, Brazil current in Atlantic ocean of Southern hemisphere with mixed climate, Mediterranean Sea of Northern hemisphere with temperate climate and Indonesian Sea in Pacific Ocean with tropical climate) clupeoids, herrings and anchovies is the most important species group in shelf catches, but other groups (mostly coastal fishes but also marine crustaceans, molluscs and miscellaneous demersal for the Indonesian Seas) also contribute significant capture production. Catch trends have been rather stable in recent years with moderate increases in total shelf catches (1999) with respect to 1990 with the exception of the Indonesian Seas where catches have been steadily increasing.

As for its catch composition, the Mediterranean Sea seems one of the most diverse and stable ecosystem in terms of species groupings, their shares in total catches and trends. Its unusual biodiversity for a temperate sea is confirmed by the fact that the Mediterranean and Black Sea together cover only 0.8 per cent of the total surface of the oceans but represent about 5.5 per cent of the total world marine fauna.

According to the productivity classification, the four ecosystems in this cluster are moderately high (Indonesian Seas), moderately (Brazil current) or low naturally productive ecosystems (Caribbean Sea and Mediterranean Sea) but the productivity of the last two ecosystems is increased by nutrient input from rivers, estuaries and human induced activities. These ecosystems have common a composite structure of environmental conditions, with local areas of upwelling, wind-driven currents, high water temperatures at least in some periods of the year, nutrient inputs from rivers or human activities.

The Patagonian Shelf in Atlantic Ocean of Southern hemisphere with mixed climate is characterized by high catches of molluscs, mostly cephalopods and Gadiformes. Cephalopod fisheries developed in the early 1980s by Distant Water Fleets, but since the early 1990s also local fleets (Argentina and Uruguay) are actively targeting these species. Following a drop in 1998, cephalopod catches in this area are still increasing. Instead, catches of Gadiformes, mostly by local fleets, increased continuously since 1970s but declining from mid–1990s.

These fisheries takes place in one of the most extensive continental shelf of the world. According to the estimates of global primary productivity the Patagonian shelf is an area of high productivity and it is influenced by intense western boundary currents and wind-and tide-driven upwelling.

Six ecosystems (Eastern Greenland Shelf in Atlantic Ocean of Northern hemisphere with subarctic climate, Iceland Shelf in Atlantic

Ocean of Northern hemisphere with subarctic climate, Barents Sea in Atlantic Ocean of Northern hemisphere with subarctic climate, Celtic-Biscay Shelf in Atlantic Ocean of Northern hemisphere with temperate climate, New Zealand shelf in Pacific ocean of Southern hemisphere with temperate climate and Faroe Plateau in Atlantic Ocean of Northern hemisphere with subarctic climate) in this cluster, with the exclusion of the New Zealand Self and the Celtic-Biscay Shelf, which are influenced also by warm currents, respectively the South Equatorial and the Gulf Currents, the other ecosystems are categorized as high latitude and extreme environments, in which temperature, currents, tides and seasonal oscillations affect productivity. The same division in two sub-groups applies also to data on primary productivity, the New Zealand Shelf and the Celtic-Biscay Shelf are considered highly productive ecosystems, and the East Greenland shelf is a low productive ecosystem.

The marine environment of the New Zealand Shelf in very diverse and includes estuaries, mudflats, mangroves, sea grass and kelp beds, reefs seamount communities and deep sea trenches. The Celtic-Biscay Shelf is characterized by strong interdependence of human impact and biological and climate cycles. The East Greenland and Iceland ecosystems are characterized by a seasonal ice cover and by marked fluctuations in salinity, temperature and phytoplankton, factors that contribute to variations of annual catches of cod and small pelagics. In the Barent Sea, the ice-coverage extends over one third to two thirds of the ecosystem and it varies considerably during the year and inter-annually. The shallow parts of the shelf in the Faroe Plateau are will mixed by extreme tidal currents and no stratification occurs during the summer.

With regard to catch composition, these ecosystems have in common high percentages of miscellaneous pelagic fishes which, for the North East Atlantic areas, are mostly due to peak catches of capelin in 1992-93, which markedly decreased in the latest years. Another fish group that shows relevant catches in all ecosystems of this cluster is cods, hakes, haddocks with the sole exception of the East Greenland Shelf that has been affected by the cod collapse of the early 1990s. In the other three northernmost Atlantic ecosystems total catches of the whole gadiform group have been rather stable during 10 years. In the two temperate ecosystems (New Zealand and the Celtic-Biscay, shelves), the second species group in terms of catches is respectively miscellaneous demersal fishes and clupeoids (herrings, sardines and anchovies). Three ecosystems (Norwegian Shelf in Atlantic Ocean of Northern hemisphere with sub arctic climate, Iberian coastal in Atlantic ocean of Northern hemisphere with temperate climate and Benguela current in Atlantic ocean of Southern hemisphere with temperate climate) in this cluster are all western boundary ecosystems. The Norwegian shelf and the Benguela current are characterized by a high productivity, where as the Iberian coastal is considered as moderately productive. The catch composition pattern is dominated by three groups, herrings, sardines and anchovies, miscellaneous pelagic fishes and cods, hakes and haddocs. Catches of Gadiformes are however very significant and important for their value, only in the Norwegian shelf and Benguela current areas.

The Norwegian Shelf ecosystem has a complex fishery history with concomitant influences of ecological anomalies, high fish mortality and early implementation of management measure. Its high productivity may be linked to the nutrient rich, cold arctic waters that characterize this ecosystem. Since early 1990s there has been a significant increase in *Clupea harengus* catches, the stock of which recovered after two decades of very low abundance.

The Iberian Coastal ecosystem's productivity is climate and upwelling driven. It is characterized by favourable factors for the production of clupeoids and other small pelagic fishes. Trends in catches by species groupings have been quite steady in recent years.

The Benguela current ecosystem is one of the most strongly wind-driven coastal upwelling systems known and it presents favourable conditions for a rich production of small pelagics of groups herrings, sardines, anchovies and miscellaneous pelagic fishes. Harvests are characterized by stock fluctuations according to the variations in the primary and secondary level productivity.

This single ecosystem cluster includes the Antarctic ecosystem which is unique both for its geographic and climatic characteristics. It is classified as a low productive ecosystem, a consequence of the extensive seasonal ice cover and extreme weather conditions. The ecological and biological characteristics of Antarctic marine species are also unique from a food-chain point of view in that it is peculiarly short and based almost entirely on krill, a key species crucial to the sustainability and production of all other fisheries.

The catches of shelf species of this ecosystem exhibits a prevalence of miscellaneous demersal catches and a much smaller percentage of coastal fishes. Catches of shelf species have been remarkably reduced in early 1990s.

Mass-Scale Fisheries

For successful development of fisheries in a given area the fishing operations must be created to fit the production system and economic conditions of that area. There are three basic principles that must be considered when such fishing operations are being created.

1. The fishery must be operated within an organically integrated net work of production, processing and distribution operations.
2. A balance must be maintained between resources, production and demand, and in the case of drastic change in any one of these three elements, the other two must be adjusted correspondingly.
3. Cold storage and/or local processing facilities must exist in order to allow the fisheries to accomodate for fluctuations either in the size of the catch or the amount of demand.

In order to establish fishery as a regional industry it is essential that the region offers a stable and mature social and economic base and that a high level of technological expertise also be available. In other words, for mass-catch fisheries, such as, surrounding net fishery, the most important requirement for a successful operation is the existence of well-established systems of raw fish transportation (distribution), processing and freezing, that will enable the fishermen to keep the pro-price level of their catches stable in times of large catches.

Surrounding net is a type of net fishing method that utilizes a purse net with a purse line or one without and the operating method can involve either one boat or two. Although a variety of fish are caught by this method including sardine family, horse mackerel and mackerel family, skip jack family, tuna family, yellow tail, dorado, Atka mackerel, it is the sardine, horse mackerel and mackerel that make up over 90 per cent of the catch while there is little difference between the types of fish caught by the different types of purse seine operations, those using 1 boat or 2 boats and those using a purse line or not using one, fishermen in different regions must choose the type of fish they will catch according to the season. At present overwhelming number of operations use a purse line.

During 1960 there was a change over from non-powered boats to powered ones in purse seine operations, and in 1970 there was change over from the 2-boat to the 1-boat fishing method. In the mean time net hauling become mechanized, synthetic net materials came to use and electronic equipment of various types became a part of fisherman's gear. As a result, first of all catching capability improved and secondly, labour need were greatly reduced. The main goal of these innovations was to cut down on labour costs and there by increase profits. For example a two boat operation using 39.9 ton boats require a work crew of 25-26 men. With new innovations, this number was soon reduced to 17-18 men. This reduction was possible primarily through the introduction of power blocks and side rollers.

There is another factor besides the desire to reduce labour, that contributed to the change in fishing methods. That was the fact that small sized fishing boats of the 5 to 9 ton classes were able to change over to fishing in off-shore waters when it became necessary due to changes in the migration patterns of fishes coming into the bay. Originally, small-sized boats were used for catching primarily horse mackerel in the central waters of the bay. When the horse mackerel schools stopped coming into the bay in sufficient numbers, the fishermen were forced to change over to catching primarily anchovy and mackerel. After five years, when spotted mackerel and sardine become the fishermen's main source of income, the fishing ground moved further off-shore.

Comparing the fish catching result and the profit rate between large (19.9 ton class boats) and small (9 ton class boats) boat operation, the former uses a large size net capable of catching 50 to 60 tonnes of fish per casting. While the later uses a much a smaller one capable of catching only 30 tonnes at best. The difference in physical productivity in shown in terms of a difference in annual catch value. But it is significant to note that the size of a boat results in little difference in profit rate. The crew can earn the same amount per annum in which ever fleet he may join.

The smaller boat can maintain the same rate of profit as the bigger one for the following reasons:

1. 19.9 ton class boats and larger are used exclusively for the purse seine fishing operation, while under 10 ton class boats can be used not only for purse seine, but also for small trawl net, boat seine etc. Therefore, an unexpected poor catch in purse seine can be made up for by the catches of other fishing nets.
2. A large size boat is operated with the main aim of increasing the amount of catch, while a small-size boat aims at catching prime fish species, rather than mere increase of whatever catch.
3. A small size boat is much more speedier (19 to 23 knots) than a large size boat (8 to 9 knots). It can go to a good fishing ground more quickly than a large size boat. It can also return port in time for the very profitable auction of the day.

In recent years, inspite of the increase of catch of mackerel and sardine, the consumer demand for these fish has failed to grow. There are several factors which can explain this lack of increase in consumer demand, such as:

- The improvement of living standards has brought a shift in tastes away from the mass catch fish.
- The growth of fish culture industry has brought an increase in demand for fish as feed material, thus upsetting the frame work of the market for mass-catch fish.
- Improved fishing methods have reduced the product value of mass-catch fish.

Marine product market is developed around the distribution of fresh fish. In fishery developed nations an increase in shipment of frozen fish and a decrease in shipments of fresh fish was noticed from mid 70s. Increased dealings in goods from out side the market, consisting mainly of imported marine frozen products and increased sales of standardized products such as, frozen processed foods lowered the demand of fresh mass-catch fish. Since the profit margin on mass-catch fish is small for the middle men and retailers, in the large city consuming area markets, where the basic marine product prices are determined, there has been a tendency to avoid mass-catch fish.

To establish a stable growing industry of mass-catch fish a few basic measures need to be taken.

- By making full use of freezing and refrigeration facilities, they will have to improve their ability to regulate their shipment of products to the market.
- They will have to send their products to smaller markets with the aid of cold-chain as a means to increase their sale.
- They will have to develop new processed foods using mass-catch fish in order to encourage the growth of the local processing industry (as sardine is used as raw material for a product like "denpu" in Japan, where powdered fish meat is boiled hard with sugar and soy.) In order to develop successful coastal fishery through the establishment of effective techniques for catching marine products with high market value, the first requirement is to develop the capability to catch sufficient amount of fish in proportion to the labour requirements of the fishing process chosen, and this process must be one that takes into consideration the traditional fishing gears of the given region and also the life cycle and movement habits of the fish to be caught.

The second requirement is that the use of this fishing gear and method be properly managed on a continuous basis by a commercial fishery structure within the fishing village and surrounding area and that the right social and economic conditions exist to provide for a sufficient income for the people involved in the fishery. For this the following social and economic conditions are necessary:

1. There must be sufficient resources of some salable type of marine product to support a given number of fishing families on a permanent basis.
2. There must be sufficiently large market for the marine products within reach of fishing village, and a marketing system or processing facilities must be available to the fishermen.

3. The fishing method must be one that is acceptable to the fishing community. For example, there must be sufficient manpower available to conduct the chosen fishing method, and a proper governing system must be created to prevent friction with fishermen involved in other types of fisheries.

Change in Mass-Catch Fish Production-Development of Commodities

The variations in boat seine fishing method are not a result solely of such natural factors as distribution of resources and geographical aspects of the fishing grounds. The variations also reflect the social and economic changes in the society supporting the fishing industry, and how the individual fisherman chooses to answer these different conditions.

Because sardine is a mass-catch type fishery it traditionally had to bear the handicap of low market prices. For this fact, the sardine fishery industry has always supported itself and continued to grow and develop by selling a major portion of the catch for non-edible use, while the minority has gone to edible use either as fresh or processed fish.

In addition, the sardine fishery has historically been the victim of fluctuations of good catch years to bad catch years. In the bad years fishermen have turned to catch a different type of fish or increased the variety of species being caught, increased the fishing grounds by moving further out to sea or introduced new fishing methods or else they reduced the scale of fishing operation to meet the change in conditions.

In order to understand the choices that the fishermen have made to change their fishing in the past, it is necessary to look at the history of the development of products for the market. In other words, we need to look at the commodity value of the marine products involved.

Although species caught under the one name “sardine” in many countries depending on the size and species there is a big differences in the way the products are used, resulting in considerable differences in whole sale prices for the various categories. While sardine are caught at all stages of their life cycle, it is the fry that are sold at the highest price, with fingerlings and half-grown sardines to be used for processed food being the next in value.

In Japan "sardines" are used in three major ways, (a) as fresh fish for human consumption, (b) as the material for processed foods, and (c) for uses other than human consumption. Of course the type of sardine that the fisherman chooses to make his primary catch or supplementary catch depends on the resource conditions and the fishing grounds available to him, but also another factor that has an extremely important impact on his decision is what sales systems or processing facilities are available to him. The business decisions that the fisherman makes are motivated on the following three factors:

1. Response to the demand.
2. Adjusting production methods to minimize costs and maximize profits.
3. Pursuit of highest overall income and lowest expense levels.

In Japan during 1960 there were two major changes in the demand for sardine family fishes.

1. With the development of yellowtail fish farming there was a sudden growth in the demand for sardine as fish feed.
2. The Japanese taste became more westernized resulting in a drop in demand of small dried anchovy for human consumption. The wide spread switch from small dried sardine as a traditional flavouring to the artificial flavouring by monosodium glutamate.

Development of Small Scale Fisheries

In the conference of the 19th FAO Indo-Pacific Fisheries Committee held in Kyoto, Japan on May 21-29, 1980, a symposium on the theme of "development of small scale fisheries" was held in which the following three topics were discussed:

1. The present condition and point at issue of small-scale fisheries, and a review of the existing plan,
2. The order of priority and methods of future actions, and
3. The role of the FAO and other corporations and foundations in development.

As marked event, many nations delivered their opinions that the promotion of fisheries and the problem of over population must be considered as social problems, and they emphasized the importance of the preparation and expansion of fundamental social overhead capital.

The symposium adopted the following suggestions to the government of member nations:

1. As a means of developing small-scale fisheries, we must give the same level of assistance to these fisheries as to other economic activities to increase income and raise standard of living.
2. In planning for development, we must give due consideration to the connection between the sector of small-scale fisheries and that of large-scale fisheries. The development of fisheries must be integrated into the development of society in general.
3. By paying attention to the results of census and various other statistics, plans must be made on the basis of careful and detailed analysis.
4. The government of each nation must remove friction within the small-scale fisheries and between small-scale and large-scale fisheries as soon as possible. We must tackle the problem of making laws on fishery rights which are appropriate from the view point of limitation of resources and conservation of natural ecosystem and acceptable to both fisheries.
5. Women are playing an important role in fisheries. Therefore, we must give thought to the organization and education of women in the society of the fishing village.
6. There is no "standard" method of solution. We hope to forward these projects by principle of solving problems case by case under the co-operation of administrative organs, research institutions and fishermen.
7. In addition to the effort to increase the standard of living, social infrastructures, social services and environment for fishermen must be set up in sufficient quantity.
8. We must pay attention to the fact that a shortage of credit has prevented the development of small-scale fisheries.

9. As regards the marketing of fishes, we must devise means for developing new co-operative associations to protect the interest of small producers while closely examining the past examples of failure in the activities of co-operative associations.
10. Coping with the increase in cost of fossil fuel, we must reconsider measures to eliminate waste in the whole process from fishing to processing and distribution.
11. We must concentrate our efforts on the development of small-scale farming.

Foods for Man, Live Stock and Industrial Materials

Sea weeds are used as foods in two different ways, they are either eaten as they are, or agar (polysaccharide) is extracted from the body of the sea weed and supplied to the food industry.

Representative of useful sea weeds used directly for food are the three kinds of green, brown and red algae. Of these, *Porphyra tenera, Undaria pinnatifida, Laminaria japonica, Enteromorpha linza* and *Hizikia fusiforme* are the ones whose production is industrialized and is distributed as commodities.

Forms used for food as well as the processing form are divided roughly into the following three types:

1. Raw sea weeds, sea weed salad or spices for dish (*e.g.*, garnishing served with sliced raw fish).
2. Dried products–cooked and seasoned food and ingredients or miso soup (the same utilization as that of vegetables).
3. Processed for food–seasoned laver, Tsukudani (food boiled in soy sauce) condiments, confectionery, etc.

Sea weeds are also used for purposes other than food.

Live Stock Feed

Dried sea weeds contain many ingredients and calory sufficient for feed. The total digestable nutrient content for live stock is estimated to be about 33-35 per cent. For dairy cattle and beef cattle a portion of non-concentrated feed is substituted with sea seeds. For pigs and chickens, concentrated feed is mixed with small amount of sea weeds. Suitable species are *Eiscnia bicyclis, Sargassum ringgoldianum, Zostera marina, Ulva pertusa, Gracilaria verrucose* etc.

Fertilizer for Agriculture

Since old times, sea weeds and water side aquatic plants have been used for fertilizer in many countries of the world. Sea weeds are of great value as a source of potassium and they are very effective as fertilizer because of nitrogen and other organic components contained in them. Besides dried sea weeds mixed with soil by ploughing have a tilling function.

Raw sea weeds are used for fertilizer by making compost, after drying or by reducing them to ashes.

Suitable species are mainly brown algae, such as, *Eisena, bicyclis, Ecklonia cava, Sargassum fulvellum.*

Industrial Paste

This is used for textile and plastering. By making the best use of the characteristics (solubility, penetrability, viscosity, adhesive power etc.) of each sea weed, the paste is used for many purposes. Mainly red algae such as *Gloipeltis tenax, Chondrus ocellatus, Neodilsia yendoana* and *Gigartina tenella* are used as raw materials.

Alginic Acid Industry

Alginic acid is a component constituting the external layer of the cell wall of brown algae and it is an elastic substance which is educed when kelps are treated with diluted, alkaline solution and acid is added to its filtrate. Alginic acid combines with metals and other various substances to form salts of different properties. Alginic acid is used for various purposes *e.g.*, as a stabilizer for food, viscosity reinforcing agent, glue, hard-water softener, purifying agent, dental molding material etc. Alginic acid is obtained mainly from *Eisenia bicyclis* and *Ecklonia cava* and various kelps.

For Medical Use

It is known that some sea weeds, such as *Digenea simplex* (red algae), Codium (green algae) and *Sargassum thunfragile* (brown algae) are efficacious as an excellent for round worms. These sea weeds are all used internally by making decoction of the dried product or by making a powdered drug.

Agar has a characteristic of solidifying at room temperature even when in low concentration. Besides, it is not dissolved by organic solvents, and it is highly resistant to bacteria unlike other carbohydrates. As suggested by these characteristics, agar is an indispensable material for making culture medium for bacteria.

Coastal Prime Fish in Relation to Urban Economy

According to the price, fishes are classified into "prime fish," "medium grade fish" and "popular fish". The criterion for a "prime fish" can not be explained simply by meat quality, flavour of the fish or the tastes of the people. In traditional cooking it is important to prepare dishes by making the most of the original taste of the material. As a main criterion in judging the value of perishable foods, the degree of freshness has been considered important. Of course the difference in taste, like the degree of freshness is also an important criterion for measuring the food value, but in the fish market, the degree of freshness of fish is considered to be the major factor in determining the price.

An experiment showed that when live fish is stored in ice after being killed by a special method (the special method to keep freshness, live fish is instantaneously killed by destroying the medulla oblongata, after which the duration of death rigor of the fish meat is extended), the time the freshness suitable for sliced raw fish (sashimi) in a fashionable restuarant can be maintained is 12 days for red sea bream and flounder, 6 days for yellowtail, 2-3 days for skip jack and 0.5 day for cod fish.

In the case of red sea bream, the commodity value in the market is clearly distinguished between live fish and ordinary fresh sea bream. Furthermore in the case of all fresh fishes, the price is strictly set according to their appearance and the degree of damage (due to fishing method). On the other hand, due to the fact that in recent years it has become possible to transport highly fresh foods to consumers by the development of freezing and cold storage techniques and the commodity value of some fish has increased with the decrease in catch, so the value of some fishes, such as horse mackerel, squid and saury pike has transferred from "popular fish" to "medium grade fish" or from "medium grade fish" to "prime fish".

The above mentioned price system for marine products has been formed based on longtime dietary habits of the Japanese people. Therefore, distribution system that is quite different from that of the advanced nations of northern Europe, which have unique processing methods for mass catch fishes such as sardine and herring has been developed in the Japanese fishing industry.

In Japan, it is a basic pattern that marine products are distributed through two markets, the market in the producing area and the whole sale market in the consuming area, where the circulating function is carried out and at the same time the appropriate price is formed by auction.

This market system has been developed in order (i) to distribute fishes in fresh form basically, (ii) to gather many kinds of small quantity catches from the various coastal areas and (iii) to make it possible to purchase those diversified catches in small quantities in the consuming area. In this way a practical and efficient market system has been completed.

As regards this market system which moves fish from the market in the producing area to the wholesale market in the consuming area, an important point to note is that the standard prices of marine products are not fixed in the market in the producing area, but in the market in consuming area. By auction at both markets, the market price is basically determined based on the daily demand–supply relation independent of production cost, and the standard price with nation wide dominating power is determined mainly by the wholesalers in the consuming area who can collect a wide variety of information and can meet the vast demand in the central wholesale markets of large cities where a large amount of diverse fishes are gathered. On the contrary, the wholesalers in the producing area decides his bidding price by judging circumstantially "what will be the price of this in the consuming area".

Generally the price of marine products fluctuate seasonally with the balance of "demand–supply" depending on the degree of fishing activity. A typical case of this fluctuation in prices is seen in mass-catch fishes, such as, sardine, saury pike and mackerel. However fishes ranked as "prime fish" show rather different patterns of fluctuation in prices.

Although drastic fluctuations caused by seasonal differences in the size of catches landed, the price of prime fresh fishes show high price stabilization" in the sellers market.

The income of the people rose along with the growth of economy, resulting in rapid improvement in the dietary life. As regards the demand for marine products too, the purchase of luxury "prime fishes" increased. Moreover, as the increase of income prevailed from large cities to local towns and farm villages, the prices of some fishes settled at higher level. This is clearly seen in the movements of the quantity and unit price of tuna, yellowtail, sea bream and shrimp purchased by the ordinary house hold.

The demand of prime fish increases sharply around the festival days, wedding season. Therefore the market price of fresh fish in the producing area shows seasonal fluctuations, showing a peak high price \at the specific demand seasons. In the markets of consuming areas, since large quantity of fish are concentratively brought in during the high price season, fluctuation in prices in rather slight throughout the year as compared with that in the markets of producing areas.

Conditions Necessary for Conducting Fisheries on a Commercial Basis

To develop commercial fishery, the following necessary conditions must be examined. First the degree of difficulty and economic value of a fishery development must be judged by considering (1) the quantity of resources and (2) location of the fishing grounds and distance to the markets. After that (3) size of fishing boats, (4) fishing methods and fishing gear and methods of storage and transportation must be concretely investigated.

For capitalized fisheries using large fishing boats require large-scale fishing grounds of the target species. This should be supported by the facts that the sea floor composition of fishing grounds are suitable for trawling or for the operation of active fishing gears together with a transporation system to markets have been made ready for use.

Next is the coastal waters if the fisheries of the target species is existing where small fishing boats are operating for the said fisheries. Even if the catch is small, but the commodity value may be high because of fixed consuming markets near the fishing ports. Fishermen engaged in the fishery can keep a steady fishery income because the management of fishery is conducted by family labour and fishing is conducted throughout the year by combining a variety of fishing methods.

The coastal fishermen should not specialize only in a single kind of fishing. Equipped with fishing gear and the techniques for conducting several types of fishing they should go fishing after selecting the type of fishing which seems to be most profitable in consideration of the migrating conditions of fishes in each season or the daily fishing conditions and market fish prices. Fishermen should not consider themselves tied to any one kind of fishing. They should always try to add profitable new types of fishing to their own repertoire, and also they will abandon types of fishing which have become unprofitable due to changes in the environment and fishing methods.

The fishery system that is to be carried on in some parts of the country will also be determined by social conditions, such as the established fisheries system and local custom. The fishing schedule of each fisherman will also depend on the condition of his local community, that is, the fishing village.

The form that coastal fishing will take in any given area is not merely the result of the isolated efforts of the individual fishermen in that area, but rather there is an interrelation between the individual fisherman and the regulatory body in his community that determines what use will be made of all the available fishery resources.

The "fishery system" of a given area means in short what fish are caught and how the catches are exchanged for money. The system comes into existence, changes and develops within a social and economic environment, with the primary objective of the individual fisherman and his local co-operative body being to ensure steady production and an economically stable income.

Characteristics of Fishery as an Industry

One of the main characteristics of fishery as an industry is the cyclical fluctuations in production as a result of alternating good and poor annual catches. Some species are particularly susceptible to overfishing that can deplete the fishery resources, while in other cases large scale changes in the sea environment can also cause drastic changes in the resources. With fishery and marine food processing developed into large industries as they are today, such changes in marine resources can have a major effect on human society.

Research in two areas of resource conservation and stabilization of fishery industry economics has taken on an extremely important role today. Depending on the species and life environment, each fishery product has its own unique resources structure and characteristics. Furthermore the amount of consumption and commercial value of marine products varies from country to country and culture to culture, meaning that problem of marine resources must be approached from a multifaceted overall view point that includes scientific, social, economic, political and cultural considerations. When this study is undertaken, it reveals a grand-scale drama of mutual interaction between productivity of nature and the productivity of man.

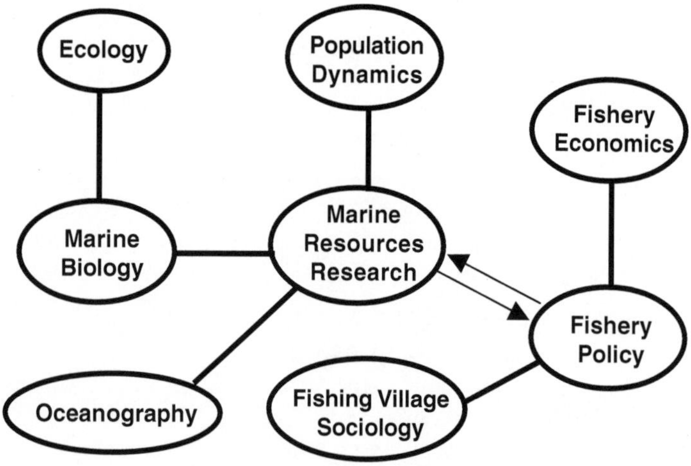

Characteristics of 3-5 Tons Class Fishing House Holds

The economic effect resulting from the increase in fishing boat size can be seen from two aspects, "productivity" and "composition of income." The labour productivity by the fishing boat tonnage class (namely, the value of annual gross catch divided by total number of workers) in Japan showed that the productivity sharply rises with the class above 3 tons. Although it is natural that the catch capacity increases as the size of the boat becomes larger, but there exists a clear difference between 3-5 ton class and 5-10 ton class. That is for 5-10 fishing boat, the required crew is 3-4 persons or more depending on the fishing type, requiring a hired labour force of at least 1-2 persons, whereas 3-5 ton fishing boats can be sufficiently operated with a labour force of 1-2 persons for all type of fishing, thus enabling them to be managed by one family. What made family operation possible was the mechanization of fishing operations and improvement of navigation by such means as remote control.

The composition of income of the 3-5 ton class fishing house hold indicates a clear qualitative difference compared to class under 3 tons. With the 3-5 ton class, the percentage that fishery income occupies in the total household income starts to exceed 50 per cent. This fact shows that the household's dependency on fisheries becomes higher and that the number of full time fishing household increases.

Fish Production

Fish production in India has, by and large, being registering a steady increase under the successive National Five Year Plans. In a period of 20 years, between 1951-70 India's fish production increased from 0.75 to 1.75 million tonnes, an increase in 113 per cent. Fish production both inland and marine sector during the year 1951, was of the order of 0.75 million tonnes. During 1951-55, the average annual production increased to 0.796 million tonnes. During the next five years (1956-60), the annual production rose further to 1.06 million tonnes. Subsequently (1961-66) there had been an increase in the average annual production to 1.126 million tonnes. During (1965-66), the annual production reached a level of 1.45 million tonnes. The production of 1985-86 was 2.87 million tonnes, while in 1990-91, it increased to 3.759 million tonnes and during 1994-95, 4.780 million tonnes.

The production of inland fish in India has nearly been doubled during the last 10 years. In 1985-86, the inland fish production had shown a small decrease of 1.1 million tonnes compared to production in 1984-85 (1.2 million tonnes) and 1.5 million tonnes in 1990-91. Inland fisheries development is, however, a slow process, compared to the capture fisheries expansion in the sea.

Marine Fish Production in India

India's contribution has been on an average 2.0 million tonnes annually. The rate of expansion of marine fisheries in India has been slow due to continued adherence to traditional fishing methods and delay in extending fishing activities to deeper waters. India contributes to more than 40 per cent of the landings of the Indian Ocean. The Indian Ocean produced 5.6 million tonnes of fish in 1988 (0.09 tonnes per square km.) as against 24.14 million tonnes from the Atlantic (0.22 tonnes per sq. km.) and 53.11 million tonnes from the Pacific Ocean (0.28 tonnes per sq. km.)

In 1990-91, India caught 2.3 million tonnes of fish which compares favourably with 1.6 million tonnes in 1983-84, 1.7 million tonnes in 1984-85 and 1.9 million tonnes in 1985-86. The West Coast accounts for 71 per cent of the total fish landings of the country and the east coast 29 per cent.

Broadly stated, the states of Kerala, Maharashtra, Tamil Nadu, Mysore, Gujrat and Andhra Pradesh contribute about 98 per cent of the total marine fish landings in the country. The composition of marine fisheries during 1989 were as below.

Name of Fish	*Percentage of Total*
Oil sardines	11.08
Prawns	10.40
Bombay ducks	6.05
Mackerel	9.90
Pomfret	2.69
Silver bellies	2.39
Ribbon fish	3.33
Sciaenids	10.62
Anchovies	3.42
Other sardines	3.42
Catfishes	3.31
Sharks and rays	2.90
Others	30.49

Production, Efficiency and Productivity of Various Fisheries and Fishing Methods

The performance of fishermen and fisheries are influenced by a number of factors; technology adopted, skill of fishermen, natural conditions like the productivity of the waters, weather conditions, accessibility of the resources, biological and ecological factors influencing the growth of the stock, fishing depth and degree of aggregation or dispersion of fish, the capital intensity and sophistication of the technology. In monetary terms efficiency and productivity are, in addition, determined by fish prices. High prices can compensate for low productivity as for example, in shrimp culture in lobster fishing.

In capture fisheries, the reproductive capacity of the fish stocks will set a limit to the long term sustainable yield and will affect the relationship between fishing effort and catch. That is to say efficiency and productivity will be affected if in the course of the development of a particular fishery more and more fishing units are added to the fleet and catches decline as more fishing units exploit the same limited resource base. This may in some cases be partly compensated by higher prices, if there is no ready alternative supplies, but there is nevertheless a size of fleet which represents the optimum investment and this is often exceeded. Over-investment in fishing vessels may increase employment opportunities for fisherman at the cost of reduced earnings. More often there may be a need to limit or even reduce the number of operational fishing units, or to divert some of the fishing units to fish on other, less heavily exploited resources.

In the culture fisheries, however, the productive capacity of the fish stocks will set a limit to the sustainable yield and will affect the relationship between carrying capacity and stocking density. The efficiency and productivity will be affected if in the course of culture more and more seeds (Fin fish and Shell fish) are added to the culture system and productions decline due to improper growth and mortality due to infection and diseases. In case of shrimp however, lesser growth and production can be partly compensated by higher price in the export market; but over-investment in seed, feed, aeration and daily water exchange, though increase subsidiary employment opportunities at the cost of reduced earning to the fish farmers. There may be a need to limit the stock according to the carrying capacity of water area for proper growth and higher production even at the cost of reduced operational costs (seed, feed, aeration).

In fisheries science, the most commonly used measure of the efficiency of the fishing operation is the catch per unit of fishing effort. "Fishing effort", in the terminology of the dynamics of exploited fish population, is what causes fishing mortality; the fraction of the fish stock removed by fishing each year. Moreover there are no simple and agreed methods of measuring the fishing effort of the types of craft-gear combinations available, except the trawlers, nor of comparing the fishing effort of different combinations fishing the same stock.

Common Property Concept

The coast, including mangroves, creeks, rivers and estuaries have hitherto considered common property of the coastal communities, who have maintained and utilised them, for years. Marine resources have also been traditionally treated as common property. Fishing communities have caught fish for themselves selling surplus catch to local and domestic markets. The FAO's definition of aquaculture, which includes the statement "Farming also implies individual or corporate ownership of the stock being cultivated" turns these common property resources into private property. However fish and shrimp, unlike live stock, are not breeding in enclosed environment strictly. Wild seed caught from breeding grounds in the mangroves, creeks and estuaries is dispersed in aquaculture ponds, which does not mean corporate or individual ownership.

Production Projections

In the case of fisheries, changing economics of operations, over exploitation of inshore stocks, availability of new stocks in off shore and deep sea areas and supply of large trawlers to operate in those areas, heavy investments involved in reclamation of water areas for inland fish culture are other factors which have their bearings on fish production. In fact, in developed countries, financial returns to fishing are not high in relation to the risky and hazardous nature of the industry and with rising living standards. It is becoming increasingly difficult to attract labour into the industry. Presuming that such situation may not come in India in the next decade, the projection of production was made for the period by Dayal (1973) as below.

Taking the average production for the years 1969 to 1971 as base, the estimates of the production projected (actual) are given below:

Year	*Actual Projected Production in Lakh Tonnes*
1973-74	21.2
1975-76	-
1978-79	28.00
1980-81	30.50
1995-96	49.50
1996-97	51.40
2001-2002	63.67

The production potential for marine fish around Indian coasts have been estimated as 38 lakh tonnes. The potential for inland fish in the country is estimated as 40 lakh tonnes. Thus the projection of production as 30.50 lakh tonnes in the year 1980-81 in only 40 per cent of the total potential of 78 lakh tonnes. There is enough scope for increasing production in subsequent years.

India may have surplus of fish meeting domestic demand in future years. In the last year of Fourth Five year Plan India produced 21.2 lakh tonnes of fish, where as the domestic requirements was only 18.7 lakh tonnes. In the last year of Fifth Five Year Plan, India produced 28 lakh tonnes of fish as against the domestic requirement of 23.5 lakh tonnes, leaving a surplus of 4.5 lakh tonnes for export or to be put to other uses like fish meal etc. New hopes of expanding export to countries like Germany, U.K., Spain and France lies before India.

Indian Institute of Foreign Trade has estimated that if planned efforts are made, India would be able to export 4.5 lakh tonnes of surplus fish after satisfying the domestic demand.

Extension: A Linkage of Increased Production

Extension is often considered as a non-productive activity. The term extension has its origin in agriculture. It was first used in connection with improvements in agriculture in the mid–nineteenth century. Organized extension activities in fisheries is a post world war II phenomenon. It started with the mechanization of traditional craft and the introduction of new harvesting technologies under bilateral or multilateral aid programmes. A rather broad and general definition of extension can be considered is "all organized communication efforts by which an individual or agency tries to bring about changes in knowledge, attitude, skill and/or behaviour of a client population, in order to reach one or more objectives that have been established within the framework of an overall development policy". In practice, extension usually refers to the provision of technical advice to the fishermen/fish farmers, either to improve productivity of existing fishing units/fish farms, or to use new production techniques. The implicit assumption in extension is that, with increased efficiency, higher levels of production can be achieved and that the incremental output can be successfully marketed, leading to better marketing incomes and therefore, a better quality of life for the fisherfolk. Less frequently extension also refers to advice on issues related to resource management and fisherfolk development.

Coasts and Returns

Extension services are provided by both public and private agencies. Examining the costs and returns of investments in extension made by the state, the study points out that the expenditure on research and development, the cost of introducing the innovations and the expenditure bourne by the fishermen while putting the innovations into practice are the main costs. The returns are higher income to the fishermen, additional tax revenues, foreign exchange earnings in the case of exportable species, better supply of protein to the people, and sometimes, more employment opportunities.

Quantification of extra returns attributed to public expenditure on extension is however difficult. Since the increase in production on account of innovation depends on technical, economic, social, cultural and political factors, it is almost impossible to isolate the role of extension in increasing production. But in certain situations, it is possible to show a positive relationship between extension and production. For example, if the gap between the actual level of production and the Maximum Sustainable Yield (MSY) is sufficiently wide, and if product and credit markets are adequately responsive, the return on investment in extension can be high. On the other hand, the relationship between extension and production can be negative if the resource conditions are unfavourable, the markets non-responsive, the fishermen indifferent and the extension network inefficient.

The catalytic role of extension has to be accepted in good faith assuming that there are direct and indirect returns to investment in extension areas where there is a positive atmosphere for such an intervention.

In situations where extension can facilitate speedier diffusion of innovations, the efficiency of its organization can have a bearing on the rate of diffusion. This would depend on;

1. The coverage of fishermen;
2. The type of fishermen reached by the extension services;
3. The sensitivity of the approach; and
4. The motivation of the extension personnel.

The intensiveness of the coverage of fishermen is important for the effective performance of the extension services. The fishermen–to–extension agent ratio has to be realistic and according to the needs of the clientele, communication, staff availability and the possibility of mass media complementing extension field activities.

Since adoption of new, capital intensive fishing techniques often involve higher levels of risk and ones who take the initiative are those who can afford to take the risk often the elite in the fishing community. Unlike in agriculture it cannot be assumed that the poor fishermen will, in the long run, adopt the innovations, because the sustainability of innovations with a strong efficiency bias is severely limited in tropical waters.

Instead of resorting to a 'top-down approach, a bottom-up one could be resorted to in situations where the fishermen themselves articulate the need for a specific type of technology, provided it is not disruptive of the labour market and destructive of the resource base. In such cases, the clientele. need not be from the village elite, they could be from the poor sections. Development and introduction of plywood boats by the Intermediate Technology Development Group and the Centre for Appropriate Technology in south-west

India, in response to a demand from *Kattumaram* fishermen for a better craft is a good example of how a client–sensitive approach can benefit poor fishermen.

The type of extension agent with whom the fishermen comes into contact is important. Extension agents should be highly motivated. They should be well paid, provided with adequate transport and should be adequately trained to propagate the desired innovations as well as to effectively understand the needs of the fishermen. The extension agent should become the effective link between fishermen and research centres, particularly in the process of making possible adaptive research that off-neglected area in technology transfer.

Four methods usually used in evaluating economic impact of extension on production:

1. Measuring the internal rate of return;
2. Correlating the expenditure on extension and the increase of production/productivity;
3. Comparing out put (between areas with or without extension input or, within the same area before and after the introduction of extension services); and
4. Assessing the microeconomic production function.

A particular extension programme considered a failure by an economist could well show success in a time frame broader than the one used by decision–makers. An American rural sociology study on hybrid corn adoption showed that the majority of adoptions took about twelve years. This is true of many technological adoptions in fisheries too. The time frame, therefore should be carefully chosen in accordance with the social and cultural characteristics of the fishermen in mind, while undertaking an impact analysis of fisheries extension.

Prospect of Aquaculture and Propagation

Aquaculture is a means to obtain a greater harvest than in natural environments, by controlled feeding in a closed ecosystem.

Recent remarkable developments in aquaculture include "sea ranching" or "culture based fisheries" beyond the traditional practice of fish culture in a closed habitat. Some of these new activities are already being practiced on a commercial basis. The activities enable increased production by stabilizing and facilitating the natural reproduction process in the ecosystem of the open sea for certain useful marine fish, resources, such as, prawns, sea breams. In other words, the aim is to influence, by means of artificial breeding and recruitment the ecosystem in a larger sea area in order to promote increased reproduction of high value fishes. Sea–ranching fisheries in coastal shallow waters may involve, as in the case of agriculture, cultivation of sea bottom, sowing seeds, elimination of predators, and fertilization or feeding to ensure higher productivity than in natural environments. In the course of time, precise plans for "sea–ranching fisheries" for warm water areas, cold water areas as well as for tropical waters will be drawn up.

Global Aquaculture Production

Aquaculture contribution to global fisheries production continues to grow, increasing from 5.3 per cent in 1970 to 32.2 per cent of total fisheries production by weight in 2000. Total aquaculture production in 2000 was 45.71 million metric tonnes (mmt) by weight and valued at 56.57 thousand million dollars, with increased production by 6.3 per cent by weight and 4.8 per cent by value since 1999. Over half of the total global aquaculture production in 2000 was in the form of finfish (23.07 mmt or 50.4 per cent of total production) followed by molluscs (10.73 mmt or 23.5 per cent), aquatic plants 10.13 mmt or 22.2 per cent) crustaceans (1.65 mmt or 3.6 per cent), amphibians and reptiles (100271 metric tonnes (mt) and miscellaneous aquatic invertebrates (36965 mt or 0.08 per cent). Although crustaceans represent only 3.6 per cent of total production by weight, they comprised 16.6 per cent of total global aquaculture by value in 2000.

In 2000, the bulk of global production was obtained from 210 different farmed aquatic animal and plant species. These include 131 finfish species, 42 molluscan, 27 crustaceans, 8 plant and 2 amphibian and reptile species. The large number of species cultivated reflects the wide range of potential candidate species available within different countries and regions of the world and the wide variety of production systems employed by farmers. Over half (54.9 per cent) of global aquaculture production originated from

Sea Food Processing Industry, Thailand

marine or brackish coastal waters in 2000, as compared with 45.1 per cent from fresh water aquaculture. The mean annual per cent rate (APR-average annual compounded growth rate in per cent) between 1970-2000 period was highest for fresh water aquaculture production (9.7 per cent) closely followed by brackish water production (8.4 per cent) and mariculture (8.3 per cent). The main species groups reared in fresh water were finfish (97.7 per cent). High value crustaceans and finfish predominated in brackish water (50.5 per cent and 42.7 per cent respectively) and molluscs and aquatic plants in marine waters (46.1 per cent and 44.0 per cent respectively).

Inland fresh water species dominated global finfish aquaculture production in 2000 (10.80 mmt or 85.8 per cent of total finfish production) followed by diadromous species (2.26 mmt or 9.8 per cent) and marine species (1.01 mmt or 4.4 per cent). Aquaculture currently provides 73.7 per cent, 65.3 per cent and 1.4 per cent of total global landings of fresh water finfish species, salmonid diadromous species and marine finfish species respectively. The observed growth rates of these different groups were very similar, the average APR (1970-2000 period) being 9.9 per cent for fresh water species, 10.6 per cent for marine species and 10.6 per cent for salmonid species.

The major finfish groups and species cultivated in 2000 is given below:

Freshwater Species

1. Cyprinids–15707109 mt., valued at 15251525100 dollars.
2. Tilapia–1265780 mt., valued at 1706538200 dollars.
3. Catfish–421709 mt., valued at 655419500 dollars.

Diadromous Species

1. Salmonids–1533824 mt., valued at 4875552400 dollars.
2. Milkfish–461857 mt., valued at 715091100 dollars.
3. Eel–232815 mt., valued at 975005700 dollars.

Marine Fishes

1009663 mt, valued at 4072151600 dollars.

It is fact that the top five cultivated species were cyprinids, representing over half of the total global finfish aquaculture production in 2000. However, it is important to note, that the production of silver carp (*Hypophthalmichthys molitrix*) and big head carp (*Aristichthys nobilis*), both key filter–feeding species has declined significantly during recent years compared with other cyprinid species, possibly due to low commercial return.

Depending on the market economy farmers in 2000 produced 62.0 per cent omnivorous/herbivorous species (94.3 per cent fresh water grass carp, common carp, crucian carp, Nile tilapia, rohu, mrigal, white Amur, bream and channel catfish); 25.0 per cent, filter feeding species (100 per cent fresh water silver carp, big head carp, and catla) and 13.0 per cent carnivorous species (68 per cent sea and brackish water species, like Atlantic salmon, rinbow trout, Japanese eel, black carp, Japanese Amber jack, coho salmon and mandarin fish). Although carnivorous species represented only 13.0 per cent of total global finfish production by weight in 2000, they comprised 34.3 per cent of total production by value, the majority of carnivorous finfish species having considerably higher unit market values than their filter-feeding or more omnivorous counterparts.

Marine shrimp production in 2000 dominated among crustacean aquaculture reaching 1087111 mt (66 per cent of global aquaculture production among crustacean and valued at

Growout Pond for Tiger Prawn, Kantai, West Bengal

Harvest ot Tiger Prawn from Pond Culture Facilities

Packing Tiger Prawn with Ice for Export

6880068900 dollars (73.4 per cent) of total value. Aquaculture currently provides just over a quarter (26.1 per cent) of total global shrimp landings. The main cultivated species are the giant tiger prawn (*Penaeus monodon*), the fleshy prawn (P. *chinensis*) and the white leg shrimp (*P. vannamei*) accounting for over 86 per cent of total shrimp aquaculture production in 2000. Although the giant tiger prawn only ranked 20th by weight in terms of global aquaculture production in 2000, but it ranked first by value at 4046751000 dollars. In terms of expansion, shrimp production has decreased to more modest levels over the last decade (averaging 5 per cent) as compared to the double digit growth rates observed during the seventies (23 per cent) and eighties (25 per cent).

Other crustaceans cultured in 2000 included fresh water crustaceans (386185 mt or 23.4 per cent of global crustacean production) and sea spiders and crabs (140256 mt or 8.5 per cent) of global production. The recent appearance and rapid growth of Chinese river crab (*Eriocheir sinensis)* with reported production increasing from zero in 1998 to 232391 mt in 2000. Impressive growth rates have also been observed for the giant river prawn (*Macrobrachium rosenbergii*) with production reaching 118501 mt in 2000.

Total global mollusc production in 2000 topped 10.7 mmt (an increase of 5.8 per cent from the previous year) and was valued at 9496615000 dollars. The Pacific cupped oyster (*Crassostrea gigas*) was the second most widely cultivated farmed aquatic species by weight at 3944042 mt and represented over 36 per cent of total mollusc aquaculture production in 2000. Other major cultivated molluscan species in 2000 included the Japanese carpet shell, (1693 thousand metric tonnes (tmt), the yesso scallop, (1132 tmt) blue mussel (458 tmt) and the blood cockle (319 tmt).

The growth of the sector has been steadily increasing, averaging 5.6 per cent per year in the seventies, 7 per cent in the eighties and 11.5 per cent in the nineties.

Farmed aquatic plant production in 2000 reached 10.1 mmt (an increase of 6.1 per cent from the previous year) valued at 5607835000 dollars. The Japanese kelp remained the top farmed aquatic species by weight at 4580056 mt and represented 45.2 per cent of total plant aquaculture production in 2000. Other major aquatic plant species produced in 2000 included laver, (1011 tmt) cotoni (605 tmt) and wakame (311 tmt). The growth of this sector is relatively steady averaging of 8.2 per cent per year from 1970 to 2000.

Regional Aquaculture Production

In 2000, 92.2 per cent of total aquaculture production was obtained from developing countries (41.68 mmt), and 83.9 per cent within Low Income Food Deficit Countries (LIFDCs, 38.35 mmt), which include in *Africa*-Burkina Faso, Burundi, Cameroon, Central African Republic, Cango Democratic Republic, Congo Republic, Core dlvoire, Egypt, Ghana, Kenya, Lesotho, Liberia, Madagascar, Malawi, Mali, Morocco, Niger, Nigeria, Rwanda, Senegal, Sierra Leone, Sudan, Swaziland, Tanzania, Togo and Zambia, *Americas*-Bolivia, Cuba, Ecudor, Guatemala, Honduras and Nicarragua; *Asia*-Armenia, Azerbaijan, Bangladesh, Bhutan, Cambodia, China, Georgia, India, Indonesia, Korea DPR, Kyrgzstan, Laos, Nepal, Pakistan, Philippines, Sri Lanka, Tajikistan, Turkmenistan and Uzbekistan; *Europe*–Albania and Macedonia; and *Ocenia*-Kiribati, Papua New Guinea and Solomon islands.

In the last decade, the aquaculture sector within LIFDCs has been growing over seven times faster (over the period 1970 to 2000) than the aquaculture sector within developed countries (total production 4.03 mmt in 2000). The bulk (93 per cent) of total finfish production within developing countries in 2000 was contributed by omnivorous, herbivorous and filter-feeding fish species. In contrast, 73.8 per cent of the total fish production within developed countries in 2000 was due to the culture of carnivorous fish species.

By region, over 91.3 per cent of the total aquaculture production by weight was obtained within the Asian region (41.72 mmt) in 2000 followed by Europe (2.03 mmt or 4.4 per cent), Latin America and the Caribbean (0.87 mmt or 1.9 per cent), North America (0.55 mmt or 1.2 per cent), Africa (0.40 mmt or 0.9 per cent) and Oceania (0.14 mmt or 0.3 per cent). The top nine aquaculture-producing countries were located within the Asian region and included China (32.44 mmt or 71.0 per cent of total global aquaculture production), India(2.09 mmt), Japan (1.29 mmt), Philippines (1.04 mmt), Indonesia (994 tmt), Thailand (707 tmt), Republic of Korea (698 tmt), Bangladesh (657 tmt) and Vietnam (525 tmt).

The second top ten country aquaculture producers by weight in 2000 included Norway (488 tmt), Korea DPR (698 tmt), the USA (428 tmt) Chile (425 tmt), Egypt (340 tmt), Spain (312 tmt), France (268 tmt), Taiwan POC (256 tmt), Italy (216 tmt), Malaysia(168 tmt) and Brazil (153 tmt).

Asia

The total reported aquaculture production within the Asian region has increased 14-fold by weight, from 2811549 mt in 1970 (78.5 per cent of total global production) to 41724469 mt in 2000 (representing 91.3 per cent of total global production). The total number of reported cultured species within the region has increased from 55 in 1970 to 107 in 2000. The top cultivated species in 2000 included Japanese kelp (4580 tmt or 11.0 per cent), Pacific cupped oyster (3741 tmt or 9.0 per cent), Silver carp (3405 tmt or 8.2 per cent), Grass carp (3379 tmt or 8.1 per cent), Common carp (2499 tmt or 6.0 per cent), Japanese carpetshell (1635 tmt or 3.9 per cent), Big head carp (1631 tmt or 3.9 per cent), Crucian carp (1379 tmt or 3.3 per cent) Yesso scallop (1133 tmt or 2.7 per cent) and Laver (1011 tmt or 2.4 per cent).

By value, aquaculture production in Asian region increased over four-fold from 9.4 thousand million dollars in 1984 to 46.3 thousand million dollars in 2000 (representing 82.1 per cent of total global aquaculture production by value). The main species group of commercially valuable in 2000 included Fin fish (23.7

thousand million dollars or 51.2 per cent), Molluscs (8.3 thousand million dollars or 18.0 per cent), Crustaceans (8.3 thousand million dollars or 17.9 per cent), Aquatic plants (5.6 thousand million dollars or 12.0 per cent), Amphibians/reptiles (0.39 thousand million dollars or 0.8 per cent) and miscellaneous invertebrates (29 milion dollars or 0.06 per cent). The top aquaculture species by value in 2000 included giant tiger prawn (4.0 thousand million dollars or 8.6 per cent), Pacific cupped oyster (3.1 thousand million dollars or 6.8 per cent), Silver carp (2.9 thousand million dollars or 6.4 per cent). Japanese kelp (2.8 thousand million dollars or 6.1 per cent), Grass carp (2.8 thousand million dollars or 6.0 per cent), Common carp (2.3 thousand million dollars or 4.8 per cent), Japanese carpet shell (2.0 thousand million dollars or 4.2 per cent) Yesso scallop (1.5 thousand million dollars or 3.3 per cent), Rohu (1.5 thousand million dollars or 3.2 per cent), Bighead carp (1.4 thousand million dollars or 3.0 per cent), Fleshy prawn (1.3 thousand million dollars or 2.8 per cent) and the Japanese amberjack (1.2 thousand million dollars or 2.7 per cent). The top ten country producers by value within Asian region in 2000 included China Mainland (28.1 thousand million dollars or 60.7 per cent), Japan (4.4 thousand million dollars or 9.6 per cent), Thailand (2.4 thousand million dollars or 5.2 per cent), Indonesia (2.3 thousand million dollars or 4.9 per cent, India (2.2 thousand milion dollars or 4.7 per cent), Bangladesh (1.2 thousand million dollars or 2.5 per cent), Vietnam (1.1 thousand million dollars or 2.4 per cent), China Taiwan (0.85 thousand million dollars or 1.8 per cent), Myanmar (0.81 thousand million dollars or 1.7 per cent) and the Philippines (0.73 thousand million dollars or 1.6 per cent).

Europe

The total aquaculture production in Europe consisting of 38 countries has increased four-fold by weight from 497898 mt in 1970 (13.9 per cent of total global production) to 2028835 mt (4.4 per cent of total global production). The annual per cent growth of the sector decreased from 4.3 per cent per year (1970-80) and 7.8 per cent per year (1980-90) to 2.3 per cent per year (1990-2000).

The total number of cultured species in the region has tripled increasing from 19 in 1970 to 60 in 2000 with the main species groups cultivated in 2000 being finfish (1.25 mt or 61.8 per cent), molluscs (769 tmt or 37.9 per cent), aquatic plants (6028 mt or 0.3 per cent) and crustaceans (209 mt). The top cultured species within the region in 2000 included Atlantic salmon (615 tmt or 30.3 per cent), blue mussel (435 tmt or 21.4 per cent), Rainbow trout (289 tmt or 14.2 per cent), Pacific cupped oysters (141 tmt or 6.9 per cent), common carp (138 tmt or 6.8 per cent), Mediterranean mussel (115 tmt or 5.7 per cent), Gilthead seabream (58041 mt or 2.9 per cent), Japanese carpet shells (55858 mt or 2.7 per cent), European seabass (41885 mt or 2.1 per cent), silver carp (37732 mt or 1.9 per cent), and the European eel (10617 mt or 0.5 per cent).

The top country producers in European region in 2000 included Norway (488 tmt or 24.0 per cent), Spain (312 timt or 15.4 per cent), France (268 tmt or 13.2 per cent), Italy (216 tmt or 10.7 per cent), United Kingdom (152 tmt or 7.5 per cent), Greece (80 tmt or 3.9 per cent), Russian Federation (77 tmt or 3.8 per cent), Netherlands (75 tmt or 3.7 per cent), Germany (60 tmt or 2.9 per cent), and Ireland (74 tmt or 2.5 per cent).

By value aquaculture production within Europe increased over three-fold from 1.42 thousand million dollars in 1984 to 4.63 thousand million dollars in 2000 (representing 8.2 per cent of the total global aquaculture production value), with the main species groups being finfish (3.79 thousand million dollars or 81.9 per cent), and molluscs (819 million dollars or 17.7 per cent)

The top aquaculture producers countries by value in 2000 in Europe were Norway (1.36 thousand million dollars or 29.3 per cent), United Kingdom (461 million dollars or 10.0 per cent), Italy (456 million dollars or 9.8 per cent), France (434 million dollars or 9.4 per cent), Spain (382 million dollars or 8.3 per cent), Greece (287 million dollars or 6.2 per cent), Russian Federation (205 million dollars or 4.4 per cent), Denmark (147 million dollars or 3.2 per cent), Germany (118 million dollars or 2.5 per cent), and the Netherlands (107 million dollars or 2.3 per cent).

Latin America and Caribbean Countries

In the region (35 countries) reported aquaculture production has increased over 714-fold by weight from 1221 mt in 1970 (0.03 per cent of total global production) to 871874 mt in 2000 (representing 1.9 per cent or total global production).

The total number of cultured species has increased from 8 in 1970 to 46 in 2000. The main cultivated species groups in 2000 was Fin fish (624 tmt or 71.6 per cent), Crustaceans (153 tmt or 17.6 per cent), Molluscs (60 tmt or 6.9 per cent), aquatic plants (34 tmt or 3.8 per cent), and Amphibians (772 mt or 0.09 per cent). The top ten cultured species by weight during 2000 were Atlantic salmon (166897 mt or 19.1 per cent), White leg shrimp (139264 mt or 16.0 per cent), Rainbow trout (94479 mt or 11.2 per cent), Coho salmon (93419 mt or 10.7 per cent), Tilapia (85246 mt or 9.8 per cent), Common carp (62241 mt or 7.1 per cent), Gracilaria seaweed (33642 mt or 3.8 per cent), Silver carp (30000 mt or 3.4 per cent), Chilean mussel (23477 mt or 2.7 per cent), and the Peruvian calico scallop (21295 mt or 2.4 per cent).

The top producer countries in the region in 2000 included Chile (425058 mt or 48.7 per cent), Brazil (153558 mt or 17.6 per cent), Ecuador (6211 mt or 7.1 per cent), Colombia (61786 mt or 7.1 per cent), Mexico (53802 mt or 6.2 per cent), Cuba (52700 mt or 6.0 per cent), Venezuela (12830 mt or 1.5 per cent), Costa Rica (9708 mt or 1.1 per cent), Honduras (8542 mt or 1.0 per cent), and Peru (6812 mt or 0.8 per cent).

In terms of value of aquaculture production in the region increased over eight–fold from 337 million dollars in 1984 to 2.98 thousand million dollars in 2000 (representing 5.3 per cent of total global aquaculture production by value). The main species groups by value in 2000 were finfish (1.89 billion dollars or 63.4 per cent), crustaceans (0.94 billion dollars or 31.5 per cent), and molluscs (128 million dollars or 4.3 per cent), with the top cultured species being white leg shrimp (848 million dollars or 28.4 per cent), Atlantic salmon (567 million dollars or 19.0 per cent), coho salmon (346 million dollars or 11.6 per cent), rainbow trout (291 million dollars or 9.7 per cent), tilapia (221 million dollars or 7.4 per cent), common carp (176 million dollars or 5.9 per cent), Peruvian calico scallop (93 million dollars or 3.1 per cent), penaeid shrimp (77 million dollars or 2.6 per cent), cachama (75 million dollars or 2.5 per cent), and silver carp (21 million dollars or 0.7 per cent). The top country producers by value within the region in 2000 include Chile (1266 million dollars or 42.5 per cent), Brazil (617 million dollars or 20.7 per cent), Ecuador (324 million dollars or 10.8 per cent), Colombia (258 million dollars or 8.6 per cent), Mexico (181 million dollars or 7.0 per cent), Honduras (59 million dollars or 2.0 per cent), Cuba (47 million dollars or 1.6 per cent), Venezuela (43 million dollars 1.1 per cent), Costa Rica (33 million dollars or 1.4 per cent), and Peru (28 million dollars or 0.9 per cent).

North America

Total aquaculture production of the region has increased over three fold by weight, from 172272 mt in 1970 (4.9 per cent of the total global production) to 551559 mt in 2000 (representing 1.2 per cent of total global production). The total number of reported cultured species with in the region has doubled, increasing from 9 in 1970 to 19 in 2000, with the main species groups cultivated being finfish (430905 mt or 78.1 per cent), molluscs (110290 mt or 20.0 per cent) and crustaceans (10364 mt or 1.9 per cent). The top cultured species within the region in 2000 included the channel catfish (269257 mt or 48.8 per cent), Atlantic salmon (90790 mt or 16.5 per cent), Pacific cupped oyster (44318 mt or 8.0 per cent), rainbow trout (32360 mt or 5.9 per cent) hard calm (23985 mt or 4.3 per cent), blue mussel (23535 mt or 4.3 per cent), American cupped oyster (14596 mt or 2.6 per cent), tilapia (8051 mt or1.4 per cent), chinook salmon (8000 mt or1.4 per cent), red swamp crawfish (7713 mt or1.4 per cent) and trouts (6407 mt or 1.25 per cent). In 2000 the total aquaculture production was 428262 mt in the USA (77.6 per cent of the regional total) and 123297 mt in Canada (22.4 per cent).

With respect to value, aquaculture production in the region has increased over two-fold from 498 million dollars in 1984 to 1.24 thousand million dollar in 2000 (representing 2.2 per cent of total global aquaculture production by value).The main species groups cultivated by value in 2000 were finfish (9.0 thousand million dollars or 85.0 per cent), molluscs (140 million dollars or 11.2 per cent) and crustaceans (46 million dollars or 3.7 per cent). The top cultivated species by value in 2000 were channel catfish (447 million dollars or 36.0 per cent), Atlantic salmon (355 million dollars or 28.5 per cent), rainbow trout (82 million dollars or 6.6 per cent), American cupped oyster (53 million dollars or 4.2 per cent), golden shiner (46 million dollars or 3.7 per cent), chinook salmon (37 million dollars or 3.0 per cent) striped bass hybrid (29 million dollars or 2.4 per cent), hard calm (28.1 million dollars or 2.2 per cent), red swamp crawfish (28 million dollars or 2.2 per cent) and Pacific cupped oyster (27 million dollars or 2.1 per cent). The total value of aquaculture production in the USA and Canada in 2000 was 870 million dollars and 373 million dollars respectively.

Africa

Total aquaculture production of African region (38 countries) has increased over 38 fold by weight from 10271 mt in 1970 (0.3 per cent of the global production) to 399390 mt in 2000 (0.9 per cent of global production by weight). The sector in the region, displayed an overall growth of 13.0 per cent per year for the period 1970-2000.

The total number of reported cultured species in Africa increased from only five in 1970 to 43 in 2000 with the main species groups cultivated in 2000 being finfish (384337 mt or 96.2 per cent), aquatic plants (7177 mt or 1.8 per cent), crustaceans (5425 mt or 1.4 per cent) and molluscs (245 mt or 0.6 per cent) The top cultivated species within the region in 2000 included Nile tilapia (161958 mt or 40.5 per cent), flat head grey mullet (80827 mt or 20.2 per cent), grass carp (66531 mt or 16.6 per cent), common carp (19590 mt or 4.9 per cent), European seabass (10483 mt or 2.6 per cent), gilthead seabream (9681 mt or 2.4 per cent), sea weeds (7000 mt or 1.7 per cent), giant tiger prawn (5225 mt or 1.3 per cent), torpedo shaped catfish (4201 mt 1.0 per cent) and tilapias (3820 mt or 0.9 per cent), excludes three spotted tilapia at 2750 mt or 0.7 per cent

The top aquaculture producer countries in African region in 2000 were Egypt (340093 mt or 85.1 per cent), Nigeria (25718 mt or 6.4 per cent), Madagascar (7280 mt or 1.8 per cent), Tanzania (7210 mt or 1.8 per cent), Zambia (4240 mt or 1.1 per cent), South Africa (4108 mt or 1.0 per cent), Morocco (1847 mt or 0.5 per cent), Tunisia (1553 mt or 0.4 per cent), Coted Ivoire (1197 mt or 0.3 per cent) and Sudan (1000 mt or 0.25 per cent).

By value aquaculture production in African region has increased over 32 fold from 29 million dollars in 1984 to 951 million dollars in 2000 (1.7 per cent of total global aquaculture production by value), with the main species groups in 2000 being finfish (911 million dollars or 95.8 per cent), crustaceans (30 million dollars or 3.2 per cent), molluscs (7.5 million dollars or 0.80 per cent) and aquatic plants (1.4 million dollars or 0.14 per cent). The top aquaculture species by value in Africa during 2000 were the flathead grey mullet (280 million dollars or 29.4 per cent), Nile tilapia (279 million dollars or 29.3 per cent), grass carp (115 million dollars or 12.1 per cent), European seabass (73 million dollars or 7.6 per cent), gilt head seabream (61 million dollars or 6.5 per cent), common carp (28 million or 3.0 per cent), giant tiger prawn (28 million dollars or 2.9 per cent), torpedo shaped catfish (12.8 million dollars or 1.3 per cent), tilapias (7.4 million dollars or 0.8 per cent), and rainbow trout (6.6 million dollars or 0.7 per cent). The top country producers by value within the region in 2000 were Egypt (815 million dollars or 85.7 per cent), Nigeria (57 million dollars or 5.9 per cent), Madagascar (28 million dollars or 2.9 per cent), South Africa (14 million dollars or 1.4 per cent), Tunisia (7.1 million dollars or 0.7 per cent), Zambia (7.0 million dollars or 0.7 per cent), Morocco (4.8 million dollars or 0.5 per cent), Seychelles (4.1 million dollars or 0.4 per cent), Cote d'lvoire (1.6 million dollars or 0.17 per cent) and Sudan (1.5 million dollars or 0.15 per cent).

Oceania

Total reported aquaculture production of ten countries in Oceania increased over 16-fold by weight from 8421 mt in 1970 (0.2 per cent of total global production) to 139432 mt in 2000 (representing 0.3 per cent of total global production by weight). An overall growth of 9.8 per cent per year for the period 1970-2000 was noticed in the region.

The total number of cultured species within the region has increased from 3 in 1970 to 30 in 2000 with the main species groups cultivated in 2000 being molluscs (95576 mt or 68.5 per cent), finfish (28763 mt or 20.6 per cent), aquatic plants (10020 mt or 7.2 per cent) and crustaceans (5073 mt or 3.6 per cent). The top cultivated species within the region during 2000 were New zealand mussel (76000 mt or 54.5 per cent), Atlantic salmon (10907 mt or 7.8 per cent), Pacific cupped oyster (10773 mt or 7.7 per cent), sea weeds (10020 mt or 7.2 per cent) southern bluefin tuna (7803 mt or 5.6 per cent), chinook salmon (6140 mt or 4.4 per cent), Sydney rock oyster (5584 mt or 4.0 per cent), giant tiger prawn (2654 mt or 1.9 per cent), rainbow trout (1949 mt or 1.4 per cent), and Australian mussel (1771 mt or 1.3 per cent).

Country wise the top aquaculture producers in 2000 included New Zealand (85640 mt or 61.4 per cent), Australia (39909 mt or 28.6 per cent), Kiribati (9509 mt or 6.8 per cent), Fiji Islands (2299 mt or 1.6 per cent), New Caledonia (1754 mt or 1.3 per cent), Guam (232 mt), French Polyneasia (53 mt), Papua New Guinea (19 mt), Solomon Islands (15 mt), and Palau (2 mt).

Total value of aquaculture production within the region has increased over 9 folds from 32 million dollars in 1984 to 319 million dollars in 2000 (representing 0.5 per cent of the total global aquaculture production by value). The main species by value in 2000 being finfish (262 million dollars or 63.2 per cent) molluscs (69 million dollars or 21.7 per cent), crustaceans (44 million dollars or 13.9 per cent) and aquatic plants (4.1 million dollar or 1.3 per cent). The top aquaculture species by value within the region in 2000 included the southern bluefin tuna (118 million dollars or 36.9 per cent), Atlantic salmon (49 million dollars or 15.5 per cent), New Zealand mussel (30 million dollars or 9.5 per cent), giant tiger prawn (22 million dollars or 7.0 per cent), Pacific cupped oyster (18 million dollars or 5.8 per cent), chinook salmon (18 million dollars or 5.8 per cent), Sedney rock oyster (17 million dollars or 5.2 per cent), penaeid shrimp (12 million dollars or 3.8 per cent), rainbow trout (7.2 million dollars or 2.3 per cent), and kuruma prawn (5.9 million dollars or 1.9 per cent).

By country, the top produces by value within the region in 2000 included Australia (246 million dollars or 77.0 per cent), New Zealand (54 million dollars or 16.9 per cent), New Caledona (12 million dollars or 3.8 per cent), Kiribati (3.8 million dollars or 1.2 per cent), and the Fiji Islands (1.8 million dollars).

Per capita Production of Farmed Aquatic Meat

The total production of farmed aquatic meat (Values calculated using mean conversion values of 1.15 for fish, 2.8 for crustaceans and 9.0 for molluscs within Asia region has increased 16 fold from 1127548 mt in 1970 (94.1 per cent finfish, 5.6 per cent molluscs and 0.3 per cent crustaceans) to 19295523 mt in 2000 (91.7 per cent finfish, 5.6 per cent molluscs and 2.7 per cent crustaceans). The calculated per capita production of farmed aquatic meat in Asia has increased nine fold from 0.54 kg in 1970 to 5.25 kg in 2000.

The total production of farmed aquatic meat in Europe has increased seven-fold from 159224 mt in 1970 (74.8 per cent finfish, 25.2 per cent molluscs) to 1175838 mt in 2000 (92.7 per cent finfish, 7.3 per cent molluscs). The calculated per capita production of farmed aquatic meat in Europe has increased three–fold from 0.35 kg in 1970 to 1.62 kg in 2000.

The total production of farmed aquatic meat in Latin America and Caribbean region has increased just under a thousand–fold from 612 mt in 1970 (66.5 per cent finfish, 3.5 per cent crustaceans and 11.5 per cent molluscs) to 604168 mt in 2000 (89.8 per cent finfish, 9.0 per cent crustaceans and 1.1 per cent molluscs). The calculated per capita production of farmed aquatic meat within the region has increased from 0.002 kg in 1970 to 1.16 kg in 2000.

In North America the total production of farmed aquatic meat has increased over eight-fold, from 47587 mt in 1970 (68.1 per cent finfish, 31.3 per cent molluscs and 0.6 per cent crustaceans) to 390655 mt in 2000 (95.9 per cent finfish, 3.1 per cent molluscs and 1.0 per cent crustaceans). The per capita production of farmed aquatic meat in the region increased over six-fold from 0.21 kg in 1970 to 1.24 kg in 2000.

In African region the total production of farmed aquatic meat has increased over 38-fold from 8834 mt in 1970 (99.8 per cent finfish, 0.2 per cent molluscs) to 336415 mt in 2000 (99.3 per cent finfish, 0.6 per cent crustaceans and 0.1 per cent molluscs). The per capita production of farmed aquatic meat in the region has increased from 0.02 kg in 1970 to 0.42 kg in 2000.

In the Oceania region, the total production of farmed aquatic meat has increased over forty-fold from 936 mt in 1970 (100 per cent molluscs) to 37442 mt in 2000 (66.8 per cent finfish, 28.4 per cent molluscs and 4.8 per cent crustaceans). The calculated per capita production of farmed meat in the region has increased from 0.05 kg in 1970 to 1.23 kg in 2000.

Aquaculture Development: Opportunities and Challenges

Aquaculture is an important provider of much needed, high quality animal protein, generally at prices affordable to the poor segments of society. It is also a valuable provider of employment, cash income and foreign exchange, with developing countries contributing over 90 per cent of the total global production. When integrated carefully, aquaculture also provides low-risk entry points for rural development and has diverse applications in both inland and coastal areas. While export-oriented industrial and commercial aquaculture practices bring much needed foreign exchange, revenue and employment, more extensive forms of aquaculture benefit the livelihoods of the poor through improved food supply, reduced vulnerability to uncontrollable natural crashes in aquatic production, employment and increased income. Fisheries enhancements using appropriate culture techniques also provide important opportunities for resource poor people to benefit from enhanced of under utilized, new or degraded resources. Such culture based fisheries have considerable potential to increase fish supplies from both fresh water and marine fisheries, with concomitant income generation in rural inland and coastal communities.

The challenge is to create an enabling environment for optimising the potential benefits and contribution that aquaculture and culture-based fisheries can make to rural development, food security and poverty alleviation. Improved participatory farming or production practices within the framework of sustainable, integrated, co-management of natural resources will improve their use. People centered development and extension management approaches, ensuring capacity building that focuses on culture systems for aquatic species feeding low in the food chain, also provide the low-cost products favoured by poor rural communities.

During the past three decades, aquaculture has expanded, diversified, intensified and advanced technologically. The Bangkok Declaration and Strategy (NACA/FAO 2000) emphasizes the need for the aquaculture sector to continue development to its full potential, making a net contribution to global food availability, domestic food security, economic growth, trade and improved living standards. In order to achieve this potential, aquaculture should be pursued as an integral component of community development, contributing to sustainable livelihoods for promoting human development and enhancing social well-being of poor sectors. Aquaculture policies and regulations should promote practical and economically viable farming and management practices that are environmentally sustainable, and socially acceptable. If aquaculture is to attain its full potential the sector may require new approaches in the comming decades. These approaches will undoubtedly vary in different regions and countries and the challenge is to develop approaches that are realistic and achievable, within each social, economic, environmental and political circumstances. In an era of globalization and trade liberalization, such approaches should not only focus on increasing production, they should also focus on producing a product that is affordable, acceptable and accessible to all sectors of society.

The major issues and concerns that need to be addressed to ensure overall sustainability of the aquaculture sector include:

1. Providing an enabling environment for sectoral sustainability with appropriate and well linked technology, policy, legal and institutional frameworks;
2. Involvement of stakeholders in the overall process of drafting and reviewing the regulatory processes which govern the aquaculture sector;
3. Involving all stakeholders in decision making, policy planning, development and management of the sector;
4. Facilitating access to key resources, including physical, monetary and information or knowledge;
5. Achieving responsible management and efficient use of common resources, such as water and land;
6. Integrating aquaculture into coastal and inland watershed management plans and adoption of integrated planning and co-management of common resources with relevent stakeholders;
7. Effectively integrating aquaculture into pro-poor national development plans;
8. Stimulating investments and private-sector participation in commercial and industrial aquaculture development, where appropriate;
9. Supplying products for specific consumer preferences and complementing the efforts of other food production sectors;
10. Promoting closer co-operation among stakeholders, countries, regions in the overall development process; and
11. Investment by both the public and private sector in aquaculture development at both small scale and commercial or industrial levels, which are aimed at sustainable development.

Aquaculture and Rural Development

Globally, aquaculture is still predominantly rural, producing species low in the food chain that require little or no inputs or capital investment (more that 80 per cent of total global finfish production is cyprinid fishes). This means aquaculture makes a significant, grass–roots contribution to improving livelihoods among poor sectors of the society. Pressure to over exploit resources under such circumstances has been as significant in aquaculture development. However it is important to examine the lessons learnt from past experience and develop strategies for improved sustainability of this important sector. Reduction of externalities and negative social and environmental impacts, through consultative planning, and dedicated co-management will ensure sustainable benefits.

Policy and Regulatory Frameworks

One of the key factors that support creation of an enabling environment is strong institutional capacity, that is the ability of countries and organizations to strengthen and implement policy and regulatory frameworks that are both transparent and enforceable. Aquaculture in the Third Millennium identified several key recommendations that would help develop conducive institutional and policy environments. These are:

1. Developing clear aquaculture policy with a clearly defined lead agency with adequate organizational stature to play a strong coordinating role;
2. Developing comprehensive and enforceable laws, regulations and administrative procedures that encourage sustainable aquaculture and promote trade in aquaculture products, with a stakeholders participatory approach;
3. Targeting organizations and institutions dealing with administration, education and research and development, that represent the private sector, non-governmental organizations (NGOs), consumers and other stakeholders, in addition to government ministries and public sector agencies;
4. Developing mechanisms and protocols for the timely collection and reporting of relevent data;
5. Sharing information on policies and legislation, rules and procedures that encompass good management practices in aquaculture;
6. Clarifying legal frameworks and policy objectives regarding access and user rights for farmers; and

7. Improving the capacity of institutions to develop and implement strategies that target the aquaculture development needs of poor communities.

Appropriate Technology

Appropriate technologies contribute to aquaculture sustainability with a variety of mechanisms that can meet the needs of the local environment. Delivery of such techniques requires effective communication networks, reliable data and a decision making process that ensures aquaculture produces choose the best production systems and species for their environment. Science and technology provide on going "new" opportunities for aquaculture development including techniques for sustainable stock enhancement, ranching programmes and open sea aquaculture; use of aquatic plants and animals for nutrient stabilisation, integrated systems to improve environmental performance, such as, recirculating systems integrated water use, artificial upwelling and ecosystem food web management. Some biotechnologies like, fertilization of ponds to increase feed availability and genetic engineering and DNA-probe development for disease diagnostics are more modern and focuses primarily on increasing growth rates, enhancement of disease resistance, production of sterile stocks and physiological tolerance of environmental extremes.

Trade of Aquaculture Products

Quality, safety and trade of aquaculture products are important aspects of sustainable industry. The importance of attaining sustainable aquaculture with negligible or minimal environmental or socio–economic impacts is forcing many exporting countries to adopt and implement more sustainable production practices. This is especially important where aquaculture is perceived to be a non–traditional food-producing sector. Safety assessments, based on risk assessment and the precautionary approach, for example, are now becoming more common, before pursuing production of new or exotic species or products from modern biotechnology.

The role of aquaculture in international trade is increasing, both in the relative and absolute sense. This is a result of increasing aquacultural production in general and of high-value commercial export-directed production in particular. As international trade statistics do not denote production methods of fishery products (capture or aquaculture), it is not possible to determine the exact share of aquaculture products in most commodity trade. However, recent legislative initiatives, such as new labeling requirements to distinguish farmed and wild products, introduced in 2002 by the European Community, coupled with increased demands for traceability of food products for food safety reasons, should improve the quality of international trade data and facilitate better and more accurate aquaculture trade analysis.

A trend towards consumer preference for organically produced aquatic products is increasing. The aquaculture sector lags behind agriculture in terms of the quantities and diversity of certified "*organic*" produce-reflecting a lack of accepted international/regional/national standards and accreditation criteria for organic aquaculture produce. Existing certifying bodies and organic aquaculturists are, primarily, restricted to a handful or organizations in developed countries of Europe, Oceania and North America, all of which contributed less that 10 per cent to the global aquaculture production in 1999. Although no official statistics are available for global production of certified organic aquaculture products, it is estimated that such production in 2000 was only about 5000 mt. primarily from European countries. This represents a mere 0.01 per cent of total global aquaculture production and 0.25 per cent of European aquaculture production. The total volume of organic aquaculture products marketed in Europe in 2000 is estimated at between 4400 and 4700 mt. Negligible production data is available for countries outside Europe. Organic certification and other "eco-certification" programmes are being discussed and established by various agencies and groups. These empowers consumers to choose aquaculture products with perceived higher quality or health attributes and grown in an environmentally sound manner. Price premiums for organically grown food products generally range between 10 to 50 per cent above conventional products. Higher prices give aquaculturists incentives to produce organic products, but incur higher production costs associated with environmental protection measures. Where certification is non-discriminatory and based on sound science-based technical standards, it can help consumers use their purchasing powers to encourage environmentally sound production practices. The issue is to ensure the process is based on sound scientific evidence, is fair and non-discriminatory. Awareness of, and sensitivity to environmental and welfare issues is increasing, particularly in developed countries where purchase decisions can be influenced by adverse publicity or a lack of information. As live stock farmers aquaculture, producers are increasingly required to act in a line with standards expected of the live stock industry. At a national level, safety and quality management systems should be put into place to ensure production distribution and sale of aquaculture products are safe and of high quality. Such measures require competent professional associations that work in close association with the legal authority in order to be successful.

Integration with Agriculture

Various types of aquaculture form an important component with agriculture and farming system development. These can contribute to the alleviation of food in security, malnutrition and poverty through the provision of food of high nutritional value, income and employment generation, decreased risk of monoculture production failure, improved access to water, enhanced aquatic resource management and increased farm sustainability.

Global aquaculture is now the fastest growing food-production sub sector in many countries. The production of all cultured aquatic organisms reached almost 43 million metric tonnes (mmt) in 1999 and it is expected that this trend will continue despite several constraints.

Starting in 70's there was a substantial assistance for developing the aquaculture sector in Latin America, Asia, and Africa. The tendency of this development, initiatives was to focus on large infrastructure development technical packages and technical training without paying sufficient attention to the role of these, often new production systems of the livelihood or farming system of the intended beneficiaries. All too often, the result was lack of adoption by one of the intended target groups, the rural poor. As a result of the apparent inability to impact the rural poor, donor support for aquaculture development has declined in the past 10 years. Paradoxically, the progress made in Asian aquaculture during this time saw a tremendous boom in commercial scale aquaculture by households with better resource bases, hand in hand with the economic expansion of the region, opening markets and increasing the flow of cash economies to rural areas.

Aquaculture in rural livelihoods

Recently there has been a revaluation of the role of small-scale aquaculture in rural livelihoods and its importance in poverty alleviation and household food security, particularly the mechanisms by which the rural poor can access and benefit from aquaculture. A recent FAO/World Bank Farming Systems study noted the importance of five major household strategies for escaping poverty for 70 farming systems across the world, intensification, diversification, increased asset base, increased off-farm income and exit from agriculture. Diversification, which includes aquaculture was judged to be the single most promising source of farm poverty reduction in the coming years.

Extensive to semi-intensive aquaculture systems still produce the bulk of aquaculture production. Extensive farming usually involves unsophisticated methods, relies on natural food and has a low input to output ratio. As production intensity increases, fish are deliberately stocked and the natural food supply is enhanced by using organic and inorganic fertilizers and low-cost supplemental feeds derived from agriculture byproducts. The system found most frequently is the farming of fish in ponds. However rice-fish farming or stocking of fish into natural or impounded water bodies are also included as aquaculture systems.

Specific examples of aquaculture activities that have positive impacts on the rural poor include, fry nursing and the development of nursing networks, the integration of fish farming with rice crops in flood plains and the more remote mountainous areas in Asia, sustaining and restoring aquatic biodiversity through simple enhancement management methods. In coastal areas, the farming of mud-crabs, oysters, mussels, cockles, shrimps, fish and seaweeds provides employment for the rural poor, mainly for direct labour inputs, as well as seed and feed collection. Intensive aquaculture systems yield more output from a given production unit, using technology and a higher degree of management control. This typically involves facilities deliberately constructed for the purpose of aquaculture, which are operated with higher stocking densities and use compound manufactured feed and chemotherapeutant intervention on a regular basis. Intensive inland and coastal cage aquaculture of high value salmonids has been encouraged and supported to develop remote areas in Europe, South and North

America. Similar systems have emerged in Asia and Australia for warm-water piscivorous fish such as, groupers, yellow tail, snappers and sea bass. Coastal shrimp farming has raised particular interest throughout the tropics because of its high value and opportunities for export and earning foreign exchange. The benefits of aquaculture in rural development relate to health and nutrition, employment, income, reduction of vulnerability and farm sustainability. It also provides this protein at prices generally affordable to the poor segments of the community. It creates "own enterprise" employment, including jobs for women and children and provides income through sale of what can be relatively high value products. Employment income opportunities are possible on large farms, in seed supply network, market chains and manufacture or repair supporting services. Indirect benefits included increased availability of fish in local rural and urban markets and possible increase in household income through sales of other income generating farm products, which will become available through increased local consumption of fish. Aquaculture can also benefit the land less from utilization of common resources, such as, finfish cage culture, culture of molluscs and sea weeds and fisheries enhancement in communal water bodies.

An important benefit which is particularly relevent for integrated agriculture–aquaculture systems is their contribution to increased farm efficiency and sustainability. Agricultural by–products, such as manure from live stock and crop residues, can serve as fertilizer and feed inputs for small-scale commercial aquaculture.

In view of all these benefits aquaculture production has grown rapidly since 1970s and has been the fastest growing food production sector in many countries for nearly two decades. The sector is exhibiting an overall growth rate of over 11.0 per cent per year since 1984.

FAO's latest studies on future demand for and supply of fish and fishery product predict a sizeable increase in demand for fish. The majority of this increase will result from expected economic development, population growth and changes in eating habits. Fish supply from capture fisheries in most countries is expected to remain constant or decline since catches have either reached or are close to maximum sustainable yield.

Contribution of Aquaculture to Rural Development

Recent meetings and consultations organized and supported by FAO and partner organizations have reached a number of conclusions and recommendations aimed at increasing the sustainable contribution of aquaculture to rural development. Land–based culture systems in inland areas have the greatest potential because aquaculture can be integrated with the existing agricultural practice of small-scale farming house holds. Coastal aquaculture also contributes to rural development by enabling diversification of subsistence fishery sectors. Differences between countries and regions, with regard to physical resources, norms and traditions, as well as economic conditions are significant, hence the development status of aquaculture differs widely.

In the past decades, there has been a move away from a predominantly top-down view, dominated by technical issues, to a more holistic perspective of improved livelihoods and greater household food security. Social, economic and institutional issues have been recognized to be the most important constraints to enhanced contributions by aquaculture to rural development. There is a need to assess the impacts of aquaculture on sustainable livelihoods and for advocating products and benefits. Advocacy issues include:

1. Rising awareness amongst policy makers of the role of small-scale rural aquaculture and aquatic resource management in rural livelihoods, including actual contributions and unfulfilled potential of aquatic resource management including aquaculture to sustainable rural development:
2. Documenting indigenous aquaculture systems and farmer-proven examples of aquaculture;
3. Developing indicators for monitoring aquatic resource management and aquaculture impacts on food security and poverty alleviation;
4. Encouraging and promoting consumption of aquaculture and inland fishery products; and
5. publicising and promoting benefits of sustainable aquaculture enterprises aquatic resource management and their products.

Government should address the design and implementation of policy, ensuring feedback mechanisms to allow the poor to influence development. This may be done through the establishment of multi-sectoral co-ordinating process both at sectoral policy formulation level and at the extension service level. Aquaculture development should complement or substitute wild fisheries as needed. Negative impacts of aquaculture projects on the food supplies of the poor should be avoided. Other recommendations aimed at improved planning and policies include:

1. Establishing national aquaculture development and inland fisheries management plans and policies in consultation with stakeholders; and
2. Integrating aquaculture planning into water resource management planning for inland areas and into coastal management planning in coastal areas, as well as into other economic and food security interventions for rural areas. Generic technologies for sound aquaculture production exist. Some of the indigenous systems require further study and more detailed documentation. More emphasis is need to:

    (*a*) Favour systems which use readily available species and local materials;

    (*b*) Decentralized seed production and seed nursing and trading networks,

    (*c*) Improving culture systems for aquatic species feeding low in the food chain and that are preferred for local consumption; and

    (*d*) Adopt and improve these systems through farmer-based learning, and promoting the results through participatory approaches.

Governments should aim to provide services and facilitate access to inputs. The rural poor need to be provided at least initially, with public sector support, while commercial aquaculture requires less intervention. In the longer term, aquaculture has to function on a self financing basis within the private sector. Necessary actions include:

1. Focussing limited public resources on strategic government infrastructure and flexible and efficient extension services that meet producers needs;
2. Promoting and facilitating the private sector production of feed and seed;
3. Encouraging credit for medium and large-scale producers;
4. Facilitating the formation of farmer's associations and encourage community production; and
5. Encouraging investment in building the institutional capacity and knowledge base concerning sustainable aquaculture practices to manage the sector.

Positive examples and case studies of traditional and other aquaculture systems that have proven to be sustainable should be promoted and disseminated. In doing so:

1. Promote collaboration, co-ordination and information exchange between national and regional aquaculture institutions and agencies; and
2. Develop strategies for an effective transfer of aquaculture know-how into areas and regions where it has no tradition.

Pre-marketing Fattening: A Modern Technology in Aquaculture

An interesting sector which has opened up in recent years is the temporary holding of bluefin tuna (*Thunnus thunnus)* to improve meat quality. The early development of this activity began in Australia with the southern bluefin tuna (*Thunnus maccoyii)* in response to falling catches from the South Australian wild fishery. Australian landings of this migratory species reached a peak of 21500 mt in 1982, but increasingly lower quotas had to be introduced, dropping to 5265 mt by 1989. The quality of the product being landed was poor which diminished export value, thus fishing and farming companies began holding 2-4 year old fish in cages for 3-5 months for conditioning. This enhanced meat quality and enabled them to sell to high-value sushi markets in Japan at prices of around 18 dollars per kg. or upto 620 dollars per fish. By 1997 tuna "fattening" had become Australia's most valuable single aquaculture sector.

Similar techniques were adopted by fishermen of Mediterranean (Malta, Croatia and Turkey) over the last few years, holding Atlantic bluefin tuna captured during the limited fishing season (May-July). The fish are on a spawning migration at this time, thus flesh quality is poor, and meat prices are depressed. The fish are held in floating cages until November or December and fed on mackerel and herring. At the end of holding period the fish improve in condition and meet high market price quality for export to Japan. The cages used to hold and transport the fish are large structures of up to 100 m. in circumference and it can take a week or more to tow them as much as 300 km. from fishing grounds to the holding area.

The capture of bluefin tuna from the declining and possibly threatened stocks in the Atlantic has caused some controversy, thus, their interest in further development of this technology for true farming and reduce reliance on wild captured stocks. In addition, there is controversy over the amount of fish required to feed the tuna during the "fattening" process, especially since these are species also used for human consumption in the Australian and Mediterranean regions. This presents a significant challenge, however for this highly piscivorous migratory species.

Chapter 3

Structure of Production–Economic Aspects

Inshore Fishing Grounds for Commercial Fisheries

There are numerous islands and straits located throughout the sea, thus providing a certain kind of marine geographical complexity. Various organic substances are constantly being supplied into the sea area through a great number of rivers, while the exchange of sea water takes place at frequent intervals in conjunction with the characteristics of tidal currents. These factors prove a great boon to the reproduction of plant plankton. In addition, the natural conditions and geographical features of the sea area provide a very inhabitable environment for various fish and shellfish species. It is estimated that more than 600 species fish and shellfish inhabit the sea area, approximately 100 of which are caught in commercial fishery operations. Fish resources are roughly divided into the following three categories:

1. Mother fish migrate from high seas into the inland sea for spawning
2. The fry which are hatched in high sea areas come to settle in the inland sea area until they become adult fish

3. Some of the common fish species remain their whole life time in the inland sea.

The above type of resource distribution has long been formed in close relation to marine geographical conditions peculiar to the sea area. An extensive tideland has been an ideal habitat for shellfish species. On the other hand, schools of fish coming from high seas pass through or settle in the channel or bay. The above factors alone, however, are enough to clarify why such sea areas have become abundant in particular fish species, thus giving a boon to the development of specific type of fisheries. It must also be taken into consideration how fishermen have dealt with these resources. In other word, they have concentrated efforts on harmonizing the exploitation of resources with progressive fishery development.

In the post war period, a great increase in fish catch amount due to the increased number of fishermen and mechanization of fishing methods and gear. Another remarkable increase in catch was due to:

1. Eutrophication of sea water–The increased amount of waste water from coastal industrial and housing zones resulted in the eutrophication of sea water. This benefited the reproduction of various fish and shellfish species.
2. Effective thinning-out–Fishing intensity was greatly increased by improved fishing methods and gear. This had a certain thinning–out effect upon the eventual reproductivity of fish and shellfish species which might have over propagated in the sea area.

In the sea area where fishermen have consistently increased their fishing intensity for the past 3 decades in an effort to boost the eventual reproductivity of fish and shell fish resources, there by increasing the amount of catch by leaps and bounds. On the other hand, however, a drastic increase in total catch amount has degraded the mean trophic level of the catch. In general, the fish of lower trophic level is lower in commodity value. The degraded trophic level of the catch has resulted in a significant impact in commercial fisheries. This is posing a crucial problem concerning an alternative decision regarding the targets of the government's fishery development policy. One is to simply increase the catch amount of middle or low classes fish species with an increase in the supply of fish protein in mind, while the other is to put greater effort in increasing the production of prime fish species. The government must choose between the two.

Built-in Fishing Grounds in Shallow Coastal Waters

Creation of fishing grounds, which is a new human endeavour is of particular significance in the shallow coastal waters of the world. This new attempt involves numerous fishing communities where hundreds of thousands fishermen and their families depend on marine fish resources. Building and development of coastal fishing grounds will no doubt increase the catch of fish and raise the standard of living for these people. It will also greatly stimulate construction of fishing harbours, processing and marketing facilities in the fishing communities which have so far often been improverished, thus enabling the people to make a greater contribution to national economy.

Shallow coastal waters are important habitats for fishes, shellfishes, seaweeds and crustaceans. These marine resources are sedentary or only slightly migratory rather than fishes, which are more migratory. The biota of shallow coastal waters is rather complex, as the waters are varied, ranging from brackish waters affected by inflow of fresh land water to high salinity waters, open to the outer sea. Fishing grounds in shallow coastal waters may be classified broadly into "inner bay fishing grounds or inshore fishing grounds" and "open sea fishing grounds". The "Inner bay fishing grounds" normally have a sea bottom consisting mainly of sand or mud, while "open sea fishing grounds" consist usually of rocky sea bottoms. The different environments result in different types of dominant marine resources inhabiting their respective fishing grounds.

To deal with the need for attaining greater self-sufficiency of food supply and also for establishing a firm policy in the face of the change of fishing environments caused by rapid industrialization an urgent task is now the consolidation and development of the coastal fishing grounds within the 200 mile limit and particularly in the 12–mile territorial waters.

Government assistance to the development of shallow coastal water fishery is required in highly affected areas in phased manner.

The following projects may first be taken up for the purpose:

1. Creation of reef areas suitable as fish habitats by placing artificial rocks or destroying unsuitable natural rocks or reefs.
2. Transplantation and release of fry and spats.
3. Provision of artificial devices to collect seeds of sea weeds.
4. Cultivation of the sea-bottom in shell fish culture grounds.

Subsequently projects for the improvement of the coastal fishery structure may be taken up in order to modernize coastal productivity and improve the level of income of coastal fishermen. These projects may include:

1. Creation of suitable habitats by placing artificial reefs and destroying unsuitable reefs.
2. Building artificial fish shelters with concrete blocks.
3. Construction of wave damping break waters and dredging or cultivation of sea bottom in aquaculture grounds.
4. Construction of artificial breeding facilities for shellfishes and sea weeds.
5. Construction of water ways to facilitate exchange of sea water in inner bays.
6. Construction of aquaculture pens in open sea areas.
7. Fertilization and disease and predator control in aquaculture grounds.

"Law for coastal fishing Ground Consolidation and Development" may be enacted. The purpose of this law will be to develop large coastal fishing grounds for the purpose of aiding the steady development of coastal fisheries and increasing the supply of high valued fishes and shell fishes to meet consumer demand in the face of the new fishery age resulting from 200 mile limits. In this phase the project may include:

1. Construction of artificial reefs and fish shelters.
2. Construction of fish nurseries.
3. Development of large scale aquaculture facilities.

4. Development of aquaculture and propagation grounds in shallow coastal waters.
5. Protection of the environment in the fishing grounds.

These projects will be unique in the history of coastal fisheries, particularly those for development of large artificial reefs and floating break waters. While artificial breeding and releasing of fish and shell fishes correspond to "sowing", the projects intend to create and consolidate the "grounds" to which fishes, shell fishes and sea weeds are attracted, settle, live and propagate.

It is expected that under these projects coastal waters can be developed in a co-ordinated manner and that the fishery productivity in shallow water areas will be greatly increased.

Artificial Tideland for Fishery Resources

The tideland presents an unique ecosystem different from those of either land or sea and is indispensable in maintaining fish resources because it plays a vital role in facilitating the growth of young fish. For example kuruma prawn (*Paenaeus japonicus*) have been decreasing in direct proportion to the amount of natural tidelands lost to land reclamation for industrial and agricultural purposes.

Fish culture centres were originally designed to make up for this loss of natural fish resources by means of stocking. The technique of stocking waters with kuruma prawn fry undergoes a two-stage crisis. First, the number of fry shows a remarkable decrease during the twenty four hour period following their release. The characteristics of the fry themselves plus the environmental conditions into which they are released greatly affect their rate of survival. Secondly, fry which manage to survive the first stage of the crisis settle on the tide land and, as they grow disperse into the area off the coast, when they fall victim to gobies and other predators resulting in a second major decrease in their numbers.

Thus, the life of artificially hatched fry between the time they settle in the tideland and the time they grow to about 3 cm in length, is extremely unstable especially when compared with the conditions over the same period for natural fry.

The mortality rate for natural fry is about 10 per cent per fifteen days, while that of artificially hatched fry is from 10 to a whopping 60 per cent, depending on the environmental conditions. Much more care should be taken to the artificially hatched fry's adaptability to environmental conditions, as their distribution is sometimes rather uneven within the breeding ground. An artificial tideland reinforces natural reproductive power. The artificial tideland duplicates a number of environmental conditions found on a natural beach at abb tide which are conducive to the growth of kuruma prawns–conditions such as the ground level relative to the tide level, the depth of tide remaining etc.

The artificial tideland is designed to control environmental conditions so that they will be the most favourable for try to grow. The new system not only means the recovery of lost natural tidelands, it also promotes natural productive power as a means to enable scheduled production of kuruma prawns. By stocking the waters with kuruma prawn fry, the system provides the following three advantages:

1. Increased prawn production.
2. Stabilized production through adjustment of the time fry released.
3. Prawns not caught are added to the natural reproduction resources.

The economic effects of artificial tideland on local fisheries can be assessed with fair certainty. The artificial tideland certainly constitutes an essential technical base for controlling the production of marine life over a large expanse of sea.

Economic Structure of Ocean Fisheries

From the view point of fishery economics, fishing operators households of South East Asian Countries are broadly classified into two groups *i.e.*, (i) those who undertake fishery on own account merely to maintain their lively hood, and (ii) those who perform fishery with the aim of gaining profit. Fishing operations of the former are done mainly with their family members, while that of the latter mainly with employed fishermen. But recent survey data indicates that employment of fishermen is practised now also by fishing operators house holds using in-board powered boat of 8 to 15 metre and 15 to 30 metre.

From the above findings, it would be appropriate to consider fishing operator's households using non-powered boat, out-board powered boat and in-board powered boat of less that 8 m. as small-scale fishermen and the remaining large scale fishing operators households as large-scale fishermen.

Owing to rapid growth of capitalistic fishery, there has been a great increase in the number of fishing labourers households in the third world countries. It is very likely that throughout South East Asian Countries, including India, the number of fishing labourer's households has exceeded that of fishing operator's households. However, the low income and poor living conditions of these fishing labourer's households are more or less equivalent to those of small-scale fishermen. It may therefore, be appropriate to consider fishing labourer's households as being small-scale fishermen, who are also in need of assistance for improvement of their income and living conditions.

Effect of Capitalization

In coastal states of India, the impact of capitalization of marine fishery has not been progressed much. In the coastal state of Kerela, where the marine fishery has been developed to a considerable extent the trawl net, hook-line, set and gill net fishery accounted for 68' per cent of the fishing trips. The trawl net, shore seine, gill net and encircling net groupings account for 67 per cent of the total man power expended during the fishing trips.

Shore seines, boat seines and hook and lines fishery in another coastal state Tamilnadu of India using non-rigid plank boats and catamarans.

Capitalization with mechanization accounts for 26 per cent and 8 per cent of total fishing effort in trawl net and large mesh gill net fishery in the state of Kerala, catching mainly prawns, perches, sciaenids, cat fish, tunnies, seer fish, pomfret, sharks and caranx.

The capitalization of Thai marine fishery, which has occurred during the past 28 years, the number of fishing labourer's households increases in relation to the development of large-scale fishery.

In Tamilnadu, India, fishing crafts are owned by one man, when there is joint ownership it is between close relatives. Though most fishermen have nets, yet the distribution is unequal, most nets are in the hand of only a few families.

Since the capitalization of fishing fleets, the productivities are poor, as too many mechanised boats share a limited resources, the artisinal fishing units on an average make a better use of invested capital than the mechanised units in most of the coastal states of India.

Extent of Dependency on Fishery

In terms of the extent of dependency on fishery, a fishing household is classified as:

1. "Full time", income derived from fishery only;
2. "Part time (major)", income derived mainly from fishery; and
3. "Part time (minor)", income derived mainly from any occupation other than fishery.

"Full time" and "Part time (major)" are fishing households, which depend on fishery for their livelyhood. The extent of dependency on fishery is, therefore, evaluated by the ratio of the sum of "Full time" and "Part time (major)" to the total and may be called "the rate of dependency, on fishery". The rate of dependency on fishery for all sizes of fishing operator's households is extremely high, being more than 90 per cent whereas the dependency rate of operator's household with non-powered boats is 100 per cent. As for fishing labourer's households, however, the rate of dependency on fishery is rather low, being a little more than 60 per cent.

Marketing Channels Mainly in Use

In Kerala, India, the pattern of marketing structures and linkages are related to the structure of the fishing fleets and their pattern of operations, that is in the scales of operation of production and marketing in different regions. In case of small scale landing between 5 to 50 kg on an average, the small-scale distribution composed of thousands of men carrying 25 to 50 kg of fish on cycles and women carrying 5-15 kg on their heads. In large production and marketing centres, lorries carry half to one tonne of fish each.

Negotiations between the producers and buyers are generally conducted through an intermediary, who is either a commission agent or an auctioner. Their function is to facilitate the exchange of fish and money at first sales and his services are paid by the fishermen.

The marketing channels as to the movement of fish from production to the consumption end are as follows:

Producer–Fishermen

↓ → Mode of selling– Bargaining–37%

Purchasers Auction–60%

Head loaders–14% Fixed price–3%

Cycle loaders–23% Others–Negligible.

Lorry merchants–40%

Exporter's Agent–13%

Middlemen–9%

Co-operative Society–Negligible

Others–1%

↓

Final market

Internal neighbouring market–30%

Internal distant market–52%

Export–18%

↓

Consumers

The women of Tamilnadu are deeply involved in marketing of fish, 60 per cent of fisher women folk have income as vendors. But not all the fish is taken to the market by women. Fish traders come to the fishing villages on bicycles; many families have borrowed money from them and the debts are usually repaid in the form of fish. Whenever auctioning is practised, fish traders have an advantage as the money normally has to be paid cash down. Since most women vendors do not have any ready cash available, this result in depriving them of opportunities to participate in marketing activities.

The marketing channels as to the movement of fish from production centres to the consumer end are:

Producer–Fisherman

↓

Purchasers

Head load–58%

Bus–16%

Head load/bus–20%

Head load/lorry–1%

Bus/lorry–2%

Head load/bus/lorry–3%

↓

Final market

Nearby villages–51%

Internal distant market–49%

↓

Consumers

Marketing of prawns is a well-organized business of private entrepreneurs. As almost all the prawn catch can be exported for high prices, there are export companies with their own freezing plants; cleaning is done either in the fishing village by the fisher women or in the factory itself. The freezing plants have contracted with local traders to collect prawns from the fisher folk. Contracted bicycle traders come to buy them. They are given relatively fixed prices by the companies but their margin of profit is higher than that of the fish traders.

In Thailand, however, of the fishing operator's households of all sizes, 54 per cent sell their catch to any fish dealer and 41 per cent sell to specific dealer. Only in 5 per cent cases, the catch is sold directly to the consumers. When the fishing operator's households sell their catch mainly to a specific dealer, it may well be assumed that they are indebted to that dealer and may be exploited by him.

Socio-economic Condition of Fishing Household

Family Size

Ninety per cent of all fishing households in Tamilnadu, India consists of nuclear families that is wife, husband and children. About

20 per cent of these families have dependent grandmother or grandfather. The average nuclear family consists of five people.

While in Thailand, the average family size of a fishing operator's household and a fishing labourer's household is 7-9 and 6-5 persons respectively. Family size of fishing operator's households with non-powered boat, in-board powered boat of less that 8 m. and 8 to 15 m. is quite high, being, 8, 8.2 and 8 persons respectively. Of the total number of family members the number of persons who actually engage in fishing is 1.5 persons for fishing operator's households and 1.8 persons for fishing labourer's households.

Employment of Fishermen

Most of the fishermen undertaking fishery do not have any side income from any other business or employment. As long as fishery is under taken by family members only, a fisherman does not have any sense of pursuing profit. However, once a fishery is carried out with hired fishermen, a fisherman runs his fishery with the aim of profit. He then become a capitalist because he must have some profit after paying wages to his hired fisherman. Thus, whether fishery is undertaken with hired fishermen or not is a criterion to distinguish between small-scale and large-scale fishery. The employment of fishermen is found for any size of fishing operator's household with in-board powered boat. But, the number of employed fishermen for fishing operator's household with in-board powered boat of less than 8 m. is very limited, compared to the employment of in-board powered boat of 8 m. above to 15 m. and 15 to 30 m.

Educational Attainment

School attendance of children is irregular and in most cases poor. This is because children have to work at home and parent's indifference to education. It is also due to the acute shortage of educational facilities, equipment and water facilities. Often teachers absent themselves resulting in the school having to be closed for those days. Because of the above situation, the rate of literacy is very low. Only 3 per cent of the fisher women claims to be able to read and write fluently. The situation however seems to be improving very slowly.

In Thailand, however, the majority of the heads of fishing households has attended primary school and above being 96 per cent for fishing operator's households and 91 per cent for fishing labourer's households. In fishing operator's households with in-board powered boats of 8 m. and above, there are some who graduated from junior high school, but there are also some who never attended school.

Fishermen's Organisation

There are generally three types of fishermen's organisation; (*a*) fishermen's association, (*b*) fishery co-operatives and (*c*) fishermen's groups. A fishermen's association is merely a non-economic organization, through which some communication is established between fishermen and government; whereas both fishery co-operatives and fishermen's groups are organizations authorized to undertake economic activities such as fishing co-operatives, service co-operatives, marketing co-operatives etc. But these activities are very limited.

Fishery Credit

Fishery credit is used mainly for the construction of fishing boats, although some fishing operator's household use such fishery credit for the purchase of fishing gear and for operational costs. Of fishing operator's household with fishery credit, however, those who were given credit from public sources accounted for about 23 per cent, whereas those who secured credit from private sources accounted for, 77 per cent. Regardless of the source of fishery credit, fishing operator's households running larger fishing boats, appear to be receiving fishery credits, which indicates that fishery credit is also required by large-scale fishery. In general size of fishery credit from public sources are larger than that from private sources.

Artisanal fisherfolk are used to taking credit from various sources. The credit is used for food items during the lean fishing season, when the catch is very low, and in cases of long-lasting sickness or disease. Credit is also taken whenever large occasional expenditures have to be incurred, as in the case of purchase of nets, gear and boats. They also barrow for family festivities and religious festivities. Taking credit is not restricted to poor families. Even those families with comparatively large assets take part in the system of borrowing for the various occasions and not only for the purchase of boats and nets.

Relatively small sums are borrowed from the head man, net-owners, local shop owners, neighbours, rice traders and fish traders. Prawn traders generally give much higher credit in order to secure their supplies of prawn.

Credits are not taken from only one source. Most of the families reported to be indebted to numerous persons. The mode of repayment will either be in cash or in fish. The rate of interest varies; prawn traders charge the highest rates–at least 100 per cent per annum.

Prawn traders give higher credit to fisherfolk, but at the same time demand a high rate of interest. The fisherfolk have to sell all their prawns to the traders as they are indebted to him, most of them will borrow cash, others rice, during the lean season. The mode of repayment is exclusively by means of prawns The traders so manages affair that the fisherfolk are still in debt at the need of the prawn season, and give him the prawn catch in the following year. As the mode of repayment is a fixed percentage of daily catch of prawn, the trader has to try and keep control over all the landings, he will be on the shore during landing so that no women can go off unseen with a catch of prawns in order to sell it on her own for a better price. Many fishermen are aware of the fact that they for are not paid a proper price for their prawns. Nevertheless they depend on the prawn traders as money lenders. The prawn traders are aware of the economic position of each fishing family and can easily judge how much money they can risk lending to them. Though the fishermen are aware of the fact that the prawn traders are making a great profit, they do not like to eliminate the trader because this would result in the loss of an important source of quick credit. As such the prawn traders rank first as a source of credit, fish traders play a smaller role in the lending.

Income of Fishing Household

Income of fishing household is the sum of income derived not only from fishery, but also, from any other source, such as, agriculture, salaries, wages and trade.

With fishing operator's house holds, the average annual income varies greatly, owing to differences in the size of fishery management. The average annual income of fishing operator's household with non-powered boat and with in-board powered boat of less than 8 m. is quite low, which is more or less equivalent to the average annual income of fishing labourer's households. The average annual income shows a considerable increase for fishing operator's households having in-board powered boats of more than 8 m. and with in-board powered boats of 15 to 30 m. being the highest.

Per capita income of fishing households was obtained by dividing annual average income by average number of family members for each size of management separately. Same trend as mentioned above, have been noticed in per capita income.

Living Conditions of Fishing Households

In terms of Engel's co-efficient, the living condition of a good number of fishing operator's household possessing in board powered boats of 8 to 15 m. and 15 to 30 m. is considered to be high. But with household's operating smaller powered boats (less that 8 m.), the living condition are low in most of the cases. Living conditions of fishing labourer's households in terms of Engel's co-efficient are also low being more that 80 per cent of the total.

A clear difference in living conditions is also seen in terms of possession of durable consumable goods. The ownership of sewing machine and automobiles (motor car) is found only for fishing operator's house holds with in-board powered boat of 8 m. and above. There is also a definite increase in ownership of a TV set, refrigerator and motor cycle in the case of fishing operator's household using in-board powered boat of 8 m. and above. In fishing labourer's households, the possession of durable consumer goods in quite low in respect of every item with the exception of a radio.

The housing conditions of fishing operators household using inboard powered boats are better than those of fishing operator's households using non-powered boats and out-board powered boats and of fishing labourer's households, since in the case of later no fishing households lives in concrete houses.

Cost and Earnings

Throughout all types of fisheries some profits are seen, to have been realised, even though fishermen complain much about the rising cost of fuel. In fact, there seems to be no fishing operator's household which suffered a deficit.

"Purse seine appears to be most profitable in terms of both total sale and net income. Thai purse seine has achieved the highest profitability, being more that 50 per cent of the total sale, being close to the best fishing ground of Indo-Pacific mackerel, which is mainly caught by Thai purse seine.

Total sale as well as net income from otter trawl is far below those of purse seine, when comparison is made for same size of boats. Poor catch due to overfishing of demersal fisheries resources may be the main reason.

Out of the total cost, material cost is generally higher than labour cost. This may be due mainly to the rising price of fuel and partly to cheaper labour cost.

Cost Component Analysis

The cost component is analysed for material cost, labour cost and maintenance cost.

Material Cost

Regardless of the type of fishery, fuel cost accounts for the major portion of total material cost, being around 70 per cent of the total. This fact clearly indicates how the rising price of fuel affects to the total operational cost, resulting in great difficulties in operating a fishery. After fuel cost, cost of ice appears to be fairly high, being around 15 per cent of the total.

Labour Cost

Labour costs consists of fixed wages, share–the amount of which is determined in accordance with the size of catch and food provided by the owner. As the amount paid for share is generally larger than paid as wages, the payments to employed crew seem to have been done mainly in the form of share system. The amount paid by a boat owner for the purchase of food for the crew would not seem to be a small expense, since it constitutes about 10 per cent of the total labour cost.

Maintenance Cost

Maintenance cost is composed of expenditure on the hull of the boat, the engine and fishing gear. Of the total maintenance cost, expenditure on the hull of the boat is the major one, generally being 40 per cent to 50 per cent of the total.

Depreciation Cost

For the computation of depreciation cost, the durability of the fishing boat and the construction cost of the boat have first to be sought. The average durability of a teak wood boat is 15 years Annual depreciation cost may be arrived by dividing the average construction cost by 15 years for all sample and type of boats separately for each size class. Even within a same size class of boat, that is, in-board powered boats of 8 to 15 m. there is bound to be a great difference in the construction of the boat due to the actual difference in the size of each boat. The use of such standard depreciation cost determined for each length of boat is, therefore, not always appropriate.

Thus it becomes clear that the problem of small-scale fishery is not the fishery itself, but a matter of fishermen who engage in this fishery. For fishery administration, therefore it becomes important to know who are the fishermen in need of assistance in order to improve their income and living conditions. With regard to the socio-economic problems of small-scale fishery, it is often said that small-scale fishermen are generally poor, meaning that small-scale fishermen include not only persons who actually engage in fishery, but also their family members who depend on, fishery for their income.

Economic Management of Small-scale Fisheries

Marine fisheries throughout the world co-exist as small scale or artisanal fisheries together with the large scale or industrial fisheries. Infact there is no standard definition of small-scale fisheries. There exist various classifications into small-scale and large-scale, into artisanal and commercial and into inshore and offshore according to vessel size, gear type, distance from shore or combination of three. Thus it is not unusual to find what is considered as small-scale fishery in one country to be classified as large-scale fishery in another. Attempting for rough categorizations of technical and socio-economic characteristics of the fishing activities of fishermen is between those who have broad spectrum of options, both in terms of fishing grounds and non-fishing investment opportunities (large-scale fishermen) and those who, by virtue of their limited fishing range and a host of related socio-economic characteristics are confined to a narrow strip of land and sea around their community, faced with a limited set of options, if any, intrinsically dependent on the local resources (small-scale fishermen).

The dualism is not confined to the scale of operation but extends to the type of technology used, the degree of capital intensity, employment generation and ownership. In contrast to large-scale fisheries, artisanal fisheries are owner operated and labour-intensive using very little capital and hardly any modern technology.

Though the small-scale fishermen landing almost half the world marine catch used for direct human consumption, yet with the exception of some motorization of canoes and the introduction of nylon nets, the fishing technology of small-scale fishermen in many parts of the world remained largely unchanged for decades. This, however, may have been a blessing for the economies of many developing countries which suffer from severe scarcity of capital and foreign exchange, ever rising fuel-imports bills and chronic underemployment. It has been estimated that the small-scale fishery uses one-fifth as much capital and one-fourth to one-fifth as much fuel per tonne of fish landed and creates a hundred times more jobs per dollar invested than the large scale fishery. Yet in many developing countries, small-scale fishermen live close to or below the subsistence level or at any rate they are among the lowest socio-economic groups.

The fundamental problem of small-scale fishermen around the developing world is their persisting absolute and relative poverty despite decades of remarkable overall fisheries development and national economic growth. They have neither adopted the advanced fishing technology nor did they find employment in the large scale fisheries or elsewhere for reasons ranging from capital market distortions and the (consequent) capital intensity of the large scale fishery to the limited mobility of the small scale fishermen or the lack of alternative employment.

Concepts of Fishery Management

Fishery management is the persuit of certain objectives through the direct or indirect control of fishing efforts or some of its components. For example, enforcement of minimum mesh size for regulating the size of fish at capture and increasing the productivity of the resources, or licensing system to control entry into fishery for maximizing the economic returns from the fishery. On the other hand fishery development, is the expansion of effective effort through a set of assistance programmes for exploiting underutilized resources and increasing fish supplies and fishermen's income. Fishery development may be defined more broadly to include, in addition to the expansion of fishing effort, improvement in post-harvest technology, marketing and transportation of fishery products as well as the provision of infrastructure and other related facilities.

Fishery management, for its 'control' feature is thought to be required once a fishery becomes "over exploited", while fishery development is thought to apply while a fishery is still under exploited'. This need not be so. One need not wait for overfishing to occur before management measures are taken. Overfishing is better avoided by judicious management measures taken along with development. Similarly the need for development is not confined to under exploited fisheries. The management of overexploited fisheries, sooner or later involves the regulation of fishing effort, development, fishery related or otherwise to absorb the surplus labour and capital.

The general objective of both management and development is the attainment of the 'optimum' rate of exploitation of the fishery. The defination of optimum depend on the policy objective. If the objective is maximum production, then the optimum rate of exploitation is defined by Maximum Sustainable Yield (MSY), that is, the maximum catch can be obtained on a sustained basis.

But if the policy objective is to maximize the economic benefit to the national economy from the fishery, the optimum rate of exploitation is defined by the Maximum Economic Yield (MEY), that is the maximum sustainable surplus of revenues over fishing costs. The fishery is said to be under–exploited in the economic sense if the actual catch falls short of MEY due to insufficient effort and require further development. On the other hand the fishery is said to be overexploited in the economic sense and call for management if the actual catch falls short of MEY due to excessive fishing effort.

In cases, where social considerations, such as, the improvement of socio-economic conditions of small-scale fishermen, generation of employment opportunities and improvement of income distribution matter, the optimum rate of exploitation is defined by a third concept, the maximum social yield. This is the level of catch and corresponding effort which provides the best possible solution to social problems. Introduction of social considerations may limit the speed with which management measures are introduced, or it may justify a more intensive rate of fishing than is justified on purely economic grounds. The concept of maximum social yields is the one most applicable to the case of small-scale fisheries in which socio-economic considerations often override both biological and strictly economic concerns.

Maximum social yield as the objective of fisheries management takes full cognizance of the likely conflict between income and employment objectives (more employment may lead to overfishing and a reduction of aggregate fishing income). Employment objectives may be weighted more heavily than income objectives only when there are no other effective means of redistributing income to lower income groups.

Biological Concepts

The basic biological concepts and relationships between fisheries development and management is the relevant relationship between 'sustainable catch' and 'fishing effort' in single species and single gear fishery. Sustainable catch is the quantity of fish in terms of weight of biomass which can theoretically be caught year after year without a change in the intensity of fishing. Fishing effort, on the other hand, is a composite index of all inputs employed for the purpose of realizing this catch. Maximum possible catch from a fishery on a sustainable basis by simultaneously adjusting the level of fishing effort, which corresponds to the highest level of sustainable yield can be achieved. It is also to be kept in mind that after a change in either the level of fishing effort or in the age at first capture, sufficient time must be allowed for the age structure of the stock to stabilize at the new conditions of exploitations.

The natural variability of the stocks as well as the possible decline in recruitment and the increase in variability associated with intensive fishing, suggest that on purely biological considerations, even maximum sustainable yield may be 'too risky' an objective for development or management.

Selection of a single optimum size at first capture is operationally difficult since what is optimum size for one species is likely to be too small or too large for some other species. In a multi-species fishery, a compromise mesh size will be required, certain species will be overexploited and others under exploited depending on the fishing technology and the biological relationships between species. The later could be particularly complex because competition and predation affect differently the various age groups of different species (*e.g.*, large predators can prey on small pelagic species which prey on the eggs of the former). These complexities obscure the effect of intensive fishing of one species on the abundance of another species.

The essentially experimental (or trial and error) approach to multi species fisheries management is probably the only option available at the current state of knowledge and management capabilities. It has the added advantage that it involves considerable learning by doing and it is amenable to modifications as new knowledge is accumulated and management capabilities improve.

Economic Aspects

Aggregate sustainable yield for a multi-species fisheries does not provide a meaningful management tool. Rather a gross economic yield or Total Revenue (TR) obtain by multiplying the catch of each species by its unit price, and expressing the resulting aggregate value or total revenue as a function of total fishing effort would be more appropriate device. The maximum value can be attained at a lower level of effort than that required for the maximum yield. At maximum yield the value of catch is less than the maximum.

Fishing costs are to be considered next. Fishing effort is an index of fishing inputs, such as, vessel, engine, crew, fuel, and other operating costs as well as fishing time. In culture fisheries sector, operational costs, such as, land rent, manures, fertilizers, seed, feed, manpower and other operational costs are taken its account. A complete bio-economic model can be obtained by putting revenues and costs together in which the net economic yield (resource rent) is obtained as the difference between revenue and costs.

The net economic yield of the fisheries is maximized at the level of effort when the total revenue equals the total costs. At this level of effort the last hour of fishing brings in a catch whose value (MR Marginal Revenue) is exactly equal to the cost of catching (MC–Marginal Cost). Each of the previous unit of effort beings in a catch whose value is in higher than its costs, while every unit beyond maximum gross value brings in a catch whose value is less than its cost. A wise owner or manager of a fishery will expand fishing effort upto that point, every additional unit of efforts adds to the profits but will not allow effort to go beyond which adds more to the cost than to the revenues and hence lowering profits.

This model is also applicable in the capital intensive culture fisheries especially in shrimp culture. Where the fishing effort is identified as inputs, such as feed, fertilizers, labour cost and rent. The farm manager or owner should be watchful not to allow further period of culture (DOC) involving costs of feed, labour and operational expenses to go beyond which adds more to the cost than to the revenues by the way of increased growth and production with the additional risks of mortalities due to overcrowding and diseases.

The maximum profit obtainable and the level of effort/inputs known as Maximum Net Economic Yield (MNEY) is the appropriate objective fisheries management as it ensures that the net benefit to the society from the fishery is maximized. In a multi-species fishery maximum net economic yield further reduces the risks of valuable species disappearing from the catch. Thus maximum net economic yield is preferable to the elusive maximum sustainable yield as the objective of fishery management, not only from economic point of view but also from the ecological point of view, since ecologically a greater diversity of species has greater chances of being achieved for lower intensities of fishing. The later has the added advantage of maintaining flexibility in the light of possible irreversibility of "extinction" of certain species about which very little is known. The absence of property rights over the resources and the presence of surplus profits (resource rents) at maximum net economic yield would encourage existing fishermen to expand their effort and others to take up fishing until all surplus profits are completely dissipated in excessive effort. Expansion of effort would only cease when total revenues just equal total cost and, hence, there are no surplus profits to attract new entrants. The level of effort (E3) which generates zero resource rents (zero net economic yield) is known as a 'bio-economic equilibrium', because at this point both the stock (bio) and the industry (economic) stabilize. From biological point of view, at E3 there is over exploitation of young fish, known as, 'growth overfishing and a risk of an over exploitation of the parent stock and chronic drop in recruitment known as 'recruitment overfishing'. At E3, the marginal returns are negative which implies that total revenues can be increased by reducing effort. But in an actual economy the potential benefits from fisheries management may even be greater since the process of entry into the fishery might not stop where all rents disappear but continue into the 'red' beyond E3, where fishing cost exceed the total value of the catch, and hence fishermen on the average do not earn even as much as they could earn from other occupations.

This could happen because the investment decisions are made on over-optimistic forecasts of yield based either on proportional extropolation of past yields or on exceptionally good fishing years. As the fishermen make their investment decision independently from each other and since the economic life of a vessels is quite long, over–investment is very likely. Once built a fishing vessel is to a large degree, a 'sunk ' cost and would keep operating, whether it covers its fixed cost (depreciation and interest on capital) or not, as long as it covers its operating cost. Another reason for the expansion and sustenances of effort beyond the bio–economic equilibrium and consequent negative resource rents is the tendency of Governments to subsidise (directly or indirectly) the industry, thereby lowering the private cost of fishing below its true social costs. Finally, fishermen may be earning incomes below their opportunity cost because of geographical and occupational immobility, itself the result of a host of socio-cultural factors.

In a fixed–price model, the price of its species is assumed to be independent of the size of the catch and this is not unrealistic assumption as majority of fishermen land only a fraction of the total catch. At national level, however, a flexible–price model need be introduced. Depending on consumer's preferences and import and export possibilities, the average price of all species combined will be relatively high at low levels of catch and relatively low at higher levels. As fishing effort expands, a number of forces are at work whose net effect may raise or lower the average unit price of the catch, at low fishing intensities; increases in catch tend to lower the average price and economies of scale from the expansion of operation also tend to reduce it, at high fishing intensities, the drop in catch tends to raise the average price, while the reduction in the size of fish caught tends to lower it.

Profit attracts new entrants, losses cause exit, and stocks are reduced with entry and increased with exit with corresponding changes in net natural growth. In multi-species fishery, changes in composition of the stock as a result of fishing take time.

An action, such as allowing of fish stock to recover from over fishing depends on whether the benefit exceed the cost of waiting. The crucial determinants of this benefits and costs are growth rate of bio-mass, the discount rate and the rate of depreciation of fishing asset. In case of multi species fisheries both the growth rates of individuals species and the rate at which certain species composition is altered or reconstituted are to be considered. The proper goal of management is the maximization of the present value of net revenues over the life of the fishery. This give rise to a dynamic maximum economic yield which is obtained by expanding or reducing effort to the point where the last unit adds to the present value of the stream of future revenues as much as it adds to the present value of the stream of cost.

In tropical multi species fisheries maximum sustainable yield is not a meaningful goal for the fisheries management. Not only does it ignore the cost of fishing effort but it may also result in a maximum catch consisting largely of trash fish at the expense of more valuable species whose relative abundance may considerably decline in the catch under the heavy fishing pressure required by maximum sustainable yield. The maximum economic yield is a more appropriate goal for fisheries management, since it results in maximization of a society's net benefit from the fishery, it keeps more options open in the light of our inadequate knowledge of ecological relationships, and it reduces the risk of collapse of certain species.

Fishery development aims at increasing the exploitation of under-utilised stocks by expanding effective effort through allocation of additional labour and capital, technological upgrading, training etc. Fishery management on the other hand calls for a reduction in fishing effort which sooner or later involves the retirement of fishermen and fishing assets. The implementation of such interventions, though justified on aggregate economic grounds, may be constrained by a variety of social considerations. Since fisheries development and management involve and affect primarily the fishermen, it is necessary to consider their values, their motivations and attitude towards the contemplated interventions and to examine the distribution of benefits from the interventions between fishermen and non-fishermen, and among fishermen themselves (small scale vs large scale, crewmen vs boat owners etc) in the light of their relative socio–economic conditions.

The introduction of social consideration into the bio-economic model leads to a new concept for the fisheries management, the maximum social yield, which is basically a modified maximum net economic yield. Introduction of social considerations may limit the speed or the extent to which management measures are introduced, or it may justify more development than is justified on purely economic grounds. This can be less illustrated by the case of an over exploited fishery in a rural economy with severe scarcity of alternative employment opportunities.

When there is considerable unemployment, fishing wages do not reflect the true opportunity cost of labour. If, because of wide spread unemployment fishermen have no alternative to fishing, their opportunity cost is close to zero and therefore, the society makes little or no sacrifice in keeping them in the fishery.

According to bio-economic model, fishermen will stay in the fishery as long as they earn an income at least as high as the opportunity cost of their labour and capital. As the fishery becomes over-crowded and profits for most fishermen disappear it is expected that those fishermen who are not able to earn from the fishery as much as they earn from other occupations to slip quietly out of the fishery, changing both occupation and location if necessary, that is, perfect mobility of labour and capital. But this does not happen.

Lack of occupational and geographical mobility may result from long isolation, low formal education, advanced age, preference for a particular way of life, cultural taboos, caste restrictions, inability to liquidate one's assets, indebtedness or just lack of knowledge and exposure to opportunities. The consequence of immobility is that fishermen may continue fishing even if they earn far less than their opportunity costs.

In fact, many of the socio–economic problems of small-scale fisheries arise from the asymmetry between entry and exit. To enter the fishery, especially in a good fishing year, is relatively easy. To leave, especially in a bad fishing year, is quite difficult, for one, fisherman might not be able to afford to spend time looking for a job or moving when income is down to subsistence level and he can hardly expect to find a buyer for his boat and gear during a bad fishing year. During a good fishing year, of course, to leave the fishery is out of the question. This entry during a good years and no exit during bad years swell the ranks of small-scale fishermen and reduce their income to subsistence levels. There is, also, the time lag between the decision to invest in fishing assets, which usually is made when fishing is quite profitable, and the actual entry which may take place when the average yield and profitability have already declined.

As per bio-economic model, the objective of every fisherman is income or profit maximization. It has often been argued that artisanal or traditional fishermen are engaged in fishing not for profit, but for subsistence, but, even substance is made possible either by consuming one's produce or by selling it for cash income. Since, fish is not a subsistence commodity (not a staple), a fisherman's subsistence depends almost entirely on his income, from sale of fish, whether as a boat-owner or labourer.

Some fishermen's objective may be to earn a certain level of income rather than maximize that income. They behave differently from the fishermen of bio-economic model, who chase every fish which has a price tag higher than the cost of catching it. Fishermen, who go after a target level of income reduce their effort when fishing is very profitable (a few fishing trips may be sufficient to meet their target) and increase their effort when fishing is poor, a behaviour with grave implications for both fisheries development and management.

In bio-economic model the purpose of fisheries management from the economic point of view was to maximize the aggregate social benefit without consideration of who gets what. But in many fisheries, small-and large-scale fishermen and boat owners and labourers in one hand, and the objectives of many governments to reduce income disparities on the other attach a bigger weight to the benefits accruing to small-scale fishermen and crew members that to large-scale fishermen and big boat owners. This would mean that social benefits would increase as a result of a change in the sharing system which increases the share of the crew, or as a result of a fisheries regulation which allocates more coastal resources to small-scale fishermen by banning trawlers close to shore even if total fishing effort is not reduced and total fishing income has not increased.

In case of small-scale coastal fishery which catches fish of small size before it is recruited into the stock exploited by off-shore large-scale fishery. The sustainable catch and its value may be raised by raising the age at first capture through the contraction of the small-scale fishery; but this by no means socially desirable if neither income redistribution nor the small-scale fishermen's participation in off-shore fishing could be made operational. Inefficiency and waste, especially of open-access resources, could be the price being paid for a tolerable distribution of wealth unattainable by other means in certain socio-political environments.

Constraints of Small-scale Fisheries

Marine fisheries, in addition to technological, capital and market constraints are still facing terminal natural resource constraints, there is a maximum quantity of fish that can be obtained from a national fishery or even from world fisheries, on a sustained basis. The new ocean regime however, brought about by the declaration of 200–mile Exclusive Economic Zones (EEZ) has increased the area over which a country has exclusive control of fisheries and hence its maximum sustainable catch under the prevailing economic and technological conditions.

In addition to resource constraint, the open-access nature of the resource is another constraint. Unlike agriculture, the sea and its living resources cannot be owned privately, because of their fluid nature and mobility. No individual operator has sufficient incentive to manage the open-access fishery, so as to maximize its productivity and economic returns. If he did, he would have to bear all the costs alone while receiving only a negligible share of the benefits. The implications of the open-access status of the resource is that even the limited productivity (maximum sustainable yield) or potential net returns (maximum economic yield) cannot be realized without some form of collective management because of the fierce competition which open access induces among the participants. However, management involves costs for research, administration, monitoring and enforcement, as well as political and social costs.

Small scale fisheries face even more severe constraints, such as, their level of technological development confines them to a much narrower area than the 200-mile EEZ; their immobility sea ward because of limited fishing range and land ward because of lack of alternative employment makes them particularly vulnerable to encroachment from both land and sea and because of relative easy and not costly entry, but difficult and painful exit due to variety of reasons ranging from chronic indebtedness to lack of better alternatives outside the fishery.

The limited fishing range of small-scale fisheries confines their area of operation in a narrow strip of sea often not exceeding a few kilometres from the coast. The occurrence and migration of fish into this area determine the resource available to the fishery. The abundance of this resource varies according to environmental conditions and off-shore fishing activity. Under constant off shore activity and stable environmental conditions there is a maximum yield that can be obtained on a sustained basis.

In tropical shallow waters, including coastal areas and reefs, there appears to be considerable consistency of yields ranging between 4 and 8 tons per square kilometre, while under estuarine conditions productivity is usually higher, excluding 10 tons/km^2 (Pauly, 1980)

Maximum Sustainable Yield per Surface Area of Selected Tropical Ecosystems

Type of Ecosystem	*Location*	*Max. Sustainable Yield (MSY)*	*Source*
Coral reef	Jamaica, Caribbean	4.0t/km^2	Munro (1975)
	West Indian Ocean	5.0t/km^2	Gulland (1979)
Shelf	Gulf of Thailand (<50 m.)	3.6t/km^2	SCSP (1978)
	San Miguel Bay (Philippines) (<15 m.)	8.0t/km^2	M. Vaxily (1980)
Estuarine conditions	Gulf of Mexico	12t/km^2	Saila (1975)
	Sakumo Lagoon (Ghana)	15t/km^2	Pauly (1976)

The majority of the fishermen complain that their catch and income have been declining in recent years despite an increase in fishing effort on their part, citing reasons, such as, increase in the number of fishermen, encroachment by trawlers, low price for trash fish whose proportion in the catch is rising, coastal pollution etc. A reduction in the average size of shrimp and of other high value species is also reported.

The resource constraint is probably binding and no further development based on the same resource appears to be possible.

A number of ways can be thought of to mitigate resource limitations. They are;

1. Control of gear selectivity and mesh-size regulation to raise unit value of catch (especially shrimp) and to return the fishery on a higher trophic level (more valuable species composition of catch);
2. Increase the value of catch through improvements in marketing efficiency and catch utilization;
3. Reduction of encroachment by trawlers and other commercial fisheries into the area reserved for the small-scale fishery;
4. Reduction of off-shore fishing to increase the abundance of the same stocks in the inshore areas, and
5. Expansion of the fishing range of the small scale fishery through technological upgrading.

In a multi species fishery an excessively small mesh size does not only prevent a fishery from attaining its maximum sustainable yield but it could also change the composition of maximum sustainable yield towards less valuable species. If selective fishing is not possible, it is necessary to chose the mesh-size, and to consider its effects on the trophic level at which the fishery should operate, that is, the mesh size which will put the fishery on highest possible total revenue.

Although long-term gains is expected from mesh size regulations, yet the fishermen are not willing to increase their mesh size because:

1. With a larger mesh size they would have no catch and no income for several months, which they cannot afford; and
2. They fear that larger shrimp will attract the trawlers, as in past, wiping out both larger and smaller shrimps and any other fish found inshore.

The catch from a multi species fishery consists of a variety of species and sizes of fish, varying commercial value according to their utilization. Some species and sizes being completely unconventional as human food are classified as trashfish and are either discarded or sold for fish meal.

Also, waste and spoilage of catch in small-scale fisheries are substantial for a variety of reasons:

1. Relatively high perishability of tropical fish;
2. Lack or, inadequate supply of ice and freezing facilities;
3. Dispersion and remoteness of small-scale fisheries from consumption centres.

These factors imply high collection, marketing and storage cost and low reservation price for fishermen (once landed fish must be disposed of immediately). The smallness and dispersion of landing points precludes both economies of scale in the collection of catch and competition among traders, while the remoteness of fishing communities, means limited knowledge of market conditions and low bargaining power for the fisherman.

It is possible to reduce the proportion of trash fish catch (including spoilage) and to raise the value of catch by increasing the supply of ice and by promoting the utilization of small sizes and unconventional species for human consumption through processing (fish paste consumption, FPC) or through induced change of consumer tastes. Similarly, the marketing margins could be reduced and the price paid to the fishermen increased through promotion of competition among traders, supply of price informations to fishermen and provision of marketing infrastructures, such as, feeder roads, landing centres and marketing facilities.

Conflicts

Not only is the resource base of the small-scale fishery limited by its fishing range and natural productivity, but often it has to compete for this limited resource with other fisheries using more advanced technology. In fact, small-scale fishermen are often more strategically positioned on the biological cycle, but this is generally more than offset by a fundamental asymmetry between the two fisheries. The large-scale fishery can operate both offshore and inshore, which the small-scale fishery is confined to inshore fishing.

Although large-scale fishing units are often prohibited by law from operating in the coastal area reserved for the small-scale fishery, the presence of high-value species, such as shrimp, and higher fish densities in shallow waters, which characterizes tropical ecosystems as well as increasing fuel costs coupled with enforcement difficulties result in encroachment and open competition between two fisheries over the same resources.

The two fisheries are also in conflict in the market to the extent that they use the same inputs or catch the same species of fish. Large-scale fishermen may bid-up the prices of fishing inputs or their massive landings may depress fish prices while this may raise the welfare of the producers of fishing inputs and the consumers of fish, small scale fishermen may become increasingly uncompetitive. A fundamental condition for fair competition is that the participants have access to the same capital market.

Small-scale fisheries are further handicapped by their dispersion and remoteness which precludes economies of scale in the marketing of catch and procurement of inputs, which may only partially be compensated by their low opportunity cost and low capital and fuel costs. Although in social terms small-scale fishermen often may be low-cost producers, in private terms their unit cost may be relatively high because of inadequate infrastructure and high cost of borrowings.

Institutional support is often skewed in favour of the large-scale fishermen because of:

1. Their apparently high efficiency and greater contribution to economic growth;
2. Their ability to concentrate in few landing areas which allow economies of scale in the provision of infrastructure, such as, landing facilities, roads etc. and the delivery of assistance programmes;
3. Their visibility, political influence and economic strength; and
4. A general bias in favour of large-scale fisheries under the open-access regime.

In contrast, small scale fishermen are geographically dispersed and isolated and politically unorganised and weak. Because of their technological stagnation and their feature as economic activity of last resort small-scale fisheries rearely seem to satisfy the conventional investment criteria.

Economic Parameters-Variations

Under constant economic and bio-ecological conditions an open access fishery will reach an equilibrium at a level of effort and corresponding catch yielding zero profit to the fishery as a whole, although fishermen with lower cost may be earning some rents of efficiency.

A rise in fishing costs would result in losses for the marginal fishing units forcing them to leave the industry. With the rising of fishing costs the total effort expended by the fishery will decline, catch would fall in short run. In the long run, catch would fall if the fishery is biologically under exploited and would rise if the fishery is biologically overexploited. Reductions in fishing costs resulting from the introduction of a new technology would have the reverse effects. However, no long term benefit for the fishermen or the society, results from technological upgrading under open access conditions.

Changes in fish prices occur as a result of shifts in the supply of or demand for fish. With a given demand, a poor catch would lead to rise in price and a rich catch to a fall in price. With a given supply, rising demand (due to population growth or increasing incomes) would lead to rising fish prices. With unchanged costs, an increase in fish prices would result in excessive profits which would induce entry into the fishery and expansion of fishing effort until all profits are dissipated and a new equilibrium is attained at higher effort. Catch rises in the short run but falls in the long run if the fishery is biologically over exploited. Daily or seasonal fluctuations or prices may or may not affect the equilibrium level of effort depending on the case of exit and re-entry (including the availability of alternative employment for labour and capital).

The effect of changes in mesh size or other regulations which improve the productivity of the resources and, thus, the total yield and, immediately, the individual catch rate, without affecting the fishing costs will be identical. On the long run, mesh size regulation will not improve individual incomes and the total economic rent, unless action is similarly taken to prevent the fishery to drift to a new equilibrium at higher effort. Therefore, mesh size and other gear selectivity regulations cannot be considered as an alternative to the regulation of the amount of fishing.

In small-scale multi species fisheries the effects of changes in prices and costs are not likely to be as clear–cut. Due to fundamental asymmetry between entry and exit (the later being more difficult than the former), favourable changes in costs or prices followed by unfavourable changes result in surplus numbers of fishermen earning incomes below their opportunity costs. By definition, lack of mobility is inability to respond to changes in economic conditions. If mobility towards the fishery is greater than mobility out of the fishery, favourable changes can hurt them. Also in multi species fisheries, changes in fishing effort resulting from a change in economic parameters may not result necessarily or only in the expected change in the quantity of catch but to a change in its composition. If the fishery supplies a large share of the market, the change in catch composition would cause a second–stage change in fish prices mitigating the effect of the original change. A fall in costs or a rise in price may make profitable (and even socially desirable) the "extinction" (economic extension in the sense of near disappearance of a certain species from the catch is not uncommon in multi species fisheries) of some species.

In the absence of restrictions on entry, reductions in the prices of fishing inputs, technological up–grading, and increase in the price of fish can benefit the fishermen temporarily. In the long-run the fishermen cannot expect to earn more than their opportunity costs. In fact, an initially beneficial change (fall in costs rise in fish prices) may ultimately hurt the fishermen if it encourages "excessive" numbers of entrants who, once in the fishery, find it difficult to leave despite possible losses.

An expansion in the resource base through manipulation of mesh size type of gear and geographical distribution of fishing operations or simply through allocation more resources to the fishery would lead to expansion in fishing effort and would have no lasting effects on fishermen's incomes, as long as, the open–access status of the resource remains unchanged. In the absence of regulations of fishing effort, improvement in the utilisation of catch, or the efficiency of the market system would have a temporary rise in incomes followed by expansion of effort and return to the previous income levels which, in the long run, are uniquely determined by opportunity costs. The only way in which a lasting increase in fishing incomes can be effected is either by limiting effort through fisheries management or by raising the fishermen's opportunity cost through non-fishery development (creation of alternative employment) Fishery development can help only if it is implemented after or concurrently with fisheries management.

Modernization of Fisheries

Modernization of fisheries in Japan since the first-nineties has proceeded in two directions. First of all modernization has taken place in the promotion of pelagic fisheries. Large fishing boats and trawling techniques were imported from Europe and America, and the technical guidance was provided to the fishermen. Fishing grounds were also enlarged from coastal to off shore waters and further to the high seas, and through these innovations, the development of large-scale commercial fisheries was promoted. Japanese fisheries began to be industrialized starting in the first–nineties, and after a lapse of about 80 years, they have developed to a constant production level of a 10 million ton annual catch.

However, about 40 per cent of the 10 million tons of fishery production consists of the production from coastal fisheries and culture fisheries. As to the size of fishing boats, 90 per cent of them are small motorized fishing boats under 5 tons.

An overwhelming majority of workers engaging in fisheries throughout the country are coastal fishermen. This fact indicates that the promotion of coastal fisheries is important as the primary means for overall development of fisheries.

Coastal fisheries are managed by families and produce a small amount of medium and prime fishes with high market value. Fishermen have continued to devise various fishing gears and methods throughout the years and have strived to elevate productivity although the form of fishery has remained as traditional small scale fishery. In the Japanese coastal fisheries, modernization of fishery production was realized by motorization of fishing boats, mechanization of fishing work and an attempt to transfer to culture fishery from capture of natural resources. However, at the same time the value of marine goods was raised by improving the methods of maintaining freshness of catches and processing techniques, and by rationalization of distribution.

Japanese coastal fisheries have demonstrated that the pursuit of ever larger fishing scale, such as, the construction of larger fishing boats and expansion of management scale is not the only means for developing fisheries. Small-scale fisheries have a direction of development of their own. There are many countries in the world that need to strive to raise their standard of living by giving employment opportunities to coastal inhabitants by promoting fisheries and by increasing fishery income. They must consider the importance of small-scale coastal fisheries from this view point. Further more one can anticipate that a smooth development of off-shore fisheries conducted by medium and large fishing vessels will occur naturally after small-scale fisheries with adequate technology and management have been developed and established as industries.

Gill Net Fishing Ensure Stable Fishing Life of Coastal Fishermen

Pelagic fishes such as, sardine, herring, horse mackerel and mackerel, migratory fishes like salmon, trout, skipjack and tuna and fresh water fishes such as, carp, crucian carp are the most familiar types of fishery resources–These three groups of fishes account for little over 65 per cent of the total catch of the world.

However, demersal fishes, such is, flounder, flat fish, cod, alaska pollack and benthic aquatic animals such as, lobsters, shrimp, crab, squid and octopus are equally important fishery resources as the above mentioned fishes. The annual catch of so-called bottom dwelling animals is over 20 million tonnes and amounts to one-third of the total catch in the world.

The fisheries catching demersal fishes, crustaceans and molluscs in greatest quantity are the various types of small, medium and large-scale trawl fisheries, and the large fishing grounds have been exploited on the continental shelves around the land masses of the world and the majority of such submarine fishery resources still remain to be surveyed, and these unused fishery resources will be further exploited in the future.

Attention is required to be given to demersal fishes, crustaceans and molluscs inhabiting coastal waters along the developing countries in abundance and to emphasize the necessity of making effective use of gill net fishing (especially bottom gill net), in addition to trawling, for the promotion of coastal fisheries. In the coastal waters along the developing countries, there seems to exist a considerable amount of demersal fishery resources which have not yet been surveyed and exploited. From technical view point catching of these resources by the gill net fishing method is relatively easy, and besides its initial cost is low, and the catching rate can usually be improved by the proper use of this fishing method.

The Japanese coastal fishermen are now enjoying a stable fishing life by mainly catching coastal demersal fishes and like using gill net.

The type of fishing conducted, as well as its operational ability and the fishery regulations involved are determined by the size of fishing boats used. In order to carry on fishing as a main occupation, an effective operation plan for fishing must be drawn up to cover the entire year. It is needless to say that a fisherman must adopt the most appropriate fishing plan according to the size of fishing boat he uses.

Utility Boat Fitted with an Outboard Motor

This is mainly used for gathering kelp along the shore. Moreover, by using this boat also for shellfish gathering and gill net fishing in shallow waters, fishermen can make a sufficient living throughout the year.

3-5 Ton Diesel Fishing Boat

Because of higher power, this boat can go to fishing grounds several kilometers off the coast. Therefore, it becomes possible to increase the catch per boat and to operate various kinds of fishing in many different ways. This makes it possible to select the fishing object and fishing method which seem to be most profitable judging from the daily fishing conditions and thus to achieve a stabilized business.

5-10 Ton Diesel Fishing Boat

This class of fishing boat is engaged in only a few kinds of fishing and fishermen are specialized in this line of fishing. Although the productivity of each fishing boat is high, in case of a poor catch, it is not so easy to convert to fishing for other fishes as in the 3–5 ton class. Therefore it is necessary for this class of fishermen to have the capability to change their fishing grounds from coastal to off shore and to have the ability to obtain outside labour when necessary.

Economics of Fishing Boat Operation

The survey team of Indian Institute of Foreign Trade reported that the cost of high speed diesel oil, crew's share and repairs and maintenance constitute the principal items of expenditure. The cost of diesel varies from place to place and ranges from 20.7 to 39.4 per cent, with an average of 28.7 per cent of expenditure for all. The share of crews worked out to between 15.4 to 36.8 per cent, with an average of 25.9 per cent for all firms. The expenditure on repair and maintenance ranges from 21.3 to 34.6 per cent, its average share being 27.8 per cent. On an average the depreciation of boats works out to 14.6 per cent, while for individual firm it ranges between 11.4 and 24.2 per cent.

Basing on the data obtained above from field survey, the Indian Institute of Foreign Trade have worked out the economics of boat operation as follows:

Manning of Vessels

Six persons in each vessel, comprising of skipper, second hand, engine driver, tindal, greaser and fisherman.

Diesel Oil

If constitutes a major portion of expenditure being of the order of 20 to 39.4 per cent of the overall annual expenditure.

Crew's Share

For 13.8 m. vessels, the crew's share has been taken 30 per cent of the value of the catch, after deducting the operational expenses. For bigger vessels, like 17.5 m., 22 m. and 27 m., it has been worked out at 20 per cent in view of increased catches anticipated from large vessels.

Depreciation

Depreciation for all the boats has been taken at, 10 per cent, normal life of vessels being 10 years.

Interest on Loan

A uniform rate of 10 per cent on the entire capital cost has been taken into consideration.

Repayment of Capital

It is a quite significant factor and has been calculated at the rate of 10 per cent, taking the life of the vessels at 10 years.

Other Expenses

Port dues, insurance, stores and over heads expenditures have been included under this head.

Gross Returns

1. A standard rate for shrimp at Rs. 4000 per tonne has been uniformly taken for all regions in view of heavy demand for processing and export market.
2. Fish prices from 13.8 m. vessels have been calculated Rs. 300 per tonne in Verabal, Goa–Mangalore and Kerala; at Rs. 400 per tonnne in Maharashtra, Visakhapatnam and Tamil Nadu.
3. The price taken for fish from 17.5 m, vessels is calculated at Rs. 300 in Goa, Mangalore and Kerala and Rs. 500 per tonne in Verabal, Vishakhapatnam and Rs. 600 per tonne in Maharashtra and Rs. 400 per tonne in Tamil Nadu.
4. Fish prices for 22 m. and 27m. Vessels have been calculated at Rs. 1000 per tonne in all regions.

Basing on the above back ground, the annual returns of the vessels of different categories in the different regions have been given below. The rate of annual returns differ to a great extent with duty-free diesel oil, when compared with diesel oils with duty.

(Profit in Rs. '000)

Region	*13.8 m.*		*17.5 m.*		*22 m.*		*27 m.*	
	Oil with Duty	*Duty free Oil*	*Oil with Duty*	*Duty free Oil*	*Oil with Duty*	*Duty free Oil*	*Oil with Duty*	*Duty free Oil*
Marashtra	56.0	82.7	130.2	169.9	109.0	177.9	89.5	175.9
Tamil Nadu–Tuticorim	"	"	66.2	105.5	"	"	"	"
Vishakhapatnam	"	"	98.2	137.5	"	"	"	"
Kerala	40.6	68.8	34.2	73.5	"	"	"	"
Verabal	"	"	98.2	137.5	"	"	"	"
Goa–Mangalore	"	"	34.2	73.5	"	"	"	"

Economics of Operation of 17.5 m. (200 BHP) in Maharashtra

(Rs.)

	With Duty	Without Duty
A. Capital Cost of Vessel and Gear		9,05,000
B. Operation cost		
(i) Fuel oil, lubricant .oil, water	79800	30700
(ii) Port dues	1200	1200
(iii) Running maintenance	6000	6000
(iv) Ice (50 per cent cost)	16500	16500
(v) Misc. and stores	3600	3600
	107000	58000
Repayment of capital at the rate of 10%	90500	
Depreciation	90500	
Average interest on capital @ 10%	59000	
	240000	
C. Gross Income		
Shrimp @ Rs. 4000/tonne for 100 tonnes		400000
Fish @ Rs. 600/tonne for 400 tonnes		240000
		640000
D. Net Income After Deducting Operational Cost	532900	582000
E. Crew's Share @ 20% of Net Income	106580	116400
	426320	465600
F. (i) Less annual repairs Rs. 18000		
(ii) Less depreciation, repayment and interest on capital	240000	
(iii) Less overheads	3600	
(iv) Less insurance	18000	
(v) Less ice cost (50 per cent)	16500	
	2,96100	
G. Profit		
	130220	169500

Economics of Fishing Vessels

The economics of operation of small sized fishing vessels (36′–38′) have been worked out by different agencies through out the country and the economic returns of those fishing vessels have been an established fact since, these small fishing boats have attracted large amount of private capital. But the range of these small vessels are only inshore waters, which according to many authorities are over-exploited areas.

Before introducing long range vessels for exploiting distant waters, it is necessary that the economics of operations and problems involved in the operations of such vessels are clearly understood.

Perumal (1973), has given some factual information regarding the operation of such long ranged vessels, attached to the Central Institute of Fisheries operatives as training vessels.

Economics of Operation of 93′ Fishing Vessels

The economics of operation has been calculated on the basis of "Blue Fin" 93 feet stern trawler. The vessel could be operated for about 230 days in a year, provided there are no major break–down.

A.	Capital Expenditures	
1.	Capital cost of the vessel	Rs. 2626500
2.	Cost of gear for a one year period including maintenance charges	Rs. 30000
B.	Operational Cost	
1.	Fuel oil, lubricant oil, Average oil consumption per hour 110 litres. For 10 hours running a day and for 230 days, 230 × 110 = 253000 litres @ 0.80 paisa per litre	Rs. 202400 or 48.26% of operational cost
2.	Cost of fresh water @ Rs. 18 per month for 12 months	Rs. 220
3.	Cost of ice @ Rs. 50 ton/month for 12 months	Rs. 27000 or 6.46% of operational cost
4.	Cost of repairs, and maintenance namely, docking charges, spare parts and labour	Rs. 38730 or 9.2% of operational cost
5.	Contingent expenditure	Rs. 5000 or 1.2% of operational cost
6.	Other expense	Rs. 15000 or 3.58% of operational cost

7. Managerial expenses	Rs. 30000 or 7.16% of operational cost
8. Pay of floating staff	Rs. 48000 or 11.47% of operational cost
9. (a) Repayment of capital at 10%	Rs. 2,62,650
(b) Depreciation at 10%	Rs. 2,62,650
10. Insurance @ 2% of the total cost	Rs. 52530
11. Crew's share for one year period	Rs. 23000 or 20% of the gross income from fish catch
C. <u>Economics of Operation</u>	
1. Total operational cost of the vessel for one year period	Rs. 419350
2. Depreciation, repayment interest on capital, insurance etc.	Rs. 705300
	Rs. 1124650

Fixing a maximum profit at 6¼ per cent, the economic operation of the vessel would be possible if the total income from the catches could be brought up to Rs. 1194940.

To achieve the above financial target, the vessel should catch on an average at least 30 tonnes of prawn and 1490 tonnes of fish and market at the following rates.

Prawn

30 tonnes @ Rs. 10000/tonne–Rs 3.00 lakh

Other Fishes

1490 tonnes @ Rs. 600/tonne–Rs 8.94 lakh

Economics of Operation of 57 Feet Vessels

These vessels have worked within 40 fathoms and could be operated for about 230 days in a year provided there are no major breakdowns.

A. Capital Expenditures	
(i) Capital cost of the vessel	Rs. 8,75000
(ii) Cost of gear (trawl net) for one year period including maintenance	Rs. 20,000
B. Operational Cost	
(i) Fuel oil, lubricant oil–Average consumption 40 litres/hour. For ten hours a day for 230 days, 92000 litres @ 0.80/litre	Rs. 73,600 or 42.63% of operational cost
(ii) Cost of water @ Rs. 15 per month for 12 months	Rs. 180
(iii) Cost of ice at the rate of 20 tons a month for 12 months and cost Rs 45/ton	Rs. 10800 or 6.23% of operational cost
(iv) Cost of repairs and maintenances *e.g.*, docking charges, spare parts, and labour charges for one year	Rs. 24000 or 13.92% of operational cost
(v) Contingent expenditure @ Rs. 300 per month for 12 months	Rs. 3600 or 2.03% operational cost
(vi) Other expenses	Rs. 12000 or 6.96% operational cost
(vii) Managerial expenses	Rs. 30000 or 17.4% operational cost
(ix) (a) Repayment of capital at 10%	Rs. 89500
(b) Depreciation at 10%	Rs. 89500
(c) Average interest on capital at 10%	Rs. 58300
(x) Insurance @ 2% of the total cost in a year	Rs. 18000
(xi) Crew's share in the form of incentives income from fish	13000 in a year or 20% of gross
C. Economics of Operation	
1. Total operational cost of the vessel for one year period	Rs. 172180
2. Depreciation, repayment, interest on capital, overhead insurance	Rs. 237300
	Rs. 409480

In order to obtain a minimum profit at 6¼ per cent, the economic operation of the vessel would be possible if the total income from fish catches could be Rs. 435072 per annum.

To achieve this, the vessel should catch at least 30 tonnes of prawn and 230 tonnes of fish *i.e.*, 130 kgs of prawn and 1 tonne of fish per fishing day and marketed at the rate below.

Prawn

@ Rs. 10000/tonne for 30 tonnes–Rs 3,00,000

Fishes

@ Rs. 600/tonne for 230 tonnes–Rs 1,38,000

Economics of Fish Processing

The freezing and canning industry has registered a phenomenal growth during last two decades. These industries have got a positive correlation with mechanised fishing and both of them depend on each other.

A survey of Indian Institute of Foreign Trade reveals that 95 per cent of raw material purchases are made through middle men, who charges commission. In Kerala, the agents charge 5 per cent of the value of purchases as commission.

The raw material cost of the industry is further increased by purchase taxes imposed by some of the state governments. Maharashtra charges 2 per cent purchase tax, while Kerala government has imposed 5 per cent purchase tax on prawn.

Cost of Production of Frozen Prawn

Nearly 80 per cent of the ex-factory cost, on an average is made up by raw material cost. As the cost per unit weight of raw material is more in the case of larger sizes of raw material which are processed into headless form, the cost of processing, which remains more or less the same in the case of all types of packs, will show itself to be relatively less than the average figure.

In 1968, average for six factories given below, shows the average expenditure for freezing and packing.

1. One pound of shrimp–Rs 0.30
2. Outward freight to USA, custom charges etc.–Rs. 0.65/lb.

As against this, the processing cost in USA and Mexico for freezing and packing as made available by the trade is on an average 7.5 per cent (equivalent to Rs. 56.25) per pound including labour, ice requirements, cost of packing materials etc. The cost of processing is much less in India. The freight costs are however, high and place a heavy burden on the Indian Industry

Cost of Production of Canned Prawns

In case of canned prawns, the expense on cans, labels, labour, transportation, fuel etc. work out to about 50 per cent of the ex-factory price of the finished goods. Of this, nearly 20 per cent is due to cans and labels, while about 11 per cent represents purchase tax, commission to middlemen or suppliers and cost of transportation to units.

Shrimp Trawling–Economic Aspects

High capital and operating costs in sea fishing mean that the operation is only profitable if revenues are high. The industry, therefore focuses on shrimp and high value fish. Retaining large bulk catches of low value fish would necessitate an early return to port or divert labour or storage from shrimp, without yielding the high revenues associated with the later.

The discard of low value fish (by catch) has fuelled a growing debate over this wasted resource and the potential of its utilisation has for enhancing fishing incomes and the nutritional status of low income groups in India.

Shrimp trawling all over the world is associated with a large by-catch. In most shrimp fisheries the by-catch comprises of 80-95 per cent of catch by volume. Most of the fish caught with the shrimp on India's west coast is landed often in poor condition. On the east coast, however, substantial quantities of by-catch are discarded at sea-a practice that results from the emphasis on voyage fishing.

The by-catch means "non target species" caught with, and incidental to, target species. In India, trash fish, low value fish and miscellaneous fish are often referred to as by-catch, and on account of species or size are usually dried for use as food or fishmeal.

It has been estimated that in 1988, approximately that 1,00,000 to 1,30,000 tonnes of shrimp, by-catch was discarded by the east coast fleet. The species and size distribution of the discarded by-catch is general rather than specific. The main selection criterion in retaining or discarding fish seems to be size, but the cut-off point is different for each type of trawler. Most of 20 m. trawlers seem to have a policy of retaining large quality fish throughout the voyage. Rest of the catch is discarded. Some trawlers will retain more of the catch in the last few days of the voyage. Every thing less than 20 cm. is discarded (80-90 per cent of the by-catch). The most abundant species discarded are jewfish, silverbellies, threadfin, sardines, ribbonfish, travellys, goatfish, Bombay duck.

Reasons for Discarding By-catch

It is often said that fish is discarded by the trawlers because its value is insignificant relative to that of shrimp. This is partly true. The average wharfside price for shrimp (head off), taking a weighted average for the three main varieties is 300 rupees/kg. The price for large good quality fish, on the other hand, is around Rs. 15-20/kg. Much of the remaining catch is sold for fishmeal at only Rs. 2/kg.

The statement on its own overlooks the question of who benefits from the sale of fish. This income may be negligible to the trawling company but is not so to the crew. In some companies, the proceed from fish sales are entirely the crew's. Even when a company take a share of this fish revenue, this is generally not as large as the shrimp revenue share.

The reality this that even if the additional income is small, it would be realistic to except the industry to realise this income, particularly with the current situation of declining net revenues in some way, or unless it is perceived as jeopardising shrimp revenue in some way, or unless there are technical or institutional reasons for not doing so.

This is, in fact, the case, as indicated by the reasons that crew give for discarding fish:

1. Limited ice or freezing (storage) capacity.
2. Lack of on-board processing on bulk reduction system (commercially viable).
3. Possible contamination of shrimp.
4. Shortage of labour for handling fish.
5. Large quantities of fish are difficult to sell and the trawler is delayed in port.
6. Transfer to other boats is risky if sea conditions are rough.
7. Insufficient space for drying or storing more fish.

In order to metigate the above mentioned constraints, there has to be a change in the value of the by-catch relative to the shrimp.

All the trawling companies and crews are of opinion that if fish prices are higher they would land more by-catch, or trawl for fish on the return journey. The solution to this is perceived, by virtually the entire industry, to be the development of "value added products from trash fish"–basically, developing an up-market product with low cost inputs, most probably in frozen form. The more ambitious technologies, however, are also the most risky, having no commercial track record in the context of tropical multi-species by-catch.

The distribution of any value-added frozen product depends on the efficient functioning of cold chains, and, in the case of India, would require significant inputs on the market development side. If these two obstacles can be overcome, India's urban middle class certainly represents a large market, and processors might best consider this market prior to venturing into the highly competitive international market.

If successful, the processor and trawler companies would have a greater incentive to land more fish, this incentive would not necessarily spill over to the trawler companies without downstream processing links. The effect here would depend on the level of demand for such products and, hence, the fish. If the development of such products brought about a real increase in wharfside prices of some fish, this would be beneficial to all fishermen, but could have negative employment, income and nutritional effects on traders, processors and consumers of fish diverted from traditional uses. Extra employment would be generated, however, in the minced fish processing industry.

Some companies are considering this type of product development in conjunction with a collector vessel system incorporating either a large "mother ship" to process by-catch from a fleet of trawlers, transferred by purpose-built collector boats, or to supply a land-based plant.

Effect on Fish Prices

Even without the impetus that new products could inject into markets for fish in India, the ratio of shrimp to fish prices is likely to diminish over time. The price of fish in India has increased relative to other food items over the last two decades while worldwide expansion in shrimp production is likely to result in stagnating, or falling shrimp prices. Any change in this ratio is likely to have an effect on the margin. For instance, the delay in port necessary to sell fish is considered costly in terms of foregoing trawling time. If no real change or increase in voyage fixed costs, and a need to spend some time in port for other reasons, then any incremental increase in the price of fish relative to shrimp should result in increased landings of by-catch if the trawling companies are maximising profits. In reality, there is likely to be a lagged response to any change in price ratios, but this scenario is consistent with the current tendency to land increasing quantities of by-catch.

Some of the problems in the marketing and distribution of fish in India (such as market development, transportation, communications, development of cold chains) will ease as India's infrastructural base is gradually developed. General population pressure and the growth of an urban middle-class will result in upward pressure on fish prices, and a change in fish handling practices.

Many trawling companies think that cold storage for fish is the solution to the delay in port, consequent on their trying to sell large quantities of frozen fish during a brief spell in port and being obliged to accept low prices. Any such facility, however, would require careful management. A build-up of supplies could be regulated by charging storage costs on a daily basis to the company concerned, and having an arrangement for auctioning fish after certain time. This would enable day to day fluctuations in supply to be evened out, and assure supplies of fish to buyers at any time. If, however, trawlers are to land substantially larger quantities of fish as a result of this facility, simultaneous improvements in marketing and distribution of frozen fish would be needed. Without this, the cold storage would just result in a higher cost product being auctioned at, or below, existing low prices.

Incentive to Land More Low-value Fish (Bycatch)

Prices of Wet Fish

Prices of wet fish are extremely variable and depend on fish size, species, quantity being sold and quality. There is also seasonal and geographical price variations. Observed wharfside prices of Vishakhpatnam during September and October, 1988 were as below.

	Rs./kg	
Tiger shrimp	270	(10% of shrimp catch)
White shrimp	160	(30% of shrimp catch)
Brown shrimp	70	(60% of shrimp catch)
Cuttle fish	30	
Pomfret	7.75	
Eels	6.5	
Perch	6	
Red snapper	6	
Shad	4	
Sharks with fins	4	
Sharks without fins	2.75	
Mackerel	3	
Mixed	1.5	

These prices were quoted for the catch of large trawlers at a time when there were many trawlers in port. The prices are considered low, but not untypical. Generally the fish listed would be the larger retained species in the by-catch, prices for such fish are generally Rs. 5-10/kg. Most of the catch of small trawlers would be smaller varieties and would sell at less than Rs. 5/kg. Very small mixed species would be sold for drying as fishmeal, and sometimes food use at Rs. 0.5 to 1/kg.

Prices of Dried Fish

Processors purchasing wet fish at wharfside obtain very low margin from dried fish. Where fish is dried on board, or by fishing families and the cost of wet fish is, therefore, not a cash out lay–it is probably the most remunerative product form of fish that is small un-iced, prone to rapid spoilage and, essentially of low value. Fish is always sorted by species and some species may be salted prior to drying. If not dried on board, it is dried at the quay or taken to nearby fishing communities.

Prices of some dried fish at Vishakhapatnam wharf in October 1988 are so low that they imply a negative return on the purchase of wet fish.

Prices of Dried Fish

Species	*Dried fish Rs/kg.*	*Break-even Price of Wet Fish Purchases (Implied)*	*Observed Wet Fish Price Rs./kg.*
Lactarius	6.25	1.56	3.00
Mackerel	5.00	1.25	3.00
Jew fish	4.30	1.08	2.00
White bait	16.00	4.00	
Ribbon fish	3.20	0.80	
Silver bellies	3.00	0.75	

Assuming the dry weight is 25 per cent of wet weight and fish takes two days to dry, and the net return on dried fish is Rs. 0.5/kg, then a processor would have to dry 250 kg of wet fish every two days, in order to earn Rs. 15 per day. This takes no account of salt costs (Rs. 0.5/kg) used in a salt fish ratio of 1:4 to 6.

Value Addition by Freezing

In 1988, the cost of sending iced fish by train to Hydrabad from Vishkhapatnam was less than Rs. 2.5/kg, though the wholesale price of wet fish in Hydrabad were Rs. 20-22/kg for pomfret and Rs. 12-13/kg for perch or eel. This appears to offer a significant margin to the broker or packer, though no mention has been made of product losses.

The profit potential for value added product (a 500 gm. cartoon of pomfret steaks', making use of large, high value fish) is indicated below. This retails in Madras of Rs. 44/kg, whereas the raw material price is about half of that.

Cost Structure for Packed Frozen Pomfret 'Steaks'

	Price Rs./kg
Pomfret, steak quality, sorted (wharf side price)	14.00
Yield for steaks at the rate of 68%, implied raw material price	20.60
Preparation/packing costs per 500 g. cartoon @ Rs. 2.60 each	5.20
Bulk packing for transfer to Tamilnadu	0.80
Transportation charges to Tamilnadu	1.50
c and f price, Tamilnadu (exclusive of capital/rent/ management costs, but inclusive of labour costs)	28.10

Distribution costs, including wholesaler's and retailer's margins, have to be added to the c&f price in addition to cold chain distribution costs and losses, likely to be high, as the facility is not well developed for retail products in India. Further such new products are aimed at the urban middle class market and consequently, require outlay on advertising.

The current interest being shown by processing companies in the development of minced fish products stems from the belief that, although the end-product would be more perishable, margin would be much higher because raw material costs would be significantly less by using low value small fish.

Voyage Profitability

An attempt has been made to model the cash flow associated with a "typical" shrimp trawling voyage, and then to compare that with a situation where more by-catch is landed. Capital costs and other fixed costs are not included since there is no change in these.

The trawler is manufactured in India, capable to stay at sea for five weeks and to return to port for five days with 4 tonnes of frozen shrimp and 4 tonnes of "quality" by-catch.

Cost Structure for Voyage by 20 m. Trawler
(35 days voyage, five days in port, operating costs only)

A. Costs	
(i) 50,000 litre of diesel at subsidised rate of 1988 at Rs. 3.40/litre	Rs. 1,70,000
(ii) 15 t of water at Rs. 37/t	Rs. 555
(iii) Salaries (4 officers and 10 crews)	Rs. 25,733
(iv) Provisions for mess	Rs. 10,544
	Rs. 2,06,832
B. Gross Revenue	
(i) 4000 kg shrimp at Rs. 117/kg.	Rs. 4,68,000
(ii) 4000 kg fish at Rs. 5/kg.	Rs. 20,000
	Rs. 4,88,000
C. Gross Margin (before crew share is paid)	Rs. 2,81,168

D. Crew Share	
(i) 20% of shrimp revenue exceeding Rs. 225000	Rs. 48,600
(ii) 40% of all fish revenues	Rs. 8,000
E. (i) Net revenue to trawler company per voyage (Net revenue to trawler company per day–Rs. 5614)	Rs. 2,24,568
F. Crew Earnings	
(i) Crew salary and share per voyage	Rs. 82,333
(ii) Crew salary and share per day	Rs. 2,058

If the trawler is to retain an additional two tonnes of by-catch valued at Rs. 10000 and shared the same at 40:60 ratio and delayed an extra, day in port for selling additional by catch and if one day's trawling is lost, then the extra by-catch results in extra gross revenue of Rs. 10000 and net revenue of slightly less when taking freezing costs into consideration. This amounts to a little less than cost of not trawling for that one day (*i.e.*, gross income foregone less fuel saved). In the case, the one day delay results in a small net loss.

It is easy to glut the market, under current marketing arrangements in landing centres, resulting in fairly volatile fish prices. It might be more appropriate to consider extra by-catch being landed at lower prices to reflect this, or a higher proportion of smaller fish in the by-catch.

The increased landings of fish, and an associated fall in fish prices would have the following effect on trawler revenues.

	4 t Fish Landed at Av. Price Rs. 5/kg.	*6 t Fish Landed at Av. Price Rs. 4/kg.*	*6 t Fish Landed at Av. Price Rs. 3/kg.*
Gross margin/voyage (before crew share paid)	28168	285168	279168
Company's net revenue/voyage	224568	226968	223368
Company's net revenue/day	5614	5674	5584
Crew salary and share/voyage	82333	83933	81533
Crew salary and share/day	2058	2098	2038

If prices falls to just over Rs. 3/kg as a result of landing more by-catch, both crew and company will loose. So Rs. 3/kg. was the lowest price at which it was worth landing fish.

The processing and trawling companies are all interested in the development of "value added products" from so-called 'trash' fish. If 20 m. trawler having adequate freezing capacity can handle (the crew) increased amounts of by-catch, there is apparently no conflict of interest; there would by no cause for delay in port if the market is assured and the economic incentive to company and crew is clear.

Collection of By-catch

Fishermen of West Bengal are engaged in collecting discarded by-catch of the trawlers. This involved a voyage of 6-7 hours, to locate double-rig trawlers at Sandheads area, take the discards by going alongside, if the skipper agreeded in lieu of supply of fresh provisions to the trawlers.

The costs and revenue from this activity is given below

Costs of Collecting Fish from Trawlers off West Bengal (operating costs only)

A. Costs	
(i) Fuel costs (150 litres)	Rs. 555
(ii) Ice 700 kg	Rs. 350
(iii) Provisions for barter trader	Rs. 150
	Rs. 1055
B. Gross Revenue	
(i) 180 kg. fish at Rs. 3/kg	Rs. 540
(ii) 1020 kg fish at Rs. 1/kg	Rs. 1020
	Rs. 1560
C. Net Revenue per Day	Rs. 505
(i) Net revenue/crew (8 crews)	Rs. 63

Income from this activity is not very great, but is on a par with normal fishing activity.

Any consideration of the economics of landing increased quantities of by-catch ultimately comes back to prices and marketing. The by-catch discarded by the east coast trawling fleet of India appear significant by comparison. India's already large population continues to grow, creating upward pressure on fish prices and an increasing fish deficit for use as food or feed. As real prices for fish are pushed higher, trawling companies may look again at ways in which by-catch can be retained. The development of high value fish products, is unlikely to influence fish consumption by low income groups positively. The distribution of employment, income and foreign exchange benefits should also be carefully appraised. Gill net fishery features a handy fishing method using a small one-man boat and net of the simplest configuration of all. The fishery with a long history is widely operated in costal and inland waters around the world.

Its fishing method is simple and outlay in fishing gear is relatively small. For this reason this fishery should be considered as a very practical method for effectual development of coastal fisheries especially in fishery developing nations.

From its catching frequency the gill net fishing gear has two different sorts of functions, enmeshing and tangling. In designing a modern gill net fishing gear, the following two points are to be taken into consideration:

1. How to increase the catching frequency, and
2. To what extent, the fish species selectivity of fishing gear is to be counted.

While the mesh size is a crucial factor in catching the fish, it also selects the fish caught by body size or shape. In designing a gill net, therefore the fisherman must select a type that would serve his particular purpose best. In addition much importance must be given to corresponding design particulars as well.

Small-scale coastal gill net fishery is conducted by single boat operations using powered boats of upto 10 tons in fishery developed nations. However the large majority of the operations use boats of 5 tons or less, which make up 90 per cent of the total in this type of fishery.

The scale of business that the coastal gill net fishing family is engaged in can be divided into three levels; those operating powered boats of one ton or less, powered boats of 1 to 3 tons and powered boats of 3 to 5 tons.

Small-scale coastal gill net fishery is conducted in every region of maritime countries, and the fish that are caught cover an amazingly wide range of species from ocean migrating surface fishes to bottom habitat fishes and crustaceans. However, the object of most fishermen involved in small scale coastal gill net fishery today is to commercialize their business by catching a particular kind of fish in particular water areas (of course, these catches still remain mixed with a considerable number of miscellaneous fishes.)

There are actually six types of nets in use in gill net fishery. Among these, the two which have proved to have the best catching capabilities and lend themselves best to a commercialized type of fishing operation are:

1. Fixed bottom gill net, and
2. Drift gill net for surface and middle depth fishery.

Two other types which must also be mentioned as note worthy fishing methods are:

1. Bottom drift gill net for prawn and
2. Excircling net.

Income and Expenses for Fixed Bottom Gill Net

The fishermen involved in both small-scale trawl and gill net fishery averages between 7.5 and 10 million yen a year in total fishery income. Both the small-scale trawl and gill net operations are performed by 2 or 3 workers, and in most cases, by family labour. With regards to fishing gear expenses, the gill net is less expensive then a set of trawling gear, but most fishermen maintain 4 or 5 nets and so the depreciation on these is a considerable amount.

Number of fishing per year	180 days
No. of workers needed	2
Gross income for fish catch Fishing expenses	7.5-10 million yen
Labour (Estimated family labour expenses)	4 million yen
Fuel expenses	800 thousand yen
Depreciation on boat/engine	1.9 million yen\
Depreciation on nets	500 thousand yen
Other fishing expenses	300 thousand yen
Total	7.5 million yen
Profit	0-2.5 million yen

Income and Expenses for Drift Type Gill Net

The fisherman in this type of fishery, with a team of his family labour seeks to earn an income of one million yen month by means of gillnet and angling. When however, their daily income fails to reach 30000 yen/day for an extended period, the male member will be engaged in dive fishing in shallow water areas near their home. This is an economically sound decision based on consideration of fuel and net depreciation costs.

Costs of Drift Type Gillnet Fishing

1. Fuel Costs

   Using 120 to 130 litres per day–5000 yen perday and 100 000 yen per month

2. Fishing Net

   Material costs–high quality netting 85000 to 86000 yen per 151 metres (1 set = 151m × 12 = about 1 million yen), low cost netting 55000 to 56000 yen per 151 metres (1 set =151m × 12 = about 700 000 yen) net framing cost–15000 yen per 151 metres (1set = 151m × 12 = 180 000 yen). Usually (151m × 3) of spare netting are kept.

3. Fishing Boat

(i) Boat	6 million yen
(ii) Engine	7 million yen
(iii) Equipment	2 million yen
Total	15 million yen

Set net fishery is considered very advantageous from the point of utilization of resources, labour and capital. It is a "wait–and–see" type fishery operated to catch only incoming fish species, with much less possibility of overfishing than trawl net or boat seine operated to go after moving fish schools.

The net is set in such a manner that it can continuously catch the fish for a long time until it is pulled up, that is, fish schools swimming with the stream are lured into the bag net by means of the tactfully set leader net. In this way fishing operation is continued day and night. Set net fishery is a very economical type fishing in terms of the amount of labour needed. The net is dragged up once a day, usually in the morning, (twice a day in some fishing grounds in the morning and in the evening) and the operation can be completed within two hours. Depending on the size of net, the number of fishermen is needed, but in case of a small–scale set net operation it does not exceed five or six.

The fishing ground for this type of fishery is usually near shore and this makes it possible to keep the catch fresh.

Different species of fish are lured into the bag net and they are caught alive. Prime fish can be sold as "live fish" bringing more profits to fishermen. When the catch amount to extremely large of schooling fishes, a part of the catch is kept in fish preserve to control the amount of fish despatch, thus preventing a drastic fall in fish prices.

Large–scale set net, which requires a great number of fishermen, takes a form of management fishery using hired fishermen in most cases. While self-employed fishermen can still make up a part of those fishing units, organizing management by co-operatives as much capital and manpower is essential to this type of fishery.

Small-scale set net can be operated by five or six fishermen even during the peak period and it takes a form of small-scale self-employed fishery using family labour alone.

In the case of large-scale set net fisheries catching mainly migrating species such as yellowtail, horse mackerel, mackerel, sardine and anchovy, the areas chosen as fishing grounds are the steep coast areas near the mouth of deep bay where a branch of the ocean current that has entered the bay, and circled it is about to leave the bay again. The reason for this is that fish species that are migrating with the circulating currents along the coast form fish routes and because each species migrates at a given depth, when they come to a place on the coast where the isobath lines are close together it is believed that the schools of fish are forced together and thus become concentrated in a small area.

In order to conduct set net fishery it is necessary to invest in basics, such as net, related fishing gear and several boats, as well as fees for setting and removing the nets, maintenence of the nets, machinery used in the everyday working of the nets and also labour expenses.

To encourage the growth of small-scale set net fishery by artisanal fishermen, a system of low-interest loans for fishery investment, financed by government funds is essential.

The characteristic of set net fishery is that a wide variety of fish are caught in the same net. A sales system should be set up to maximize the freshness of the fish when they reach the market.

1. Small-catch fishes can be sold to local tourist inns and hotels.
2. Home processing of catch for sale as processed foods.
3. Extra fish from large catches can be used as material for aquaculture feed or fish meal.
4. More than 160 species of fish are caught by set net fishery.

Large scale set net fishery adopts five different types of management, each of which is capitalized by fishermen alone. In this fishery operation each fishing unit formerly required about 150 fishermen. But the requirement has been reduced to about 30 fishermen due to the introduction of labour saving machinery, equipments and improved nets.

Small scale set net fishery is mostly conducted by self-employed fishermen. In most cases it is operated by a fisherman with his family labour. Even in case no family labour is available, hired labour is kept to a minimum.

Set net fishing is carried out roughly 150 days of the year, with one day's catch equalling 20 to 50 kilograms or 10 to 15 thousand yen worth of fish. The income from these set net operations varies greatly from one fishing unit to the other, but each fishing household earns at least 5 to 6 million yen a year from their fishery businesses. Labour costs, fishing gear maintenence and fuel costs total between 3 and 4 million yen in every case.

Pot fishing is a relatively simple fishing method which has been used traditionally by fishermen all over the world to lure and catch marine animals. The gear used in this type of fishery include a wide variety of baskets, cylinders, boxes and jars, all of which for the sake of simplicity is given one name "pots"

When primitive trap-type fishing gear are used by the fishermen, fishing efficiency is too low to enable them to constitute a commercial fishery operation. In the pursuit of technical advancements that will enable the higher catching capability, required in a commercial operation, alternatives, such as larger scale fish wires, keddle nets and set nets come first to mind.

However, with the introduction of modern technology and adaptive measures suited to the local marine and labour resources, pot fishing can still be made to function as a significant part of a modern fishery industry.

The basic innovations that have taken place over the years in Japanese pot fishing are:

1. Improvement of materials used fishing gear construction (use of synthetic nets, steel frames and plastics) resulting in lighter, longer lasting gear.
2. Expansion of fishing grounds due to the introduction of motorized fishing boats, and the increased mechanization of fishery equipment.
3. Introduction of long-lined type fishing methods due to introduction of motorized line haulers.

As a result of these innovations, coastal fishermen have been able to easily include pot fishing as a part of this commercial fishing operations, and also it has enabled them to exploit particular marine resources in deep off shore sea areas.

Angling is the fishing methods requiring the smallest capital investment, and has been a primary fishing method used in all parts of the world. However, since the number of hooks which can be employed in one operation is very, small, and since the catching efficiency depends largely on the individual skill of the fishermen the catch per boat per day is generally small. In order to continue angling as a successful commercial fishery method, in recent year angling fishery has focused on the catching of high market value fish, such as, large deep sea fish or other high priced fish and has sought to preserve their market value even further by outfitting the fishing boats with fish-reserve tanks that enable the fish to be landed alive, or facilities which enable the fish to be preserved in ice on board and landed within a short period with a high degree of freshness.

The amount of catch by species for coastal long line fishery in Japan (1983) is as follows:

Species	*Amount of Catch in Tons*	*% of Total Catch*
Tuna	140	
Marline and Sword fish	63	
Shark	1491	1.5
Salmon and trout	2898	3.0
Mackerel	124	
Yellow tail	140	
Flounder and flat fish	1465	1.5
Cod	49963	52.0
Atka mackerel	1107	
Rock fish	435	
Drums and Croaker	648	
Lizard fish	91	
Pike conger	350	
Cutlass fish (ribbon fish)	997	
Ray	324	
Sea bream	4128	5.0
Dorado	1378	1.5
Flying fish	153	
Sea bass	322	
Other spp	23788	25.0
Octopus	5493	6.0
Other marine animals	25	
Total	95536	100.0

Shrimp Farming: A Commercial Activity

Generating a huge amounts of foreign exchange employing millions of people and cultivating vast areas of previously unused land, shrimp farming continues its relentless march towards dominance of world shrimp markets. Starting from nowhere in mid 1970's the industry now produces 28 per cent of shrimp placed on world markets. In Japan and United States, the number one and number two shrimp consuming nations, farm raised shrimp often captures 50 per cent of the market.

In 1991 world shrimp, farmers celebrated record production of 690100 metric tons up 9 per cent from 632900 metric tons in 1990. About a million hectare ponds yielded 694 kg/ha. Aquaculture Digest estimates that the world has 4708 shrimp hatcheries and 35895 farms. The western hemisphere produces approximately 20 per cent of the farm-raised shrimp, the eastern hemisphere 80 per cent.

Western Hesmisphere

With 94.8 per cent of production in Latin America, the western hemisphere produced an estimated 20 per cent of the world farm-raised shrimp in 1991. Ecuador, the leading producer with 75 per cent production exported 400 million dollars worth farm-raised shrimp. Colombia which quadrupled production in 1989 and 1990, moved into second position and Mexico with lots of new activity moved into third position. Honduras occupied second place in 1990, but fell to fourth place in 1991 on slightly improved production.

Peru, Panama, Brazil and Guatemala have small shrimp farming industries and scattered farms exist in Venezuela, Nicaragua, El Salvador Belize, Costa Rica and the Carabbean. Most new farms adopt scientific semi-intensive strategies. Shrimp farms in Latin America market most of their shrimp in the United States, while developing receptive markets in Western Europe, particularly France and Spain which purchase raw frozen whole animals. In the United States, a five-pound box of raw, frozen shell-on tails is the most popular products.

Ecuador with one of the most organized and efficient shrimp farming industries in the world has been the consistent production leader in the western hemisphere for more than two decades. In 1991, a year of declining shrimp prices, it increased production by 35 per cent United States Corporation operating in Ecuador reported mixed results with improving numbers towards the end of the year. In some farms viral disease reduced shrimp growth. So farmers produced more crops of smaller shrimp. Ecuador exports 70 per cent of its production to the United States (frozen tails) and 30 per cent to Europe (frozen whole).

From March to May, Ecuadorean coastal waters produce abundant supplies of wild seed stock and prices for hatchery produced seed stock drop into year low. Some hatcheries had to dump their surplus production of seed stock. During Ecuadorean winter (June-September) when wild seed stock is not available, hatcheries come to the rescue. While large-scale hatcheries continue to prosper, an increasing number of small to medium-scale hatcheries handle the industry's seed stock requirements.

Ecuador prefers warm, wet El nino. That's when there is plenty of wild seed stock, plenty of warm weather and plenty of rain to flush the estuaries and ponds. During the El nino production of farm-raised shrimp sky-rocketed.

In Ecuador 145000 ha is used to produce 100000 metric tons (live weight) of shrimps (*Penaeus vannamei*–90 per cent and *P. stylirostris*–10 per cent) in 1700 farms of which 60 per cent extensive, 25 per cent semi-intensive and 15 per cent intensive. Among these farms 25 per cent are large scale, 50 per cent medium scale and 25 per cent small-scale in size. 150 hatcheries produce shrimp seed to meet the requirement of these farms and the production rate is 690 kg/ha (live weight).

During recent years Colombia's shrimp farming industry has grown leaps and bounds. It produced 9000 mt in 4000 ha in 30 farms (5 per cent extensive, 90 per cent semi-intensive and 5 per cent intensive) by culturing three species (*P. vannamei*–85 per cent, *P. stylirostris*–10 per cent and *P. schmitti*–5 per cent). Twenty hatcheries (small scale–5 per cent, medium scale–90 per cent and large scale–5 per cent) produce shrimp seed to meet the culture requirement. But still it is a distant second to Ecuador in the production of farm raised shrimp in the western hemisphere. The Colombian Industry is dominated by a small number of companies that operate most of the farms. Some of the companies are incorporated, most of the rest are limited partnerships.

Farmers argue that shrimp farming conditions among the Caribbean coast of Colombia are better than those along the Pacific coast of Ecuador. Sea surface temperatures are stable at 27 to 29°C and the moderate rainy season (May to November) rarely exceeds 15 cm per month; so there are few sharp swings in salinity. In addition, the relatively modest tidal amplitude simplifies daily operations (pumping, water exchange, harvesting, transportation). Pond construction is easy because there are fewer mangroves. Typically the Caribbean coast has beach and bay habitats with tidal ranges averaging about 0.6 metres.

A United States consulting farm, Tropical Mariculture Technology, has worked for many of the large shrimp farms in Colombia. It advocates large high-yielding, semi-intensive farms. Judging from the jump in Colombia's shrimp production, this approach has been very successful.

Private sector shrimp farming is a priority of Mexico's administration. In late December, 1989 the Mexican congress approved 14 amendment's to its fishery laws, allowing for the first time, private sector shrimp farming and foreign investment. Formerly shrimp farming was reserved for fishing co-operatives and agrarian reform communes. The regulations implementing the amendment clearly permit private as well as social groups (co-operatives and communes) to use their own land to farm shrimp. Fishery co-operatives and communes retain the exclusive right to collect postlarvae and gravid females.

Co-operatives produce shrimp in large low-lying extensive ponds. They control the best shrimp farming sites, some of them along estuaries with the mid-range salinities that favour shrimp farming. But the co-operatives cannot use their land as collateral, they can not borrow money to develop the sites. There are some co-operative/private sector joint ventures. The co-operatives supply the land and labour and the private sector provides the capital, technology and management. They split the profits 50:50.

In late 1991, the private sector reported severe seed stock shortages, primarily because it was not allowed to harvest the abundant supplies of wild seed stock in local estuaries. Now 18 hatcheries (50 per cent small-scale, and 50 per cent medium–scale) produce seed of *P. vannamei* (80 per cent) and *P. stylirostris* (15 per cent) to cater the need to 100 shrimp farms (40 per cent extensive, 55 per cent semi-intensive and 5 per cent intensive) to produce 5000 mt of shrimp from 5000 ha (1000 kg/ha.).

In Honduras, farmers raised 4500 mt of *P. vannamei* (70 per cent) and *P. stylirostris* (30 per cent) from 7000 ha. (643 kg/ha) in 25 farms of which 40 per cent are extensive and 60 per cent semi-intensive. Two hatcheries, (one small scale and the other medium-scale) produce shrimp seeds. Yields of *P. stylirostris* are low averaging 425 kg headless per hectare. Some farmers import better feeds to meet the higher protein requirement of *P. stylirostris*. The main product is shell-on frozen tails exported to United States.

Ample supplies of wild seed stock, which fluctuate seasonally between *P. vannamei* and *P. stylirostris* limit hatchery development. In the past, however, seed stock shortages have disrupted the industry. So there was lots of hatchery talk, but few sites on the Pacific coast which offer the high quality water necessary for a hatchery.

Panama has been raising shrimp since the mid-1970s. Agromarina-de Panama, owned by a United States company is the largest farm in Panama and operates a hatchery which supply seed stock world-wide. This farm/hatchery is reported to be very successful and expanded in a big way. In addition Panama has several other world class hatcheries (6 hatcheries of which 3 medium-scale, 2 large-scale and 1 small-scale) which produces shrimp seeds (*P. vannamei*–75 per cent and *P. stylirostris*–25 per cent) and supply seed stock to farms in north, centraland south America.

Overall shrimp farming has progressed slowly in Panama. In 1991 it produced 4000 mt of shrimp in 40 farms (15 per cent extensive, and 85 per cent semi-intensive) covering an area of 4000 ha (1000 kg/ha). Many farms lose money, others are only marginally profitable. Some farms started with no plan, no practical experience and no general understanding of the complexities of shrimp farming. Other farms paid little attention to site selection and farm design. Small farms were unable to take advantage of the economics of scale. Other problems which hindered the industry included poor commercial feeds, a long dry season and high energy costs. These problems were compounded in the early 1980s by the regional debt crisis and in the late 1980s by the economic and political instability. Few investors were willing to commit funds in the chaotic Panamanian economy.

Since fall of Noriega Government several small farms have gone out of business. New farms were better designed, larger and succeeded. Consequently the outlook for Panamanian shrimp farming improved. Eventually the hatcheries have eclipsed the farms in economic importance.

Shrimp farming sites are limited in Peru. In 1991, it produced 3500 mt of shrimps (*P. vannamei*–99 per cent and *P. stylrostris*–1 per cent) in 60 farms (95 per cent semi-intensive and 5 per cent intensive) from 4000 ha (875 kg/ha). But there is plenty of room for increased production from existing farms. Most shrimp farms harvested

12 to 14 gms. (31/35 count), although this depends very much on market conditions. Some of the major producers use aeration and high quality feeds, hoping to produce yields of 7000 kgs per hectare per year.

The demand for seed stock creates high prices for postlarvae, which are consistently one to two dollars more per thousand than in neighbouring Ecuador. Total hatchery production averages 20 to 30 million post larvae per month, compared to demand which averages 40 to 50 million per month. Despite seasonal red-tides and a difficult economic climate the Peruvian hatchery industry is in a growth phase for 3 medium–scale hatcheries in 1991. Intensification and further development of shrimp farming will lead to even greater demand for postlarvae. Farmers get their seed stock from hatcheries, from sporadic catches of wild postlarvae and from illegal shipments from Ecuador. The construction and development of well managed hatcheries and the application of effective maturation technology should lead to self sufficiency in seed stock production.

Four Peruvian feed manufacturers supply the industry. Feeds are expensive (700 to 760 dollars a ton in 1991), approximately twice the cost of Ecuadorean feeds. Six plants process shrimp in Peru.

The United States plays a major role in World's shrimp farming, not because of its production which is insignificant, but because of its role as a supplier of capital, feeds, equipments, research, education, information, research and technology to shrimp farmers in fifty countries. In addition, the United States Department of commerce supports shrimp farming research through the National Sea Grant College Programme, which funds applied and basic research at Texas A and M University and several other Sea Grant institutions. The United States Department of Agriculture supports the industry through its co-operative State Research Service which funds programme at Mississippi's gulf coast Research laboratory, which heads a consortium of U. S. research facilities that are attempting to determine the feasibility of shrimp farming in the United States. At the University of Arizona researchers study shrimp diseases.

Twenty-five shrimp farms (95 per cent semi-intensive and 5 per cent intensive) located in Hawaii, South California, Texas and to a lesser extent, Puerto Rico. In Hawaii with its high land and labour costs, the Oceanic Institute backs an intensive round pond production system and conducts research on maturation, nutrition, equipment, economics and diseases. In South Carolina, the Waddell Mariculture Centre, with about half its budget devoted to shrimp farming encourages semi-intensive and intensive shrimp farming in earthen ponds. Texas, the leading producers in the United States supports semi-intensive and intensive shrimp farming. Good support from the state and an influx of Taiwanese capital and farms have revived the Texas shrimp farming industry.

In 1991, the United States produced 1500 mt shrimps (*P. vannamei*–99.9 per cent and *P. setiferus*–0.1 per cent) from 450 hectares (3556 kg/ha). Three hatcheries (two medium-scale and one large-scale) supplied the seed stock to the industry.

Eastern Hemisphere

With most producing countries located in South East Asia, the eastern hemisphere produced an estimated 80 per cent of the World's farm raised shrimp. China was leading the pack in 1991, but Indonesia a close second to China expanded its vast potential. Thailand, India, the Philippines, Vietnam, Taiwan and Bangladesh also produced world class crops of farm raised shrimp. Japan, Australia and Malaysia have small shrimp farming industries. The big market is Japan, but countries throughout South East Asia ship processed and frozen raw product to the United States and Europe.

Intensive farms in South East Asia require high-quality feeds, but the right feeds are not always available and feed prices are high (about twice as much as in Latin America). At some farms feed represents 50 per cent of operational expenses.

Shrimp farming walks a tightrope in China, balancing jobs and foreign exchange against pollution and shrimp diseases. Production peaked in 1988 and has been slipping ever since. The gulf of Bohal, the very heart of the industry, handles a tremendous load of human and animal sewage, enriched even more by the fertile effluent from shrimp farms. Water quality problems and toxic accumulations on pond bottoms seem to be the industry's biggest problems. Uneven pond bottoms, poorly designed water control structures and the low-lying nature of the ponds make it very difficult to flush out the toxins. Many ponds have been in production for over a decade without being renovated.

In northern China, where most production takes place, large government run semi-intensive farms produce one gigantic crop of Chinese white shrimp (*P. chinensis*). They stock in May and harvest in October. The industry relies on live feeds (primarily crushed calms and mussels) supplemented with pelleted feeds. Some shrimp feed mills existed and more developed. In Southern China, extensive, semi-intensive and intensive government–run and private sector farms produce two crops a year of *P. monodon* and some other penaeid species.

In 1991 China produced 145000 mt in 2000 farms (5 per cent extensive, 90 per cent semi-intensive, and 5 per cent intensive) of shrimps (*P. chinensis*–85 per cent, *P. monodon*–10 per cent and *P. penicillatus*–5 per cent) covering an area of 140000 hectare (1036 kg/ha). There were one thousand hatcheries of which 80 per cent were of large-scale and 20 per cent medium scale.

Heavy flooding hit much of the Chinese shrimp farming industry, lowering production for the year. But increasing it in the long run because ponds, bays and estuaries receive a good flushing. China faces chronic threats to its "most favoured–nation" trading status with the United States. If she loses this status, it could mean high tariffs on Chinese farm-raised shrimps.

With its long coastline, Warm temperatures, freedom from hurricanes, good supplies of seed stock, entrepreneurial spirit, Japanese capital and government support Indonesia stood second to China in the production of farm-raised shrimp in 1991. Due to some big projects, the production was doubled.

As a result of its increasing production, Indonesia is a major consumer of shrimp feed. Feed is available from at least 10 domestic and 19 foreign feed companies (primarily Taiwan but also China and Germany). Most of the imported feeds are relatively expensive, high quality diet for intensive farms. They cost about 0.99 to 1.36 dollar a kilo. Intensive farmers feel that Taiwanese feeds are somewhat superior to domestic feeds. A few semi intensive feeds are sold for 0.66 to 0.68 dollar per kilo, and some relatively crude semi-intensive feeds are being manufactured in central Java for only 0.31 dollar per kilo. Competition has forced the feed companies to offer special services to shrimp farmers like free delivery and financial services.

In 1991 Indonesia produced 140000 mt shrimps (*P. monodon*–60 per cent, *P. merguiensis*–20 per cent and others–20 per cent) in 20000 farms (50 per cent extensive, 40 per cent semi-extensive and 10 per cent intensive) from 200000 hectares (700 kg/ha). 250 hatcheries (55 per cent small-scale, 35 per cent medium-scale and 10 per cent large-scale) supplied seed stock to these farms.

United States, shrimp imports from Indonesia have jumped dramatically in last few years, from 9 million dollars in 1980 to 12 million dollars in 1987 and 83 million dollars in 1990.

In 1989 and 1990, the unwise wild expansions, overstocking, poor drainage and pollution exacted a heavy toll on Thai shrimp farming. In 1991, the industry and government attacked these problems. New licensing requirements and stricter pollution controls were instituted. Charoen Pokphand (the CP group, a major international commodity trader, a wholesale distributor of food stuffs and Kentucky Fried Chicken franchisee in Bangkok and China) spearheaded the attack. It encouraged small intensive shrimp farms in Southern Thailand to implement pond management practices which reduced pollution.

In the same year the country produced 110000 mt shrimps (*P. monodon*–85 per cent, *P. merguiensis*–10 per cent and others–5 per cent) in 3000 farms (50 per cent extensive, 25 per cent semi-intensive and 25 per cent intensive) from 80000 hectares (1375 kg/ha). 2000 hatcheries (60 per cent–small-scale, 30 per cent medium-scale and 10 per cent large-scale) supplied seed stock to the industry.

In January, 1991, as a result of unacceptable antibiotic (oxytetracycline) levels in farm raised shrimp from Thailand, the Japanese health authority instituted a compulsory inspection programme for all shrimp from Thailand and other countries of Southeast Asia. In a speedy response to the antibiotic scare Thailand's health officials recommended:

1. All shrimp farms register with fisheries department;
2. The fisheries department to inspect all shrimp farms,
3. Farmers to stop all chemical treatments 14 days before harvest; and
4. All farm-raised shrimp be inspected before export.

The United States imported 279 million dollars worth of shrimp from Thailand in 1990. Unlike other shrimp farming countries, where the main product is raw, shell-on, frozen tails, Thailand also exports processed products, like cooked and cleaned, individually quick frozen tails, just perfect for the cocktail and catering markets.

The Marine Products Export Development Authority (MPEDA) with in the Ministry of Commerce has taken the lead in the promotion of shrimp farming in India. On the West coast it established a research complex near Kochi to train managers and technicians who assist farmers with site selection, financing, pond construction, seed procurement, stocking, management and harvests. MPEDA has also established two large commercial hatcheries one in the state of Orissa and the other in the state of Andhra Pradesh. MPEDA also subsidised hatchery development and feeds.

A large number of companies have set up modern semi-intensive and intensive shrimp farms. About 5000 hectares of these new farms were on the line since 1991.

In 1991, the country produced 35000 mt of shrimps (*P. monodon*–45 per cent, *P. indicus*–30 per cent and others 25 per cent) in 2500 farms (80 per cent extensive, 15 per cent semi-intensive and 5 per cent intensive) from 65000 hectares (538 kg/ha). 16 hatcheries (65 per cent small-scale 25 per cent medium-scale and 10 per cent large-scale) met only a part of seed stock requirement. The bulk seed stock came from wild.

To increase the volumes of shrimp export, the government has identified 100000 hectares of poorly utilized coastal lands, mostly in the states of Andhra Pradesh, West Bengal, Kerala and Goa for the development of shrimp farming. These lands are being released at low annual rents to fishing communities and the poor (60 per cent), entrepreneurs (20 per cent) and self-employed technocrats (20 per cent). The first groups received help with pond construction, pumps, power and for the first year free seed and feed.

Some of India's largest corporations like Indian Tobacco Company, Water base Ltd, Tata (oil seeds/steel) and Hindustan Lever (consumer products) have invested in shrimp farming.

In November, 1990 barely four months after an intensity seven earthquake disabled shrimp farms on Luzon, a super typhoon devasted shrimp farms in Central Philippines with 240 km/hour winds and heavy rains. Incurring the greatest damage were the Western Visayan Islands of Cebu, Negros and Panay. Low-lying extensive ponds experienced the greatest damage. Farms lost electricity and were forced to sell their crops because they could no longer aerate their ponds. Consequently the after typhoon, shrimp prices in the area fell by 15 per cent to 20 per cent. Farm gate prices of medium-sized shrimp dropped by about a dollar from 6.80 to 5.80 a kilo.

In 1991 the country produced 30000 mt shrimps (*P. monodon*–90 per cent, *P. merguiensis*–5 per cent and others 5 per cent) in 3000 farms (40 per cent extensive, 40 per cent semi-intensive and 20 per cent intensive) from 50000 hectares (600 kg/ha). 250 hatcheries (80 per cent–small-scale, 10 per cent medium-scale and 10 per cent large scale) produced seed stock for the industry.

21 shrimp feed mills in Philippines have a combined capacity of 100 000 metric tons a year. Led by San Miguel Corporation, they supply 85 per cent of the industry's feed requirements. Feed accounts for 60 to 70 per cent of production costs. Tariffs of 30 per cent on imported feeds and taxes of 10 per cent on locally produced feeds push feed costs upto 1.20 dollars per kilogram. In Thailand shrimp feeds averages 0.95 dollars per kilogram and in Indonesia 1.05 dollars/kg. 30000 metric tons of shrimps (*P. monodon*–75 per cent, *P. merguiensis* 20 per cent and *P. indicus* 5 per cent) were produced annually in 1000 farms (90 per cent extensive, and 10 per cent semi-intensive) from 160000 hectares (188 kg/ha). 120 hatcheries (93 per cent small-scale, 5 per cent medium-scale and 2 per cent large scale) cater the seed stock requirements in Vietnam.

In the country ponds of family operations are typically 1 to 3 hectares, while those of large co-operatives and state-run-farms range from 100 to 300 hectares. Some large farms utilize several thousand hectares of tidal areas. Production per hectare is low by international standards, but increased stocking densities, pest eradication and increased water exchange help produce larger yields.

Due to recent changes in economic policy provided more incentives for free enterprise, some hatcheries and farms have developed in the private sector. While Vietnam tests several methods of producing post larvae most shrimp farmers still rely on the seed stock that arrives with incoming tide. Spring is generally the main period for impounding wild post larvae because that is the beginning of the best grow out period. Hatcheries, focus on *P. monodon* and to a lesser extent *P. merguiensis*. A few utilize imported technology and consultants from Japan. Most hatcheries are located in Central Vietnam, near Nha Trang and Cam Ranh, where there is a reliable supply of wild. *P. monodon*.

Normalization of trading relations with the United States would have been a great aid to the shrimp farming industry in Vietnam. There are Vietnamese business people in the United States who are eager to develop the industry.

In 1987 Taiwanese shrimp farmers, relying on seed stock from a couple of thousand small hatcheries produced approximately 100000 metric tons of farm raised shrimp (*P. monodon*) from 10,000 hectares of small family-owned intensive ponds. In 1988 production dropped to 30,000 metric tons because of mortality and stunting problem. In 1989 production dropped to 20,000 metric tons. In 1990, the industry showed its first signs of recovery, as farmers lowered densities and switched to other species like red-tail shrimp (*P. penicillatus)* and kuruma shrimp (*P. japonicus*).

In 1991 the country produced 30000 mt of shrimps (*P. mondon* 65 per cent *P. japonicus* 30 per cent and others 5 per cent) in 2000 farms (60 per cent semi-intensive and 40 per cent intensive) covering 8000 hectares (3750 kg/ha). 800 hatcheries (80 per cent small-scale, and 20 per cent medium scale) catered the seed stock requirement for these farms.

The concentration of intensive farms in the South West part of the country has broken up as farms relocated in the North West and North East.

Taiwan has high production costs making it difficult for it to compete with countries in Southeast Asia, like the Philippines, Thailand and Indonesia that have lower production costs and longer growing seasons. Taiwanese shrimp farmers concentrate on home markets, while supplying speciality products, like live kuruma shrimp to Japan.

The newest sporting craze among office workers in Taipei is shrimp fishing. They are paying by the hour to jiggle their hooks over the city's new spate of urban high-rise shrimp ponds, which are stocked daily with jumbo shrimp. The workers can either take their catch home or grill it over barbecues placed around the ponds.

In April 1991 a powerful cyclone roared up the tunnel shaped Bay of Bengal smashing ashore in Chittagong, a province with 30000 hectares of shrimp ponds. Officials of the Bay of Bengal Programme think the destruction of mangroves by shrimp farmers and deterioration of coastal embankments amplified the effects of the cyclone. With most shrimp farms in the area badly damaged or destroyed, it is an opportune time for Bangladesh to study the balance between mangroves and shrimp farming–so that the industry can rebuild in harmony with the environment.

In the year Bangladesh produced 25000 mt of shrimps (*P. monodon* 60 per cent *P. merguiensis* 20 per cent and others 20 per cent) in 1000 farms (90 per cent extensive and 10 per cent semi intensive) covering an area of 100000 hectares (250 kg/ha). The industry depend only on wild seed stock as no hatchery was developed yet.

An estimated 30 per cent of Bangladesh's farm raised shrimp is produced in Chittagong/Cox's Bazar area, the area hit by the cyclone, and 70 per cent, in the Khulna area which appears to have been unaffected by the cyclone. The Bay of Bengal Programme estimates that the cyclone caused a loss of 4000 metric tons of farm raised shrimp valued at 20 million dollars. With ponds destroyed and the infrastructure in a mess, recovery will be slow. The Bay of Bengal programme lost its shrimp hatchery in Chokoria but constructed a new one again.

Abundant land, water and labour allow Bangladesh to produce shrimp at very low costs. Seed stock arrives with the incoming tide, and some farmers stock. *P. monodon* fry. Stocking rates remain typically low 10000 to 30000 fry per hectare and most farmers do not use supplemental feed. Shrimp hatchery technology has been slow to develop in Bangladesh due mainly to the abundant availability of wild shrimp fry and the absence of good hatchery sites.

In the early 1960's, the Western part of the Seto Inland Sea was the starting point of Japan's shrimp farming industry. In the Amakusa area of Kyushu and around the Seto Inland Sea and the Southern islands shrimp farmers use two types of ponds; partial embankment and complete embankment. The partial embankment approach takes advantage of the area's large tides. Farmers built embankments to a height half way between the high and low tides and then construct a fence on the top of the embankment to enclose the shrimp. Other farmers use typical semi-intensive ponds that are completely enclosed with embankments.

In Kagoshima area of Kyushu and no where else in Japan farmers use large round land-based tanks to raise shrimp. Seawater is pumped through the tanks which produce between 1.5 to 2.0 kilograms of shrimp per sq. metre. That is between 15000 and 20000 kilograms per hectare. The incoming water creates a circular flow which concentrates wastes in the centre of the pond where they can be drained off. A layer of stone covered with a layer of sand lines the pond bottom. Water and air are circulated through the bottom layer. Shrimp burrow in the sand during the day and come out at night to feed.

Japan is the world's largest consumer of shrimps and prawns. With a per capita consumption of over 2 kg annually, it, far exceeds the 1 kg and 0.5 kg consumption of the U.S. and the E.C. countries respectively. For this reason, Japan presently imports over 200000 tons of prawns annually from more than 50 countries around the world. In addition, Japan produces between 40 to 60 thousand tons (*P. japonicus*–95 per cent, others–5 per cent) domestically. Of this production, however, only 3000 tons or so are kuruma prawns (*P. japonicus*) from culture fisheries. By focusing their culture operations on the market for live kuruma prawns, which command exceptionally high prices, culture fishery operators are able to secure for themselves a sound commercial base.

In the Southern islands, attempts at shrimp farming got started around 1970, and by 1979 the industry hit full stride. The warm-climate permits year-round production. Air transport supplies markets on Honshu.

World Shrimp Production by Species

Giant tiger shrimp (*Penaeus monodon*) dominates production (43 per cent) every-where in Southeast Asia except China. Reaching a maximum length of 356 mm is the largest and fastest growing of the farm-raised shrimp, but shortages of wild brood stock often exist and captive breeding is difficult. Japan consumes huge quantities of tigers. They are also marketed in the Western hemisphere frequently as pre-cooked Individually Quick Frozen (IQF) tails.

Chinese white shrimp (*Penaeus chinensis*) dominate production (18 per cent world production) in China. This species has many positive characteristics. It is second to *P. monodon* in growth rate (reaching more than 25 grams in less than five months), it grows at lower water temperatures (16 to 26 degrees Celsius) than *P. vannamei* and *P. monodon*, it likes muddy bottom, and it matures and spawn in captivity. On the negative side it has a high protein requirement (40-60 per cent), it is small (maximum length 183 mm) and its "meat" yield (57 per cent) is less than *P. vannamei* (65 per cent) and *P. monodon* (65 per cent). Japan and the United States are the biggest markets for Chinese white shrimp.

Western white shrimp (*Penaeus vannamei*) is the leading species (17 per cent of world's production) in Ecuador and the Western hemisphere. It can be stocked at small sizes and has a uniform growth rate, reaching a maximum length of 230 mm. It's protein requirement is lower (20-25 per cent) than *P. chinensis* and *P. monodon*. It breeds in captivity better than *P. monodon* but not as well as *P. chinensis*. During grow out it has a reputation as a "tough" animal. Markets include the United States (70 per cent raw frozen tail) and Europe (30 per cent whole frozen).

Other important species (constituting 22 per cent of world production) are *P. syslirostris* (The Pacific coast of Latin America), *P. japonicus* (Japan and Taiwan). *P. penicillatus* (Taiwan and China) and *P. merguiensis* and *P. indicus* (extensive farms throughout Southeast Asia). Researchers and farmers also work with a wide range of other penaeid species in the Western hemisphere. These include, *P. subtilis, P. paulensis, P. setiferus, P. brasilensis, P. duorarum, P. occidentalis, P. schmitti* and *P. calitomeinsis* and in the eastern hemisphere, *P. semisulcatus, P. latisulcatus, Metapenacus monoceros, M. dobsoni, M. affinis* and *M. brevicornis.*

Economic Consideration for Shrimp Farming in Countries of Eastern Hemisphere

To day the culture of shrimps and prawns is spreading at a tremendous rate all over the world. In particular, the development of a culture industry for Penaeidae of the kuruma shrimp family has contributed greatly to the increased in shrimp production.

Demand and Supply

The demand in kuruma shrimp (*Penaeus japonicus*) is very strong. So, either the price will keep the same level despite a large production increase or shrimp culturists can expect the trend of higher price to continue inspite of production increases.

Total production of kuruma shrimp are 3000 tonnes from culture fisheries. They are shipped to the market alive. However, because of its extravagant price, kuruma shrimp is hardly consumed at home.

Since the oil crisis, increase in consumer's expenditure per capita in real terms has become stable averaging 2-3 per cent and the value of the consumption became stable during the same time. The reason why the demand for cultured shrimp a luxury food, has been increasing in this economically stagnant period is complex.

The situation as to the supply and demand of imported frozen shrimp is different. Total imports of Japan increased from 20000 tonnes in 1965 to 164000 tonnes in 1979. After that period they became stable at about 1,60000 to 1,70000 tonnes. On the other hand, the price of imported shrimp increased upto 1972 not withstanding the increase of imports, but dropped considerably from 1973 to 1975 and since then fluctuated severely.

Consumption of imported frozen shrimp is divided into two markets, (1) out-of-home and the other at-home. Out-of-home consumption seems to have reached its quantitative limit recently. Thus, there may be little possibility of increasing consumption in quantity at the present price and although at-home demand is strong, it is not expected to expand further due to the present high price compared to other foods. The income elasticity of shrimp is very high compared to other animal protein foods. If the income goes up 10 per cent, consumption of food will increase by 10 per cent in value at home. But the quantity of home consumption has been fairly stable since 1980. So shrimp price must be lowered in order to enlarge its consumption in the present food market in Japan. The intake of animal protein per capita in Japan has reached its ceiling. So, an increase in consumption price of one food will be at the expense of another. At present the competition between animal protein foods is very severe. In these market conditions, price becomes very important in promoting marketing of shrimps against other foods. Considering the high price elasticity of shrimp at-home, consumption is sure to increase with a little drop of price.

In contrast to the stagnation of the Japanese market, the shrimp market in the United States has shown rapid expansion since 1982, due to drop in domestic shrimp production (capture fishery) and the increase of exports from Ecuador and Mexico to the United States.

The per capita intake of shrimp has been increasing recently. So exports to the United States is expected to increase compared to that of Japan. The intake of fish and shrimp in the United States increased in the latter half of the 1960s because of advertisement that fish is better. The demand for fish, however, has become stagnant since 1977 due to their high prices. At present, the price of many fishes are higher than beef. On the other hand, the intake of chicken has continued to increase at its stable price. So even in the United States, price competition in the animal protein market has been severe.

In Europe and other areas, the market for shrimp is expanding, but the size of these markets is still small. Despite the probable decrease in the production of shrimp from the sea, the total supply may increase owing to the world wide development of shrimp farming.

The favourable economic situation in which the consumption and price of shrimp increased simultaneously will disappear in the near future. There will be a seller's market for shrimp for so long many people will think that the price of shrimp will go up forever as if the demand is limitless. As a result, the recent severe fluctuations of price in Japan caused by the unbalance of supply and demand during periods of low price, has given rise to the wrong idea that these fluctuations are caused by the intrigues of traders and intermediaries in the distribution channels. In fact, the severe fluctuations shows that the market is beginning to saturate, when only a shift of supply will cause a severe drop in price. Considering the situation, in order to increase export to Japan and other markets, it is most important to cut the production cost, as the demand based on the present price is now nearing a stable state.

Production Cost

In pond culture, the use of natural productivity is very important in minimizing production cost.

There are three main production method for kuruma shrimp in Japan. In each the method of culture is different. This results from differences in the growing condition of shrimp and in economic differences between culture firms in each area. The ponds in the Seto Inland sea area (pool type on land) have been converted mainly from unused salt fields. The ponds in the Amakusa area were constructed in the sea with a concrete dyke topped by standing wire-nets. The ponds in the Kagoshima area are of the circular type on land with water changed by pumps.

The former two types of culture ponds are semi-intensive and the later intensive. The productivity of former two types is 400 to 700 grams per square metre. The change of water in these types of ponds is effected by tidal movements at about 20 to 90 per cent daily. On the other hand water in intensive type is completely changed 3 or 4 times a day. With the large supply of oxygen high density shrimp culture becomes possible. With the development of special artificial feeds about three decade ago super-intensive culture has been possible.

The main characteristic of pool type ponds on land is that the culture technique is aimed at diatom production. Since diatoms do not grow well with either too much or too little water entering the pond, optimum growth is achieved by regulating the water change. In a pond with good diatom growth, the water colour is dark brown. Production of 40 gram shrimps (2 pieces/m^2) without any supplementary feed is possible. This is because diatoms supply oxygen, and animal plankton and detritus (feed for shrimp) prevent the growth of harmful algae in the pond. Good production conditions for shrimp are thus maintained. That is why, the food conversion ratio from feed to shrimp is very low compared to circular type pond.

Table 1: Different Types of Kuruma Shrimp Culture in Japan Productivity (gm/m^2)

Type of Pond	*Area (ha)*	*Rate of Water Exchange (% per day)*	*Present Level*	*Present Target*	*Feed Conversion ratio*
Pool on land	1–5	0.25	300–500	600–800	10–12
	0.5–1	0.5	400–600	800–1000	13–15
Pool pond in the sea	0.5–1	0.9	300–500	600–800	13–15
Circular pond on the land	0.1	4.0	1500–2000	3000–3500	17–20

Pool pond in the sea is built along the shore, where tidal movement is wide (5-6 m). The construction cost of a pool type pond with high concrete dike is prohibitive. Culturists, therefore, make low concrete dikes and set iron piles on them with a wire-net spread between the piles to hold the shrimp. At high and low tide water flows freely in and out of the pond through the nets. The rate of water change is much higher than on land type, but regulation of water is difficult. The supply of oxygen to this pond is more than land type due to large amount of inflowing water. But diatom growth is not as good due to limited water control. The total supply of oxygen is about the same as land-type and as such, the productivity per square metre is also similar. As the amount of natural food (diatoms) in this pond is low, the conversion ratio of feed is high in this pond and the production cost is higher to that on land type.

Due to geographical features and to cope with high land prices and wages, high intensive and labour saving facilities were designed in Kagoshima area, Japan. All knowledge and techniques in aquaculture were utilised in designing these circular ponds on land for shrimp culture. The culture method adopted in this type ponds leads to high production cost per kg. compared to other two types of ponds because of high priced feed and high feed conversion ratio. In the circular pond the use of high quality feed makes high density culture possible, but high rate water change limits diatom growth. In addition feed conversion is not good at high density. Circular ponds neglect natural productivity and is dependent on artificial feeds. This results in high production cost.

It is a common belief that the high production cost of this type is due to the high cost of pond construction and other facilities. But this is not true. Even if the fixed cost per unit area is very high, the fixed cost per kilogram of shrimp is lower in circular pond than in other two types due to higher productivity per square metre in circular pond. The high production cost mainly comes from the high conversion ratio and the high price of feed per unit and high electricity. This costs are a consequence of highly artificial techniques ignoring the benefit of natural productivity.

The use of natural productivity in aquaculture in temperate areas (Japan) is important, especially important in aquaculture in Asia. This is because the natural productivity of ponds in tropical and sub-tropical areas is higher than in temperate areas.

Table 2: Cost per Kilogram of Kuruma Shrimp in Different Culture Methods in Japan

	Pool on Land	*Pool Pond in Sea*	*Circular Pond*
Selling price (yen)	6000	6500	7000
Cost (yen)	4500	5000	5600
Feed	1540	1800	2400
Seedling	60	100	250
Wages	1000	1000	500
Shipment	410	420	450
Repair	300	500	600
Electricity	100	100	670
Rent	50	10	10
Depreciation	200	300	150
Management and others	840	470	570
Productivity (g/m^2)	600	600	3000
Conversion ratio	12	14	18

As regards to the existence of different types of culture methods in Japan, a firm operating with a high production cost cannot stay in business in the long run. However shrimps in ponds in the Seto Inland Sea cannot be overwintered because of low water temperature so the culturists in the area have to harvest and market shrimp at the end of the year when the price of shrimp is rather low.

The water temperature in Amakusa area is a little higher than in the Seto Inland sea. Here shrimp can survive the winter in a state of hybernation. Culturists are able to ship shrimps from January to March, having waited for the price recovery after shipments from the Seto Inland Sea. But the physical strength of these shrimps is poor as the shrimp fast until the middle of March. In addition the shipment of live shrimp requires the shrimp hybernate again. This results in lower survival rates of shrimps transported during this period. So Amakusa farmers complete most of their shipments between January to March.

The ponds in Kagoshima area have a warmer water temperature due to the warm Kuroshio current that follows the coast. So shrimp in circular ponds are healthy even in winter since they do not stop feeding. From April to May, the price of cultured shrimp is highest, as this time of the year there are no marketable shrimp from Seto Inland Sea or Amakusa area and there are no live shrimp caught by fishing boats. The demand for luxury foods becomes high in April and May. The supply of live shrimp from Kagoshima fills this market. The circular farms are able to concentrate all their shipments aimed at this period of extra-high price.

Accordingly circular ponds can co-exist with pool on land and pool pond in sea, because they enjoy a higher sale price that compensate their higher costs. There is a trend toward constructing new culture ponds due to high demand for live kuruma shrimp. Due to high land costs and inefficient circular pond from economic point of view the new ponds proposed to be constructed may be a combination pool type pond in sea. This new pond will have a lower land cost due to construction in the sea, But it will have a high dike and will not use wire netting. Complete control of water flow will result in high utilization of natural productivity, but the production cost per kilogram will be lower than pond in sea. There are many suitable places for the new ponds as they can be located in the sea along the shore.

Cost Study and Earning from Shrimp Culture

Many types of culture systems and many kinds of cultured shrimps are available in Asian region. Differences in farm scale can be very great even in one area. Between shrimp culturists in Taiwan and Japan there is not much difference either in productivity or technical level. It is difficult to set a standard scale or average size of farm due to small and large farms in many areas. There are limited statistics on economics of shrimp culture in all countries except Thailand, so that the comparative analysis of productivity, cost and earnings between different culture methods is very difficult to obtain. Also there are many species of shrimps cultured in various regions. Black tiger shrimp. *P. monodon* is the main species of one country, while white shrimp, *P. merguiensis* or *P. indicus* may be that of another.

This situation makes it difficult to clarify by economic analysis general trends in shrimp culture in the countries in Asia. To overcome the difficulty, the Black Tiger Conversion (B.T.C.) method is adopted.

Table 3 shows a model of calculating B.T.C from the production of cultured shrimps and fishes using price per weight. The method can be applied to shrimp culture as the price of cultured shrimps is decided mainly by the international market. A common international producer price throughout the world even in different areas and countries is thus possible. In other cultured species, the domestic market is usually larger than the export one so that the price of the cultured product is decided by the demand and supply relationship in the country. As such, it is not possible to do comparative analysis on production and cost. In shrimp culture, however, the cases are suitable for analysis on a world-wide scale.

Table 3: Calculation of Black Tiger Conversion (B.T.C.)

	Production (kg) A	*Price Yen/kg* B	*B.T.C (Yen/kg)* $A \times B$
Total	3500	707	
Tiger shrimp	100	1800(100%)	100
White shrimp			
Large	300	1300(72.2%)	217
Medium	500	1000 (55.6%)	278
Other shrimps			
Medium	600	200 (11.1%)	67
Small	1000	50 (2.8%)	28
Fish	1000	30 (1.7%)	17

Figures in parentheses in column B refer to price of commodity expressed as per cent of price of tiger shrimp.

Table 4, shows results of the BTC method using examples different types of shrimp culture in Asia. These examples were obtained from limited field survey. In the table, the production figure does not include the small-sized shrimps and trash fishes which become feed for large shrimp or are given to the workers free of cost. Examples of kuruma shrimp culture are excluded from the table, as live kuruma shrimps have a special domestic market in Japan and imported quantities from outside are few.

The price of this shrimp is decided by the domestic production. The producer price (BTC) is from 1800-2000 yen/kg excluding the examples from India and Malysia. The price is comparatively low in India due to low wages and inconvenient transport and in case of Malaysia due to large number of shrimp damaged or killed by high water pressure during harvest making a large part of total production.

Table 4: Comparison of Productivity and Costs for Each Type of Shrimp Culture Based on Field Survey in 1981, 1982, 1983

	Intensive	*Semi-intensive*		*Extensive*			
	Tung-kang	*Ponte-vedra (1)*	*Ponte-vedra (2)*	*South Bangkok*	*South India*	*Ganges*	*Johor Bahru*
Production (kg/ha)	15000	1065	520	500	240	315	250
Tiger shrimp	15000	525	310	_	10	5	6
White shrimp	_	_	_	300	80	160	57
Small-sized shrimp	_	_	10	200	150	150	187
Milk fish	_	540	200	_	_	_	_
Others	_	_	_	_	_	_	_
BTC production	15000	580	340	280	110	155	115
BTC price (Yen/kg)	1800	1950	1950	1900	2630	1710	1530
BTC cost (Yen/kg)	1620	1258	1320	900	2100	1000	1140
Fixed cost	450	650	760	600	990	880	940
Variable cost	1170	608	560	300	110	120	200
Seedling	230	260	200	_	_	_	_
Feed	730	184	160	_	_	_	_
Others	210	164	200	300	110	120	200

Generally in Asia, the culturists sell their products to merchants by the pond side, so that the price of shrimp is the producer's price. Wide range of production costs prevail in different economic situations and types of culture in each country. The production cost of extensive culture is less than that of semi-intensive and intensive culture in proportion to the degree of intensity. The production cost is highest in Taiwan and lowest in South Bangkok. The ponds in the South Bangkok area are so well prepared that the natural productivity is high enough to produce many small low quality shrimps and Acetes spp, which make good feed for white shrimp. Also small white shrimp are returned to the ponds, so that the productivity is high compared to other extensive culture systems.

The examples of semi-intensive culture in table-4 were taken from the Pontevedra area, near ~Roxas city, Panay Island in the Philippines. The area has more semi-intensive ponds than other countries and the production cost is roughly half between extensive and intensive cultures.

When the production cost is splitted into two parts, fixed and variable cost and shifting from extensive to intensive culture, the fixed cost tends to fall. The fixed cost per hectare in the intensive culture is much higher compared to extensive but the fixed cost per kilogram is lower because of high productivity. As the fixed cost is composed of depreciation, interest and fixed wages etc. the rate of decrease is very distinct in proportion to the increase in productivity.

On the other hand, the variable cost per kg becomes higher when proceeding form extensive to intensive culture. So the high level of variable costs keeps the production cost at higher side. In the manufacturing industry the variable cost per unit product may have only small differences even if the productivity is different between factories. In factories with high efficiency, the production cost per unit product becomes low in proportion to the decrease in fixed cost, and the variable cost per unit will remain at the same level.

The wide differences in the variable costs in shrimp culture, come mainly from the cost of seedlings and feeds. The seedling cost in kuruma shrimp culture in Japan is very low, compared to other area, as it is easy to get enough mature shrimps from the sea and recently many shrimp culturists have built their own hateheries to supply seedlings whenever they are needed. On the other hand, the seedlings cost per kilogram in semi-intensive culture in Asia is very high, since mature tiger shrimp are few and hatehery technology has not yet reached a high level, except for Taiwan.

The large differences in the feed cost are pre-eminent, and the feeding cost will increase at a higher rate than the decrease in fixed cost. Intensive culturists must turn to more feed to compensate for the loss in natural productivity. In Taiwan, shrimp culture from 1978 to 1982, the feed conversion ratio became higher in proportion to the increase in productivity. This is tied to high feeding cost. In addition to this, a high quality and high priced feed is needed in super-intensive shrimp culture.

The other main costs included in variable cost (Table 4) are temporary workers wages shipments, repair, electricity etc. Besides semi-intensive culture, the extensive culture in South Bangkok shows the highest variable cost compared to other countries. This is due to culturist's effort into preparation of pond, such as levelling, the pond bottom and digging a ditch along the earthen dike to make a nice habitat for the shrimp. In addition, they remove the mud that settles at the bottom. If they did not do this work, the pond would become silted wild field within 2-3 years. Moreover, they use electricity to control the flow of water into the pond. In fact the culture method in South Bangkok is between extensive to semi-intensive.

It is possible to ascertain two important economic trends from Table 4.

1. The fixed cost per hectare is very high in intensive culture, but fixed cost per kilogram is very small compared to semi-intensive and extensive culture.
2. The percentage of variable cost out of total cost per kilogram is very high in intensive culture. This is due to the high cost of feed and seedlings.

The productivities of all extensive culture were above 100 kg/ha (Table 4). Usually the productivity of these extensive ponds are under 100 kg./ha. The cost per kilogram in the extensive system having productivity of 50-60 kg/ha is higher than that of the intensive systems presently used in shrimp culture.

Although a general trend of the economics of shrimp culture is understood, it is very difficult to say exactly the true production costs. In Asia, the condition of lease, land price and interests often vary individually. So it is not easy to determine standard costs for each item.

Inspite of these difficulties, the cost per kilogram for each culture system can be compared with different levels of productivity.

Theoretically, production cost per kilogram will eventually approach, variable cost per kilogram, if productivity is increased to infinity. It is possible to raise the productivity of semi-intensive ponds from the level of 400 kg/ha to 1000kg/ha. On the other hand, it would be difficult to push up the present level of 15-20 tonnes/ha in intensive culture ponds, since it has already attained a high productivity.

The range of productivity of semi-intensive culture in the Philippines in about 300-500 kg/ha and that of intensive culture in Taiwan is about 15-20 tonnes/ha. At present there are many culturists in Taiwan whose productivity level is over 15 tonnes/ha. So the semi-intensive culturists having under 400 kg/ha. productivity cannot survive in the future when the total production of cultured shrimp from intensive culture continues to increase.

While it is almost certain that the production cost line of extensive culture will be positioned to the lower left part of the semi-intensive line when its productivity is beyond 100 kg/ha, the present level of extensive culture having under 100 kg/ha productivity will be higher than that of intensive culture. These low productivity ponds cannot survive when production from intensive and semi-intensive culture increases.

Costs per kilogram for intensive culture are now lower then those for a large number of the existing semi-intensive and extensive culturists. But this situation may easily reverse when the productivities of the latter increase a little.

That the variable cost per kilogram does not change as the productivity level changes is not always the case. Feed cost will increase following the increase in productivity. An increase in the productivity per hectare of high intensive culture is followed by an increase in the food conversion rate. The food conversion ratio during initial years was 7-8 and yellow tail grew about 1 kg during 8 months rearing. After five years the conversion is 8-9 and fish only grow to 800 g in the same rearing period. The reason for slower growth is due to the deterioration of the habit at bottom caused by left–over feeds.

In the case of shrimp culture good bottom condition in ponds can be maintained if construction allows total water drainage and drying subsequent to the removal of accumulated black soil. However, in high density culture it is difficult to prevent the increase in the feed conversion ratio. On the other hand, in semi-intensive and extensive culture it is possible to grow adequate feed plants provided the ponds are properly fertilized after the pond bed is completely dried and bottom soil turned over. In these ponds the growth of natural food is rather good and an increase in seedling number upto a certain point will not cause an increase in the food conversion ratio.

The results of high density tiger shrimp (*P. monodon*) farming beyond the intertidal zone of Bay of Bengal coast in West Bengal, India during 1995 to 1997 revealed that with a stocking density of 17 seedling/m^2 unsatisfactory growth and weight gain (average size 13 cm. weighing 15 g) has been observed after 120 days of culture with high FCR value. Production cost was higher (Rs. 138/kg) than the return from the sale of prawn (Rs. 87/kg).

Decreasing the stocking density to 10 seedling/m^2 and switching on to demand feeding along with plankton based natural food, phenomenal improvement in growth reaching an average weight of 30 g in 85 days was observed. The food conversion ratio was simultaneously lowered. The yield of tiger shrimp fetched Rs. 275/kg against the production cost of Rs. 162/kg (Biwas, 2000). The slight increase in production cost (due to increased water exchange by pumping) was recouped from the higher sale price.

In kuruma shrimp culture, the colour of ponds in which diatoms have grown very well becomes dark brown and the transparency of water falls below 30 cm. High intensive culture in such a pond show the possibility of rearing large numbers of shrimp at a faster growth rate than in other ponds. Culturists are now trying to produce the dark brown colour of water. Thus with proper management it is possible to enhance the natural productivity of ponds in Asia.

The cost curves of semi-intensive may in reality rise a little at high levels of productivity because of the drop in feed efficiency. On the other hand the cost curve in intensive culture may descend more rapidly at initial production levels with increased natural productivity. The result of increasing productivity can be great in semi-intensive and extensive culture.

Intensive culture firms can remain in business only if they have a good selling price to compensate for high cost. Even if the profit per kilogram is small, the intensive culture will still get adequate profit per hectare from the high intensive ponds due to high production of shrimp.

The object of culturists has always been to gain large profit and not to gain high profit rate per kg. The intensive firms are capable of achieving such profit by the sheer quantity of production even if the profit per kilogram becomes small. Due to these conditions, culturists are eager to improve the culture methods from extensive to semi-intensive and from semi-intensive to intensive.

As long as the selling price has exceeded the cost the intensive culturists will continue the culture. The situation will change as soon as the price of shrimp decreases due to an increase in production. When this happens, the intensive culture firms will not stay in business, because there are no way to cut down the production cost per kg in intensive culture due to high percentage of variable cost. It is clear that the profit in intensive culture will soon disappear with a small reduction in selling price. On the other hand, semi-intensive culture can remain profitable with a little effort to raise productivity.

Table 5: Economic Estimation of White and Tiger Shrimp Culture in Thailand

	White Shrimp	*Tiger Shrimp*
Size at harvest (g) A	13	35
Rearing period (days) B	70	120
Growth rate (g/day)	0.19	0.29
Rearing days/year D	240	240
Number of harvests E = $^D/_B$	3.4	2.0
Number of seedling stocked F	20000	10000
Survival rate (per cent) G	80	80
Production of one harvest (kg)	208	280
H = A × F × G		
One year harvest (kg) I	707	560
Price (Y/kg) J	1080	1800
One year revenue (X 10^3 Y)	764	1008
K = I × J		

Table 5 shows a rough economic estimation in culturing both species (white and tiger shrimp) for 240 days. An assumption is that one rearing period for white shrimp is 70 days giving 3-4 harvests and for tiger shrimp is 120 days giving two harvests. The number of white shrimp seedling per hectare stocked into ponds is assumed to be twice that of tiger shrimp due to the availability of seed. Assuming the price of white shrimp is 60 per cent that of tiger shrimp (in Thailand) revenues per hectare are about 76 per cent that of tiger shrimp. If costs of seedlings are considered, profit will be equal or larger in white shrimp culture. The ease of rearing white shrimp seedling may provide young shrimp for the main growing pond at the beginning of the culture year. This means that it is possible to harvest over four times a year and to achieve about the same sales as for tiger shrimp culture. Thus profit will surely be high in white shrimp culture.

There are many adult white shrimp in coastal sea zones and estuaries. They lay eggs in ponds and canals naturally. Artificial hatching can be done by lay men. At present in Philippines after picking out tiger shrimp seedling, the remaining seedling are discarded. The tiger shrimp seedling only make up about 10 per cent of the total seedling catch, the rest being mainly white shrimp. In Thailand there has been a 50 year history of shrimp culture centred around the white shrimp. The inner part of the Gulf of Thailand is a main production centre of culture. However new ponds have been sited in the Southern part of the bay recently. While many old ponds were converted from salt fields, most of the new ponds are constructed from mangrove areas.

Table 6: Economic Status of White Shrimp Culture in Thailand (1984)

Area (ha)

	0.8–4.6	*4.8–9.4*	*9.6–15.8*	*16*
1. Productivity (kg/ha)	532	506	472	454
2. Producer price (y/kg)	648	670	620	765
3. Production cost (y/kg)	297	269	375	365
4. Net profit (y/kg)	351	401	245	400
5. Return rate (%)	54.2	59.8	39.5	52.3
Net profit × 100				
Producer price				
6. Total net profit	187	203	116	182
7. Number of samples	11	10	6	7

Table 6 shows some results of extensive culture in the inner part of Thailand. The family farm having 4.8–9.4 ha pond is the most profitable size based on the profit rate.

The future of shrimp culture will be decided by the contribution of each type of culture method to the total production change. Intensive culture will be forced to drop out when the production of extensive and semi-intensive culture become dominant. After the dropping out of intensive culture, semi–intensive culture firms may have severe competition from extensive culture. However semi–intensive firms can survive because there is plenty of room for them to raise productivity: On the other hand, production from extensive culture may not be dominant in the market, even if many new areas suitable for culture are found, because of low productivity.

Economic Considerations for Shrimp Farming in Countries of Western Hemisphere

Investment and Operating Costs

To determine the economics of size associated with total size of the farm and size of pond, investment and operating costs have been developed in the analysis using the Generalized Budget Simulation Model (Griffin et al. 1983) for semi-intensive shrimp farms in United States using *P. vannamei* as the cultured species.

The Table 7 shows the total investment of major items for a semi-intensive 200 surface ha shrimp farm using 20 ha ponds. This is similar to a typical large farm being constructed in Ecuador today. Land and construction costs are the major investment items. Land is 43 per cent of total investment cost. Pond construction includes earth moving, pipes, gate valves and engineering fees etc. and is 40 per cent of total investment which for the 200-ha facility is slightly under 2 million dollars.

Table 7: Investment Cost in a Semi-intensive 200 Surface ha Shrimp Farm Using 20-ha Ponds Producing *Penaeus vannamei* Located in the Southern United States, 1984

Item	*Cost (US$)*
Land	828000
Pond Construction	764232
Building construction	64155
Equipment	183529
Machinery	74724

For the 200 ha system, the total investment would increase from 1.9 million to 2.2, 2.7 and 3.4 million dollars as the size of the grow-out pond is decreased from 20 to 10, 4 and 2 ha respectively. Investment cost increases because it requires more land, earth moving and inflow and outflow equipment to maintain 200 surface ha of production as the size of ponds decreases. With the increase in size of grow out pond, investment cost per surface hectare will decrease for any given size farm.

Assuming a constant production of 1159 kg/ha (based on data at Corpus Christi, Texas), investment cost per kg of annual production (single crop per year) in the United States would be 8.30 dollar for a 200-ha farm with 20-ha ponds. For the same farm size the investment cost would be 6 dollar higher (14.60 dollar) if 2-ha ponds are constructed. There are significant economics of size to be captured relative to investment cost both by increasing the size of farm and ponds. This analysis is consistent with other aquaculture systems studied (Giachelli *et al.*, 1982).

The operating costs represents the total annual cost of producing shrimp in ponds under conditions of certainty (no risk).

Table 8, presents the annual variable and fixed costs of operating a 200-ha shrimp farm using 20-ha grow-out ponds. Farms in this analysis are assumed to stock 150000 post-larvae/ha (15/m^2) at a cost of 12 dollars per thousand. Only one crop is produced during the growing season of 185 days. After stocking, water is exchanged in the pond from 3 to 5 per cent daily until harvest. Shrimp are fed 3 to 18 per cent of their body weight depending on the average size of animals in the pond. The average food conversion ratio is 2.5 : 1.

Feed which costs 440 dollars/mt in the United States is the most expensive item and represents 36 per cent of variable costs. Post-larvae is second to feed at 32 per cent of variable cost. Labour is the next highest (12 per cent) followed by the harvesting cost (10 per cent). Total variable cost is in excess of 1 million dollars and is 83 per cent of total cost.

Depreciation is more than half of total fixed cost (53 per cent) and overhead which includes a manger and an assistant manager's salary is 36 per cent fixed cost. Total annual cost for producing one crop of shrimp per year is 1.3 million dollars. Cash operating expenses are 1.2 million dollars per year.

Table 8: Annual Cost of Operating a Semi-intensive 200 Surface ha Shrimp Farm Using 20-ha Ponds Producing *Penaeus vannamei* Located in the Southern United States, 1984

Item	*Cost (US$)*
Variable Cost	
Post larvae	360000
Repairs	27729
Fuel	45093
Feed	408000
Fertilizer	10845
Labour	132640
Utilities	3912
Harvest	109545
Pay roll taxes	20185
Total	1117949
Fixed Cost	
Overhead	80975
Depreciation	118944
Insurance and taxes	11023
Taxes	14131
Total	224893
Total cost	1342842

As the size of the system becomes larger costs (variable, fixed and total) increases. The difference in total cost for using different size grow-out ponds is almost the exclusive result of fixed cost. Thus once a system is built, it takes basically the same amount of variable cost to operate the system, regardless of size of grow-out pond used. This is because post larvae, fuel, fertilizer and harvest cost per hectare are constant across all size facilities. Some small economies of size are available for repairs, labour, utilities and payroll taxes.

A 400-ha system using 20-ha ponds can produce shrimp for 5.50 dollars/kg (heads-off) where a 20 ha system using 2 ha ponds cost almost twice as much. Increasing the pond size from 2 to 20 ha for a given size system reduces the cost of production by almost 0.70 dollar/kg.

As in Ecuador farmer would purchase his post larvae from outside source. But if a firm is not able to stock its ponds at the beginning of the season a significant portion of the limited growing season is lost by the time post larvae are acquired. The break-even price per thousand for producing *P. vannamei* in an outside farm reproduction unit is estimated between 11 to 12 dollars/thousand in a production system of 100 ha or greater. Shrimp are assumed to spawn at 5 per cent per night with eggs having a 50 per cent hatch rate. Survival in the hatchery is assumed to be 40 per cent.

For farms less than 100 ha in size, the farmer would benefit by purchasing post-larvae if the market price was 12 dollars/thousand. The price per thousand for producing post larvae for these smaller farms increases as pond size increases due to restrictions on how fast a pond or hatchery tank must be stocked. A 20-ha farm using a 20-ha pond requires the pond to be filled in one hatchery run causing it to have the highest unit cost for post–larvae production. In countries with year-round growing season, the size of the reproduction unit can be reduced substantially since it could be operated year-round to stock ponds, thus reducing fixed costs. The need for reproduction units in Latin American countries is based more on shortage of post-larvae rather than a high market price.

Although there are several differences between Ecuador and the United States in their ability to produce shrimp in ponds, the two most important differences are availability of post larvae and length of growing season. Ecuador has wild post larvae available through fishermen and a year-round growing season. The United States, on the other hand, does not have a ready source of post–larvae and the growing season is limited to 180 to 240 days per year.

The United States farm is based on data from research ponds of Corpus Cristi, Taxes. Ecuadorian farms are based on actual farms as described by Hirino (1983).

Stocking density of the semi-intensive US farm is triple that of the semi-intensive Ecuador farm. Stocking density decreases as the production decreases for the Ecuadorian farm. Percentage survival generally increases as the stocking density decreases.

A 19 gram shrimp is produced in approximately 190 days on the semi-intensive US farm, whereas a 21 count animal is produced in approximately 175 days (45 days in nursery and 130 days in grow-out ponds) on the semi intensive Ecuador farm. Only one crop is produced per year on the semi-intensive US farm. As the firms in Ecuador become more intense in their operation, the number of crops produced per year increases.

The total kilogram produced with one crop on semi-intensive U.S farm is only a little less than the semi-intensive Ecuador farm that produces 2-4 crops per year. The annual production decreases substantially as farms become less intensive in their operations.

Table 8: Production Details of Semi-intensive Shrimp Firm Located in the United States and Semi-intensive, and Extensive Farms Located in Ecuador

Item	*United States*		*Ecuador*	
	Semi-Intensive	*Intensive*	*Semi-Extensive*	*Extensive*
Stocking density/ha.	150 000	50 000	25000	12000
Survival (per cent)	65	75	80	85
Harvest size (g)	19	21	22	25
Number of crops	1	2-4	1.8	1.3
Total Production (heads off) (kg/ha/yr)	1159	1323.1	554.4	232.1
Food conversion ratio	2.5:1	2.5:1	1:1	–
Fertilizer	yes	yes	yes	yes
Water exchange	Continuous	Continuous	Continuous	Minimum
Nursery ponds	No	Yes	Yes	No

The results of economic analysis are presented in Table 9.

Production per hectare is greatest for the two semi-intensive systems. The semi-intensive Ecuador farm produces 14 per cent more kg shrimp per hectare than the semi-intensive US farm. The reason being that semi-intensive Ecuador farm produces 2-4 crops through a full year of production using nursery pond in the system, whereas the semi–intensive US farm produces only one crop per year without a nursery pond in a 185–day growing season. Production per crop is much greater in the semi-intensive US, since stocking density is three times larger and the crop is growing in the pond almost 50 per cent longer. Production on the semi-extensive and extensive farms in Ecuador are only 42 per cent and 18 per cent respectively of production on the semi-intensive Ecuador farm.

Table 9: Economic Comparison (per-hectare) of a 200-ha Shrimp Farm Using 20 ha Ponds in United States and Ecuador,1984

Item	*United States*		*Ecuador*	
	Semi-Intensive	*Intensive*	*Semi-Extensive*	*Extensive*
Kg/ha/yr.(heads off)	1159	1323	554	232
Dollars/kg	8.47	9.00	10.00	11.00
Value/ha ($)	9798	11908	5544	2553
Total variable cost ($)				
Post larvae	1800	480	180	62
Wages	663	317	190	78
Fuel	225	106	75	40
Feed	2040	1995	334	0
Fertilizer	54	269	269	0
Repairs	138	311	234	179
Packing	548	448	188	79
Misc.	120	687	339	129
Total	5588	4613	1809	567
Total fixed cost ($)				
Overhead	404	230	130	100
Depreciation	595	396	268	192
Misc.	175	91	57	50
Total	1174	717	455	342
Total cost ($)	6762	5330	2264	909
Revenue before taxes ($)	3036	6578	3280	1644
Taxes ($)	1518	3289	1640	822
Revenue after taxes ($)	1518	3289	1640	822
B-E price/kg (heads off) ($)	5.83	4.03	4.09	3.91
IRR (per cent)	21	59	39	25
Total investment	1915	1243	937	715

The differences in prices received by each type of farm is a result of the different sizes of shrimp produced. As the size of shrimps increases the price increases. Prices received by Ecuadorian farmers for a given size shrimp are only slightly lower than those received in the United States. The value of the annual production of the semi-intensive Ecuador farm is 22 per cent greater than that of the semi-intensive US farm. The production value of the semi-extensive Ecuador and extensive Ecuador farms are only 47 per cent and 21 per cent respectively of the semi-intensive Ecuador farm.

The most significant variable cost item for the two semi-intensive systems is feed. It is 37 per cent of variable cost on the semi-intensive US farm and 43 per cent of variable cost on the semi-intensive Ecuador farm. The unit cost of feed was estimated at 18 per cent higher in the United States than in Ecuador. Post-larvae cost is the second most significant variable cost in the semi-intensive U.S. farm and ranked third for semi-intensive Ecuador farm. Post-larvae cost per thousand was three times greater (12 dollar vs 4 dollar) in the United States than Ecuador and the total shrimp stocked in one year is 25 per cent greater for the semi-intensive US farm than the semi-intensive Ecuador farm.

The second most important cost for the semi-intensive Ecuador farm is miscellaneous, which is composed of miscellaneous, pay roll tax (40 per cent of wages) and meals. Wages are third most important item for the semi-intensive US farm, but rank fifth for the semi-intensive Ecuador farm. Even though wages are much higher for the semi-intensive US farm, it has only 11 employees compared to 30 for the semi-intensive Ecuador farm.

Table 10: Total Variable Cost, Total Fixed Cost and Total Cost as a Per cent of the Value of the Crop Produced per Year

Cost	*United States*		*Ecuador*	
	Semi-Intensive	*Intensive*	*Semi-Extensive*	*Extensive*
Total variable	57	39	33	23
Total fixed	12	6	8	13
Total	69	45	41	36

Table 10 shows the present value of the crop produced for variable cost, fixed cost and total cost for each type of farm. Cost per value of crop produced is approximately 50 per cent higher for the semi-intensive US farm compared to the Ecuadorian farms.

The cost to produce one kilogram of shrimp (heads off) is greatest for the semi-intensive US farm and least for the semi-intensive Ecuador farm (Table 9). For Ecuador, the less intensive the farm operation, the higher the cost per kilogram to produce the product. The two of the major cost items for the semi-extensive and extensive Ecuador farms are repairs and miscellaneous. The cost of maintenance in Ecuador would be greater than the United States because of low availability of replacement parts and skilled labour. If these cost were reduced by half, than the cost to produce shrimp for semi-extensive and extensive Ecuador farms would be approximately the same as semi-intensive Ecuador farm.

The after tax internal rate of return (IRR) based on a 10 year planning horizon is attractive for all farms discussed. The IRR is much greater for most of the Ecuadorian farms which explains the rapid size of shrimp culture in Ecuador. Also the significant increase in the IRR as the intensity of the farm increases explains why investors are putting more semi-intensive systems.

Risk and Gestation Period

In earlier discussions, no consideration was given for risk and time consideration. It was assumed that production, prices and unit cost were known with certainty and they did not vary from year to year. In addition it was assumed that in the year the initial investment was made, the farm would be in full production. When these assumptions are made, the results can lead to over-confidence in the economic feasibility of the investment.

Large shrimp farms are usually built in stages. The first year will, more than likely, not have production. The second year will partially produce while in the third year full production could be realized.

There are many factors that investors will not know with certainty and that will vary over time. Price received, inflation and interest rates will vary and can be rather volatile at times. Production can vary from pond to pond through growth rates and mortality. Temperature variation can effect the shrimp. Environmental conditions such as cyclones and heavy rainfall can cause damage and loss of production.

A firm level simulation model (MARSIM) was developed to simulate the annual activities of a shrimp farm taking into account timing and risk. A firm is replicated 50 times over a 10-year planning horizon. Random values of pond growth, pond survival, temperature, cyclones and prices received in each of 10 years are generated from empirical profitability density function for those variables.

When all timing and risk are incorporated into the analysis it can have a substantial impact on the IRR. Table 11 shows when producing *P. stylirostris* on a 200-surface hectare farm using 20-ha ponds, the IRR is less than half when risk and timing of production are considered. The high IRR in Ecuador allows for a larger margin of error when an investor is performing a feasibility analysis. Many investors do not have the luxury of error through overly high returns.

Table 11: Comparison of After Tax Internal Rate of Return (IRR) for Producing *Penaeus stylirostris* in Semi-intensive Ponds and Operating a Post Larvae Reproduction Unit in the Southern United States, 1984

Total Surface Hectare	*Surface ha. per pond*	*IRR No Risk and Full Production Since 1st Year (%)*	*IRR Risk and Production Development Developed Over 2–3 Years + (%)*
40 (developed in 2 yrs.)	4	7.3	1.56
	10	9.1	2.19
	20	10.2	3.44
100 (developed in 2 yrs.)	4	15.9	9.69
	10	19.3	11.41
	20	20.8	15.31
200 (developed in 3 yrs.)	4	20.1	9.65
	10	23.8	10.68
	20	26.8	11.80
400 (developed in 3 yrs.)	4	24.7	13.20
	10	28.0	13.98
	20	31.6	14.06

Chapter 4

Profit and Loss

Economic Feasibility of Fishery Project

The appraisal of the economic feasibility of an investment in a fishery project involves three commonly used methods, the payback period method, the average rate of return method and the discounting method. Of these, the discounting method, which is also called the present value method is considered the most useful and when used by the private sector becomes the basis for financial analysis. It primarily determines the potential profitability levels that the project would generate for the project operating entity in the entire course of operation.

The discounting method is further composed of the net present value, the internal-rate-of return and the benefit-cost ratio methods.

Payback Period Method

This method estimates the number of years required to recover the initial in investment out of the expected earnings from the investment before any allowance for depreciation and is represented by the formula.

where,

A = Payback period in years,

B = Initial investment cost and

C = average annual profit expected before depreciation, such as, if Rs. 50000 is the initial investment cost (B) and Rs. 10000 the average annual profit expected before depreciation, (C)

then,

$$A = {}^{50000}/_{10000} = 5 \text{ years (pay back period).}$$

The payback period method, however, is a weak indicator of investment feasibility in the sense that (i) it does not account for profits realized after the recovery period, (ii) it fails to consider the timing of expenditures and incomes and (iii) it neither measure the profitability of an investment, nor does it provide a correct measure for ranking it among other investment projects.

Average Rate of Return Method

The average rate of return is measured as the ratio of the average annual profits expected after depreciation to the project's initial investment as is represented by the formula:

$$D = {}^{C-E}/_{B'}$$

where,

D = Average annual rate of return

E = Annual depreciation

B = Initial investment cost, and

C = Average annual profit expected before depreciation.

B = Rs. 50000, C–Rs 10000 and E = Rs. 200

then,

$$D = {}^{10000-200}/_{50000} = 0.196 \text{ or } 19.6 \text{ per cent.}$$

Like payback period method, the average rate of return method also fails to consider the crucial timing of earnings and expenditures. It is, however, a good preliminary indicator of investment returns and a useful tool for comparing projects whose time scale profits of expenditures and earnings are similar.

Discounting Method

The weakness of the payback period and average rate of return methods of not taking into consideration the timing of expected earnings on outlays necessitates the use of discounting method for financial analysis. This method considers the concept of "time value of money" which means that money received or consumed at a particular time has greater value than the same money received or consumed at some future time. A rupee received today can be more valuable to the recipient than the same rupee received a year from now.

When a person lend Rs. 80 to day and charges 10 per cent annual interest, would earn Rs. 88 after a year. (Rs. 80+80´0.1 = Rs. 88).

If the interest is represented by r and the sum loaned is P_o.

then at the end of first year,

$$P_1 = P_o(1+r) \qquad \text{.............. (1)}$$

If he lends again the whole sum P_1 for the second year, the amount he will receive at the end of second year would be,

$$P_2 = P_1 (1+r) \qquad \text{.................. (2)}$$

Substituting equation (1) into equation (2),

$$P_2 = P_o(1+r)(1+r) = P_o(1+r)^2 \qquad \text{................(3)}$$

So, a sum P_o loaned at an interest r for n years is expected to be

$$P_n = P_o (1+r)^n \qquad \text{.......................... (4)}$$

At the end of n years, thus finding today's value (P_o) of the sum P_n gets

$$P_o = P_n/(1+r)^n \qquad \text{...................... (5) or,}$$

$$P_o = p_i/(1+r)^i \qquad \text{....................... (6)}$$

where,

i = the number of years in operation.

The present value of any series of future cash inflows and outflows can be calculated by equation (5). This discounting process can also be simplified considerably with the use of a pre-calculated discounting in Table 1.

As an illustration of the discounting process and the use of discounting table, consider the example of a fisheries project that yields profits from 0 to 4 year in the amount presented in Table 2.

Table 1: Present Value of Re 1 Received After n Years

n years hence	*Discounting Rates*						
	8%	*10%*	*12%*	*15%*	*18%*	*20%*	*25%*
1	0.926	0.909	0.893	0.870	0.847	0.835	0.800
2	0.857	0.86	0.797	0.756	0.718	0.694	0.640
3	0.794	0.751	0.712	0.658	0.609	0.579	0.512
4	0.735	0.683	0.636	0.572	0.516	0.482	0.410
5	0.721	0.621	0.567	0.497	0.437	0.402	0.328
10	0.463	0.386	0.322	0.247	0.191	0.162	0.107
15	0.315	0.239	0.233	0.123	0.084	0.084	0.035
20	0.215	0.149	0.101	0.061	0.037	0.026	0.012
22	0.184	0.123	0.083	0.046	0.026	0.018	0.007
25	0.146	0.092	0.059	0.030	0.016	0.010	0.004
30	0.099	0.057	0.033	0.015	0.007	0.004	0.001
40	0.046	0.022	0.011	0.004	0.001	0.001	—
50	0.021	0.009	0.003	0.001	—	—	—

Assuming a discount rate of ten per cent, derivation of the present value of profits for each year and for the entire period yields the following results.

Table 2: Present Value of Profits of a Hypothetical Fisheries Project

	t_0	t_1	t_2	t_3	t_4	t_5
Profits undiscounted	2500	4000	6000	7500	8500	28500
Discount factor	1.0	0.909	0.826	0.75	0.683	0.621
Present value of profits	2500	3636	4956	5633	5086	22500

The Net Present Value Method

As the first of three profitability indicators the discounting technique discussed earlier, the net present value is defined as the difference between the present value of project and the present value of project costs. It is shown by the equation.

$$NPV = \frac{A_1}{(1+r)^1} + \frac{A_2}{(1+r)^2} + \ldots\ldots\ldots\ldots \frac{A_n}{(1+r)^n} + \frac{S}{(1+r)^n}$$

where,

NPV = Net present value

$A_1, A_2, \ldots\ldots A_n$ = Net benefit of individual year (difference between total revenue and total cost)

r = Discount rate

n = Number of years in operation, and

S = Salvage value of assets in year n.

Another calculation of NPV is by independently discounting the stream of annual cash inflow (B) and cash outflow (C) and subtracting the sum of the latter from that of the former.

$$NPV = \sum_{i=1}^{n} \frac{B_i}{(1+r)^i} - \sum_{i=1}^{n} \frac{C_i}{(1+r)^i}$$

The decision rule for the NPV criterion is that;

If NPV > 0, investment would be profitable

NPV < 0, investment would not be profitable

NPV = 0, break-even situation.

The Benefit Cost Ratio Method

This ratio is measure as the total present value of benefits over total present value of costs. It is represented as;

$$\text{Net Benefit Cost Ratio} = \frac{\sum_{i=1}^{n} \frac{B_i}{(1+r)}}{\sum_{i=1}^{n} \frac{K_i}{(1+r)^i}}$$

or

$$\text{Gross Benefit Cost Ratio} = \frac{\sum_{i=1}^{n} \frac{R_i}{(1+r)}}{\sum_{i=1}^{n} \frac{K_i + C_i}{(1+r)^i}}$$

where,

B = The net annual benefit

K = The capital outlay for assets

R = The gross annual benefit or revenue, and

C = The annual operating cost.

The decision rule is to accept investment projects with a benefit cost ratio of greater than 1 and reject otherwise. In case of competing projects, select the project with the highest benefit–cost ratio.

The Internal Rate of Return Method

The internal rate of return of a project is defined as the discount rate which equates the present values of the project benefits and costs, so that the net present value is zero and the benefit–cost ratio is one. It can be solved by setting the left hand side of the net present value formula equal to zero

$$\sum_{i=1}^{n} \frac{A_i}{(1+K)^i} = 0$$

The internal rate of an investment (K) represents the average earning power of money used in the project over the project's life. If K is greater than the appropriate opportunity cost of capital, the investment is feasible.

Finding the appropriate rate of discount basically involves a trial-and-error process. A higher discount rate should be used when the result of the discounted net-income stream is positive, while a lower discount rate should be used when the result is negative. An approximate method can then be applied to shorten the trial and error exercise. To find the first approximate discount rate, divide the capital cost by the average expected annual net income.

Production Economics and Feasibility of Aquaculture Projects

In the fast developing aquaculture industry, the production economics and economic feasibility of aquaculture project is much needed especially to the culturists who like to arrive at sound investment decisions.

The basic factors affecting public and private investment decisions are more or less the same, though the primary goals may be different. While government funded projects are more designed to achieve developmental objectives, privately funded projects are basically inclined toward the satisfaction of the private motive (profit). In the aquaculture industry, therefore, one may find that privately owned fish pond operations are more profit generating than their research oriented government–run–counterparts.

In a capitalistic environment, this is only rational. The enterprising fish farmer chooses investment opportunities he believes would maximize return from his business efforts. When faced with several alternatives offering different cost requirements and profit potentials, he logically selects the one he thinks would provide the highest profits given from limited available resources. With the employment of a well founded economic feasibility analysis, this crucial process of decision making proceeds with justified confidence. In the absence of an economic analysis on investment potentials, however, the decision making may turn out to be an arbitrary process devoid of sound economic basis.

Socio-economic Contribution of Aquaculture Projects

Fish culturists operate aquaculture projects primarily along profit oriented lines. But they contributes to the socio-economic development for the country besides their own families and communities. Their aquaculture projects contribute towards the improvement of standard of living, community development, foreign exchange earnings, lowering of prices, increased foreign exchange reserves and the utilization of the locally available materials.

The demand for material inputs (backward linkages) and the supply of commodity out put (forward linkages) could be made possibly by the aquaculture project, which trigger a chain of economic activities that provide further growth to the economy. The demand for material inputs, like feeds, fertilizers etc. increases income of local producers of these inputs. With more income in the hands of producers the production of more producer's inputs be demanded, stimulating the generation of further employment and income.

Coupled with this, the supply of fish produce into the market further increases economic activity by,

1. Depressing fish prices to the benefit of the consuming public;
3. Increasing foreign exchange reserves through export of fish to other countries; and
3. Inducing savings in foreign exchange with the increased consumption of locally produced fish in lieu of imported ones.

Income taxes paid by the aquaculture project also promotes socio-economic development with the construction of public goods, like roads, market, buildings and other public facilities and the provision of public services such as subsidized education, health care and other freely acquired forms of government assistance.

As a labour intensive operation, the aquaculture project further promotes the alleviation of unemployment and poverty in the community with the hiring of otherwise unemployed workers, providing them an additional source of income and improving standard of living.

In retrospect, the pursuance of a privately funded aquaculture project serves both the private motive of profit generation and the higher social goal of attaining socio-economic progress. So an aquaculture project financed by the private sector may be economically feasible and would eventually benefit to all other sectors like private fish culturists, the community, the aquaculture industry and the national economy as a whole.

Consideration for Investment Decision in Aquaculture Project

For an investment decision, besides economic feasibility, the other the following major factors need consideration.

Resource Availability

1. Availability of suitable site for culture pond construction and the site is requited to be close to natural resources.
2. Availability of land at suitable elevation to enable drainage at highest tide level. This will reduce the water exchange cost of culture ponds.
3. Reasonable price of land, not too high to increase fixed cost.

4. Suitable topography of the site for economic digging and filling.
5. Availability of abundant good quality water to be used as culture media.
6. Soil and water of high natural productivity to facilitate ample quantity of natural food for the cultured stock
7. Optimal tidal amplitude–not too high for over flooding the ponds and not too low to pump for filling ponds and thus increasing the variable costs.
8. Availability of productive soil like clay loam, silty clay, silty loam and sandy loam in pond bottom.
9. Availability of inputs, like, seed stock, feed and fertilizers at reasonable price.
10. Availability of management personnel and labour.
11. Availability of electricity and generators.
12. Infrastructural facilities.
13. Easy accessible for transport by land or water of equipments, supplies, materials and produce during construction, operational and harvest phases.
14. Proximate processing plant and ice plant.
15. Easy access to a local market or product outlet or to a harbour or airport.

Minimizing Production Cost and Maximizing Profits with a Family Run Fishing Business

Changes have taken place with the number of groups operating sardine family fishing throughout the Japan between 1955 and 1975. The changes occured due to:

1. The inefficient passive or waiting type fishing methods (gill net, set net) were replaced by active off-shore fishing methods (purse seine, boat seine).
2. Large-scale net fishing methods requiring a large number of workers were improved by the introduction of labour saving operating methods.
3. Instead of relying solely on sardine catches, fishermen increased their overall catch by including a number of other species of fish.

Another noteworthy change was the switch-over by many fishermen in shallow inland sea areas from purse seine and shore seine to boat seine fishing method. Catching sardine family fishes by purse seine brought a decrease in the sardine resources and at the same time continued low market prices. These two factors along with a sharp rise in labour costs combined to force many fishing operations to extreme financial difficulties. With rapid industrial development, there was a rapid shift in the labour force from primary industries to secondary industries, making it impossible to maintain a cheap labour force for the fishing industry. The sardine fishing industry responded to these changes by shifting their operations from the traditional purse seine method, which required 40 to 50 labourers per fishing unit to boat seine method requiring 6 to 10 labourers per fishing unit.

Concerning the problem of securing sufficient labour force, since the boat seine fishing method only requires 6 to 10 labourers per unit, they were able to make up the major part of the labour necessary from their own family and relatives. For the labour involved in the processing part of the business they found that they could rely mainly on the wives of fishing households working on a part time basis. By reducing the scale of their operations they were able to achieve a stable income from the fishing and processing businesses, and by making every effort to introduce labour saving methods they were able to reduce their expenses, and in doing so raise their income to a sufficient level to support the enterprise.

Price Gap Between Places of Production and Consumption

The price difference between large urban centres and the areas of production is not attributable to the supply-demand gap alone. Rather the greatest contributing factor is the tendency of the people to give priority to sea food freshness in their dietary life.

According to the statistics of fishery developed nations (Japan), the variation in price between places of production and central wholesale markets in consumption centres is no more than 1.5 times in favour of the place of consumption, with respect to salted fish, but is more than four times, on the same basis of comparison, with regard to fresh fish. The difference stems from the high cost of distribution, that includes the expenses incurred in maintaining product freshness and also transportation's risks, in turn, are reflected in central wholesale market prices in the urban consumer markets.

The price difference in consumer areas between live fish and fresh (dead) fish is even more conspicuous. In the producing fishing villages, where dead fish is consumed with the same degree of freshness as live fish, the price difference is only 20 to 30 per cent. Conversely the difference in retail prices between the two types of products in the consumer cities is often two to three times more for live fish.

Selection Between Production of International Commodities or Development of Home Industries

With the "practical" beginning of 200 mile era, coastal countries began to have jurisdictions over their own coastal fishing grounds. This has forced, both coastal countries and those engaging in pelagic fishery to cope with new situations. Within the fisheries industry complicated international relations, such as, interdependence between the developed nations and the developing nations have begun to appear.

Hereafter, a marked tendency for marine products to be produced as international commodities is expected. Also many countries will be striving to promote fisheries as a major means of national development. Countries that are thinking of starting fisheries and making good use of the unused aquatic resources for their economic development, must choose between one of the courses of development, either to obtain foreign currency by export or to improve the fishing village economy by supplying sea foods to the people.

Needless to say, it is impossible to put these two courses into practice separately. One of the two must be chosen as the primary objective, based on conditions, such as the biological characteristics of the resources intended for development, catching method and funds that will be involved, and international commodity value of the fish caught.

To illustrate this problem, the state of Japan's crab fishery may be mentioned. Japan was once the largest crab fishing country in the world, but now Japan is the largest crab importing country. Moreover, in the coastal waters throughout Japan many small fishing boats are engaged in commercial crab fishing. These phenomenon seem to have aroused the interest of crab producing countries around the world, because in Japan there exist both cold sea crabs in comparatively large quantities and warm sea crabs which are not so abundant, but are omnipresent over a large area; and both types of crabs have been utilized highly refined forms, establishing their value as a priority catch.

In the coastal waters of Japan's northern islands, crab fishing has been carried on for many generations. But it was done only as a side job, for the fisherman's private consumption, and the fishing method used was primitive long-lining. Since the beginning of twenteeth century, crab fishing has rapidly developed as a commercial industry. The development was caused by two important facts. One was the development of a "crab factory ship" on which the crabs caught could be canned and the other was the success in using sea water for washing crab meat.

Owing to these two successive technical innovations, the Japanese crab industry came to include a canning industry by means of which it developed into a commercial capital initiative type industry which extended its market overseas where the United States is the main buyer.

Economics of Carp Culture

Fish farming seem to be fast expanding on account of successful demonstrations on cultivation techniques. The length of time required to produce a marketable crop and the varied species combination possible in the pond have been successfully shown to the farmers which provides the major economic incentive in fish culture. There is also a big supply and demand gap for fish in the country, which offers the farmer the challenge to produce more and more fish in the shortest possible time. Prices of fish fluctuate, but this has no serious draw back to the farmer as the catch of fish from the ponds can be scheduled during the high price period.

Though the industry is favoured by some powerful economic incentive for extensive development, there are problems which confront the fish farmers. One of the most pressing problem is the high cost of feed supplement and fertilizers (inorganic and organic). In India however, the cost of organic fertilizers has not yet posed a problem.

Another problem is the availability of quality fish seed. The collection of fish seed from the natural sources, (rivers and estuaries) and the production from artificial spawning need be intensified.

The limitations on fish pond production lies on the management skill of the farmers and high cost of feed. Most fish farm operators rely on traditional techniques rather than improved methods.

Despite these problems, fish ponds continue to expand in the area, which indicates that profits are obtained even on average management level.

Economics of Fish Seed (Major Crap) Production

Cost of production of major carps spawn fluctuates, not only from place to place, but varies at the same place as well. The cost of production of one lakh spawn (5-8 mm) from riverine source in Gujrat is Rs. 100 (Amon, 1966) and in Orissa is about Rs. 40 (Patro *et al.*, 1968). However, when the expenditure on transport is not taken into account, it works out to only Rs. 26.

The cost of spawn produced in bundhs in Madhya Pradesh varies from Rs. 70 to Rs. 80 per lakh (Amon, 1966). Dubey and Singh (1968) reported the cost of one lakh of spawn as Rs. 63.5 in a dry bundh of 0.5 to 1.5 ha, taking into consideration both the recurring and non-recurring expenditures. The details of expenditure on various items of production for a 0.5 ha. bund are given below (1968 price level)

(A)	Non-recurring Expenditure	
(i)	Cost of construction of a bundh–	Rs. 6500
(ii)	Cost of construction of an observation tower-cum-store	Rs. 3500
(iii)	Construction of two storage ponds (65m. × 30 m. × 2 m.)	Rs. 5000
		Rs. 15000
(B)	Recurring Expenditure	
(i)	Cost of equipment for collection and transport of breeders (400 kgs for four crops)	Rs. 400
(ii)	Watch and ward wages for one man for 4 months	Rs. 240
(iii)	Labour charges for collection of eggs and hatchlings operations for four crops	Rs. 100

(iv) Equipment for collection of eggs and hatchlings	Rs. 1660
(v) Misc. expenses	Rs. 200
(vi) Annual maintenance charges of the bundh, storage ponds etc. (5 per cent of the construction cost)	Rs. 750
	Rs. 3350
(C) Spawn Production in One Crop from 50 kg of Female Breeders	
(i) Presuming 100 per cent breeding (fecundity 0.2 million per kg. body weight of female breeders)	10 million
(ii) 90 per cent fertilisation of eggs	9 million
(iii) 60 per cent recovery	5.4 million
(iv) 50 per cent hatching and survival upto the spawn stage say 2.5 million	2.7 million
(v) Spawn production in 4 crops @ 2.5 million per crop	10 million
(D) Production Cost	
(i) One fifth of the non recurring expenditure of Rs. 15000	Rs. 3000
(ii) Recurring expenditure	Rs. 3350
	Rs. 6350
(iii) Cost of production of 10 million spawn–Rs. 6350	
(iv) Cost of production of 1 million spawn–Rs. 635	

Based on the above calculations, the coast of spawn production from Sonar Dry Bundh, Madhya Pradesh works out to Rs. 661.4 to Rs. 2270 between 1956 to 1965. Besides spawn, fingerlings from the left over eggs of bundh were also collected bringing down the cost of spawn.

Cost of Spawn Production through Induced Breeding

The cost of production of spawn has been worked out taking into account, the recurring and non-recurring expenditures, which include, cost of clearing and manuring of ponds, collection of brood fishes, artificial freeding, cost of breeding and hatching hapas, pituitary glands, drag nets, other accessories and labour charges. Salary of technical staff and miscellaneous expenditure have also been taken into account. Cost of breeders have not been included, because after breeding the fishes could be sold at cost price and the amount involved may be recouped. Only 50 per cent cost of clearing and manuring has been considered as the pond would be utilised for rearing other fishes for the next six months after the breeding season. Fifty per cent cost of outer hapa has been accounted since the half of the total number of hapa can be used in the following year. Inner hatching hapa, which are more durable, are usable, for three year. As such, one-third of their total cost has been included. Depreciation value has been calculated for the store items, which can be used for more than one year.

The estimate of expenditure for the production of 10 million spawn are given below.

A. Non-recurring expenditure	Rs. 95.50
B. Recurring expenditure	Rs. 2395.00
C. Labour charges	Rs. 1320.00
D. Miscellaneous	Rs. 50.00
	Rs. 3860.50

Therefore, the estimated cost of production of one lakh of spawn was Rs. 38.60 as per 1969 price structure.

Economics of Spawn Production through Eco-hatchery (Chinese Circular Hatchery)

During the major carps breeding season lasting for about 120 days in a year, the breeding and hatching operations can be carried out into about 30 batches, each batch of 4 days. About 10 million eggs can be hatched in one batch (spawning pool 8 m dia and hatching pool 3.6 m. dia) and with 95 per cent hatching success, 285 millions spawn can be produced of about 6 mm. size. With the sale price of Rs. 10 per thousand, the total income of Rs. 28.50 lakhs could be derived during the season. Deducting about 50 per cent cost per annum on an average capital investment, maintenance, supply of brood stock and other miscellaneous expenses, net profit of about Rs. 14.00 lakh could be obtained during one season subject to the condition that proper arrangements are made to carry out uninterrupted operation of circular hatchery during the breeding season. But assuming that about 15 batches may be normally handled on an average due to non-availability of matured brood stock regularly and other unforeseen interruptions, the net profits would be Rs. 7.00 lakh per breeding season per circular hatchery.

Approximate Cost Estimates of a Chinese Circular Hatchery Unit (1986-87 Rates)

Sl.No.	Items of Work	Cost
1.	One spawing pool, brick-cum-RCC pond of 8 m dia and 1.4 m average depth	Rs. 50,000
2.	Two hatching pool, brick-cum-RCC pond of 4 m dia having another inside chamber with screen, depth 1.2 m.	Rs. 20,000
3.	Hatchling receiving pond, brick-cum-RCC walls 4 m. × 2.5 m × 1.2 m (one number)	Rs. 12,000
4.	Overhead water storage tank, bottom 2.6 m above ground level and capacity, 30000 litres	Rs. 1,00,000
5.	Tubewell with water supply lines	Rs. 90,000
6.	Screen for spawing pool with removable M. S. pipes	Rs. 6000
7.	Operational shed	Rs. 20,000
8.	Miscellaneous, such as, barbed wire forcing, paved area, electric service lines, site clearance area 25 m × 25 m, drainage, approach road etc.	Rs. 80,000
9.	Contingencies at the rate of 5 per cent and establishment at the rate of 2 per cent.	Rs. 22,000
		Rs. 4,00,000

A smaller unit, at a much lower cost (Rs. 66,000) in Ganjam district of Orissa during 1986-87 was able to product. 70 million of spawn in a breeding season deriving a net profit of 2 lakh rupees. Such units are suitable and economically viable for the private sector in rural areas. The details of cost of such hatching unit is given below.

Sl.No.	Items of Work	Cost
1.	Spawing pool, brick cum–RCC base of 4 m dia and 1.4 m average depth	Rs. 10,000
2.	Two hatching pools, with brick masonary work of 2 m dia with 1.2 m depth	Rs. 6,000
3.	Overhead storage tank, bottom 2.6 m above ground level, 15000 litres capacity	Rs. 10,000
4.	Open well with 4 m dia for water supply	Rs. 25,000
5.	Water pump, pipe lines and other misc. expenditure	Rs. 15,000
		Rs. 66,000

Carp seed production through multiple breeding technology has been reported to be a commercial venture (Gupta *et al.*, 1998), which involve recruitment of professional brood fish, partial replenishment of water in the brood pond, economic feed dispersion, instant attempt on induced breeding on brood maturation, minimizing spawning stress and improved hatchery management.

A hatchery with 1 ha water area can yield to the tune of 80 million spawn in a year. For such hatchery complex, the total fixed capital investment would be Rs. 5 lakh. Recurring expenditure including bank interest on fixed capital would be Rs. 2.02 lakh. Gross annual income will be Rs. 4.4 lakh. Net return of Rs. 2.2 lakh, which is 103 per cent over variable cost and 45 per cent on fixed capital. In this system cost of production of spawn comes to Rs. 357/lakh. The same infrastructure facilities, if utilised for multiple breeding can produce 240 million spawn. Considering 30 per cent risk factor on production, 160 million spawn can be produced with an additional recurring expenditure of Rs. 1.2 lakh. As such the net return will be Rs. 4.4 lakh and production cost of spawn will be reduced to Rs. 254/lakh.

Carp Seed Trade at Kakdwip, South 24 Parganas, West Bengal

Fishermen community, start trading carp seeds between June to September at the on set of monsoon every year. Carp seeds are brought from riverine collection and from hatcheries of private sector. At the selling point they construct small huts (6 m × 2.4 m) at their own cost paying a rent of Rs. 75-100 per season, inside which a rectangular earthen pit is excavated, partitioned into two halves for keeping fries and fingerlings separately. A piece of net cloth (*gamcha*) is used to collect seed from the pit for sale or transfer.

The size range and species combination of seeds are as follows:

Species	*Size range (mm)*	*Average size (mm)*	*Percentage (%)*	
Mrigal	13-30	19.7	70	Fries
Rahu	12-22	16.8	20	
Catla	12-20	15.5	9	
Bata	12-18	15.0	1	

Species	Size range (mm)	Average size (mm)	Percentage (%)	
Mrigal	33-68	50.5	60	Fingerlings
Rahu	48-65	55.5	27	
Catla	50-70	60.0	13	

The traders supply seed to the farmers locally. They purchase and sale fingerlings by weight and fries in volume. The fingerlings are purchased Rs. 40-45 and sold at Rs. 70-75 per kg. While fries are purchased at Rs. 150-200 per *Kunka* containing 10000-12000 fries and sell them at Rs. 250-300 per *Kunka*.

Floating Cage Rearing of *P. monodon* Larvae: An Economic Activities

Shrimp farming in Asian countries has been enjoying a boom these past 15–20 years. Of all the different species, it is *Penaeus monodon* or Tiger prawn, which has been the prime target of farmers, due to its high market value and its export and, consequent foreign exchange potential.

Some shrimp–producing countries have developed their hatchery business to such an extent, that they can even meet all the needs of their farms for shrimp fry. On the other hand, countries like Bangladesh and India still rely mainly on seed from natural sources. The huge delta of Bangladesh and neighbouring West Bengal, with its numerous rivers and channels, offers an excellent brackish water environment for post-larvae of *P. monodon*. Collection of seed of tiger prawn (*P. monodon*) has consequently, developed as a major income generating activity for thousands of people living in these areas of Bangladesh and West Bengal.

Most of the people engaged in shrimp fry catching in these parts of Bangladesh and West Bengal are landless peasants and poor fishermen, who often have very limited alternatives for subsistence income. Equipped with scoop nets, shooting nets and set bag nets they collect the tiny and fragile post larvae of *P. monodon*, from brackish water sources and sell their catch to middlemen at prices determined by the later. Needless to say, the fry catcher's desperate need for money makes them an easy target of exploitation.

In coastal areas of Kontai, Midnapore district of West Bengal the middlemen collect 10 to 15 post larvae of *P. monodon* from children shrimp fry catcher in exchange of one 'coke' worth Re 1/–and subsequently sell them to shrimp grower at Re 1/–per post larvae.

Before the post-larvae are stocked in the grow-out farms, they usually go through the hands of one or two more middlemen, are stored for 6-12 hours and counted for a second time before releasing in the shrimp ponds. Hence mortality, due to stress and release into an unprotected environment, often soars to 50-70 per cent. The loss can only be compensated by putting additional fry, and this eventually puts pressure on the natural resource and the few existing hatcheries.

Technology of Shrimp Nursery Cage Rearing

The economic viability of cage rearing of shrimp post larvae have been first tried in 1989 in Ramnagar area of Midnapore district as a means of improving the income of the shrimp fry catcher of the region.

The idea behind cage culture of shrimp larvae came for the reasons:

1. Nursery rearing of post-larvae of juvenile size greatly increases their viability in the grow-out ponds. The farmer would not require such a large number of fry to compensate the loss, compared to post-larvae, extra crops could be harvested within the same season.
2. Instead of selling the fry right away to middlemen at a small price, the nursery would give the fry-catchers the opportunity to raise post-larvae up to juvenile size in floating cages, before selling at a price that is reasonable. At present juveniles fetch a price 5-6 times more that of post larvae.
3. Nursery culturing provides a profitable link between upcoming hatcheries and shrimp farmers. Hatcheries could sell young post-larvae to nursery operators, which would improve hatchery efficiency.

On evaluating the socio-economic status and experience in natural shrimp seed collection twelve participants were provided with material and equipment for the construction of floating cages-one for each family. On-the-job training in cage construction, nursery rearing and management were also provided.

The floating cage consists of a frame of four narrow bamboo platforms. The sides (measuring 0.45 × 6 m) and the end pieces (0.45 × 3 m) are attached to one another by rope to form a rectangular frame with an inside dimension 2 × 5 m. Four to six 100 litre plastic barrels are attached under the raft as floats. Two nylon happas (net enclosures) or 1 mm mesh size and measuring 2 × 1 × 1 m and one happa of 2 mm mesh size and dimension 3 × 2 × 1 m are fixed inside the frame. To prevent folding during tide water movements, the happas are stretched out by frames of galvanised wire and stone sinkers attached to the bottom line. Feeding nets are suspended vertically in the happas and small shrimp, trash fish are minced and applied on them once or twice a day. The entire rearing volume is 8 cubic metre approximately when the upper 20 cm portion of the happas are above the water line. Stocking capacity is upto 20,000 post larvae, distributed according to size in the three happas.

Economics

The cost of one cage at 1990 price was Rs. 3200. Several variations in design and choice of materials exist

Price at stocking is the price the fry catcher would get for one thousand larvae if he chose to sell them at the time of catching, and price at harvest is what he obtains for the bigger size post larvae or juveniles after a certain culture period. Revenue is earnings from the sale of juveniles, less the cost of post larvae.

In terms of economic gain, the outcome demonstrates that the faster turnover rate of post larvae is more profitable, for example, the average gain after 40 days of culture is Rs. 54, for 21 days of culture Rs. 133 and for 10 days culture is Rs. 120.

Though the fry reared for 10-20 days may mot have reached juvenile size, it would appear that such bigger size post-larvae are desired and valued by the farmers.

Period of culture–10 days

No. stocked–4000

Survival rate–80 per cent

Price at stocking (Rs./1000)–50

Price at harvest (Rs./1000)–100

On the basis of the above economics of culture the beneficiary will be able to repay the cost of cage within 3 months at his present 10–day turnover rate.

Economics of Fish Culture

In the United States, it has been shown that the farm size affects the cost of production of channel cat fish. Travis. E. Mitchel and Meda. J. Usry indicated that the size of the farm upto 400 acres, shows an economical level of production cost and capital investment, but beyond this, it does not prove efficient to put additional capital and inputs. In the case of channel cat fish farming, the level of production is most economical upto 100 acre farm.

In Taiwan, the economical size of farm is about 50 hectares. S. Y. Lin stated that in Taiwan, a margin of profit of 21 per cent over the cost of production is good enough. But when the cost of the salary of the manager or owner is included, there is practically no profit in small farm operations. A farm size of less than 10 hectares is considered a subsistence level farm operation.

Owing to the high profitability of catfish culture and the improvement of irrigation, the number of culture ponds increased rapidly in Thailand. In order to obtain more income these farmers stock the fry for rearing catfish more intensely, as far as three times a year the catfish are harvested.

From the economic view point, the following four types of management prevailed in Thailand.

1. An independent farmer, who is able to manage every things by himself	10%
2. A pond is dug by a merchant who supply fry and feed to the farmers	3%
3. A merchant, who sells the fry and feed to the farmers on credit until harvest	6%
4. A merchant, who sells only feed to the farmers on credit until harvest	81%

Because most of the fish farmers, lack capital to run their own business, most of the fish farming business are controlled by a group of merchants.

The cost of feed is the most important factor in the culture. It amounts to 67 per cent of the total cost.

In Philippines, the fish produced from fish ponds contribute about 8.5 per cent to the national fish production. The average level of production of most milk fish ponds in Philippines is only 500 kg. per hectare per year. It has been demonstrated, however, that this can be increased 500 per cent more than the national average production. This is equivalent to 2500 kg. per hectare per year or about 8.5 kgs. per hectare per day (Tang, 1967).

On the other hand, the level of production of fresh water fish ponds is higher even with less intensive management applications (500-800 kgs/ha/year in case of low land fish ponds).

In India economics of composite fish culture with three (Catla, Rahu and Mrigal) species and seven species (Catla, Rahu, Mrigal, Silver carp, grass carp, common carp, calbasu) done at Cuttack. The combined culture of three and seven species yielded a production varying from 1500 to 2975 (average about 2000 kg) and 2228 to 4210 kg (average about 3065 kg) per hectare per year respectively. By adopting modern technique, average yields of about 2000 kg and 3000 kg per hectare per year could be achieved with composite culture of three and seven species combination respectively in moderately fertile ponds.

Sl.No.	*Items*	*7 Species Combination (Rs.)*	*3 Species Combination (Rs.)*
A.	<u>Expenditure per Hectare per annum</u>		
1.	Preparation of pond (including poisoning with mahua oil cake calculated for a pond of average depth 1.5 m. and @ Rs. 27 per 100 kg	910.00	910.00
2.	Cost of fingerlings (100-150 mm @ Rs. 70/1000)	350.00	262.50
3.	Fertilizers (RCD 25 tons @ Rs. 10/ton, Ammon.	250.00	250.00
	Sulph. 1125 kg at the rate of Rs. 43/100 kg.,	483.75	483.75
	Single Super Phos. 500 kg. @ Rs. 26/100 kg)	130.00	130.00
4.	Supplementary feed (MOC + Rice bran mixture @ Rs. 45/100 kg)	1215.00	810.00
5.	Charges for feeding grass carp with weeds	100.00	—

Sl.No.	Items	7 Species Combination (Rs.)	3 Species Combination (Rs.)
6.	Contingencies for labour, transport, netting	250.00	225.00
7.	Nets (Depreciation charges only)	100.00	100.00
		3788.75	3071.25
	or	3800.00	3100.00
B.	Income per Hectare per annum		
1.	Gross weight of fish expected from the ponds with 7 and 3 species combination–3000 kg and 2000 kg. respectively and the rate of fish calculated at the rate of Rs. 4 per kg.	12000.00	8000.00
Expenditure		3800.00	3100.00
Net income		8200.00	4900.00

The above economics of fish culture was estimated by the research-demonstration unit of ICAR at Cuttack, Orissa. It is not known, how far this economics will be feasible in private sector where the rent of the land, capital investment on land and technical supervision of expertise will be involved. Assuming the rent of land in the area is Rs. 1000 and other fixed costs as Rs. 1000, the net profit would come to Rs. 6000.

Though the intensification of management through increase of inputs result in the increase of production, the point of maximum inputs, which produce a maximum economic out put, need further investigations to provide the farmer with the optimum input levels that will produce maximum production per unit measure. The various factors of input–output relationship should be worked out with more details to determine the exact limit where the farmer should continue to add input to get the best profit attainable without wasting input materials applied into the fish pond system.

Fish Culture-Economic Study

So far, the economic study of fish culture operation has been done in the world only occasionally. Yamamoto (1970) has given some concept of economic study of fish culture as below:

Type of Survey Unit

The type of survey unit for the economic study of fish culture is an establishment in which an integrated fish culture operation, from the stocking of fry or fingerlings to harvesting cultured fishes is undertaken in a same compound.

Survey Period to be Covered

It would be advisable that an economic study of a fish culture establishment covers, in principle, a specific calender year, regardless of whether a cycle of culture operation is done.

Type of Data to be Sought

Types of data to be sought are listed below together with some of the reasons for each.

General Indices

For the purpose of sorting an individual case study data into various different types of fish culture, the following indices may first have to be surveyed for each culture establishment selected.

Varieties of Species Under Culture

Type of Fish Culture

Each culture establishment should be categorized into one of the following:

1. Culture in pond impoundment
   (*a*) Intensified culture
   (*b*) Non-intensified culture
2. Culture in cage and pen.
3. Culture in paddy fields.

Purpose of Fish Culture

1. To produce fry and fingerlings.
2. To rear fry and fingerlings into commercial size.

Size of Fish Ponds

1. Total area of land in use at a compound of fish farm.
2. Area of surface water.
3. Depth of water at the centre of pond.

Type of Management

1. Individual management.

2. Partnership.
3. Company and Corporation.
4. State ownership (U.S.S.R).

Operational Status

1. Number of days for which fishes were stocked.
2. Yield in number or weight.
3. Number of persons engaged.
4. Number of man days spent for fish culture.

Capital Invested

1. Coverage of fixed capital–(a) land, (b) pond, (c) buildings, (d) pumping machine, (e) mixer, chopper etc.
2. Assessment of fixed capital–(a) Land itself and pond, (b) Fixed assets other than land and pond.

The amount of depreciation per year will be calculated as:

$$\frac{\text{New Purchased} - \text{Scrap Value}}{\text{Economic Life}}$$

Current value at the beginning of the survey will be calculated as:

Newly Purchased Value – Depreciation × No. of Years Used

Earnings

Operating Expenditure

1. Purchase of materials: The expenditures refer to the cost of materials, inputs for the operation of fish culture. Such materials are in principle, worn off during operation of fish culture.

Grouping items relating material input may greatly benefit the evaluation of national income in terms of gross net profit which is generated from a sector of fish culture. This is measured by subtracting the cost spent for material input from the earnings.

The following items have been covered under this heading:

1. Stocking materials (including transportation cost)
2. Water fee, if any.
3. Fish feed by kind.
4. Fertilizers by kind.
5. Weedicides by kind.
6. Materials used for predator control, disease control.
7. Fuel and lubricants for pumping, generating electricity.
8. Others.

Maintenance and Repair of Fixed Assets

Remuneration

1. Wages or salary paid to employee in cash.
2. Wage or salary paid to employee in kind.
3. Social security premium paid by an operator.
4. Estimated salary or wage of an operator.

Management Cost

1. Sales charge.
2. Salary or wages.
3. Incidental expenses.

Depreciation

Analysis of Data

Assessment of Capital Invested

1. Fixed capital.
2. Floating capital.

Returns (Profit)

Nominal Returns

Earnings – Operating Expenditures

Net Returns

Earnings (Operating Expenditures + Depreciation costs)

Rate of Net Return per Sale (%)

$$\frac{\text{Net Returns}}{\text{Earnings}} \times 100$$

Rate of Net Return per Capital (%)

$$\frac{\text{Net Returns}}{\text{Capital Invested}} \times 100$$

Returns per Unit Area

Capital Invested per Unit Area (Area of surface of water in ha.)

Yield per Unit Area

$$\frac{\text{Annual Yield in Kg}}{\text{Area of Surface of Water in ha.}}$$

Gross Receipt per Unit Area

$$\frac{\text{Earnings (Gross Receipts)}}{\text{Area of Surface of Water in ha.}}$$

Net Returns per Unit Area

$$\frac{\text{Net Returns}}{\text{Area of Surface of Water in ha.}}$$

Labour Productivity

The following productivities may have to be studied:

Labour's Productivity in Terms of Man–year

Yield per Man-year

$$\frac{\text{Annual Yield in Kg}}{\text{No. of Permanent Workers}}$$

Gross Return/Man-year

$$\frac{\text{Remuneration} + \text{Net Return}}{\text{No. of Permanent Workers}}$$

Labour's Productivity in Terms of Man–days

Yield per Man-day

$$\frac{\text{Annual Yield in Kg}}{\text{No. of Man Days Spent for Fish Culture}}$$

Gross Return/Man-day

$$\frac{\text{Remuneration} + \text{Net Return}}{\text{No. of Man Days Spent for Fish Culture}}$$

Production Cost

Cost of production of pond fish is defined as a cost to produce a unit weight of pond fish, and is obtained by the following formula:

$$\frac{\text{Operating Expenditure} + \text{Depreciation Cost}}{\text{Annual Yield in kg or tonne}}$$

Economic Analysis of Fish Culture

For self supporting economic fish culture units, fish should be produced at a price competitive with fish or other animal products. The fixed costs in fish culture are so high that they can be covered only by high fish yields. The invested capital for construction of ponds and the water system, capital invested in gear and costs of water fertilizers etc. are almost the same, whether the fish grows and gain weight or not. Only and costly recurring item which is directly related to the growth and fish yield is food (natural food available) and fish feeding, means feeding of warm water fishes (mainly carps) cultured in ponds for human consumption on a commercial scale, and that is why, an analysis of the economic aspects of fish feeding is essential before lunching fish farming on commercial scale.

The first question to be answered when considering feeding fish in ponds is whether it is economic. This question may be put in a more specific form, is fish culture economic without feeding? that is, would fish yield from fish farms based on the pond's natural food alone (with or without fertilization) cover both capital and running costs, and still leave some profit for the owner? From the analysis it seems that, except for a very few fertile ponds, the answer is usually negative.

Fish feeding can be economical only when special care is taken to ensure the constant supply of feed and its daily routine application to the ponds. This is only feasible when the culture facilities (fish farm) is above a minimum size, which seems to the about 20 hectares. It is possible of course, that in some cases a smaller pond area may be successfully incorporated into a larger mixed agricultural farm where the pond management is efficient and organized well enough so that the fish will be regularly fed with good results. These cases, however, are few and exceptional.

In this connection it may be remembered that the general trend in agricultural techniques today is toward bigger farm units, where modern methods and equipment can be better utilized.

The fixed costs in fish culture remain the same whether the fish are fed or not. There is some difference, in labour costs, which are higher when fish are fed. This difference however, is not proportional to the increase in yield obtained by feeding. The result is that the costs per ton of fish yield are higher without feeding than with feeding of the fish.

The average carp yield in India, in ponds where fish are fed, is 2500 kg/ha., where as in ponds without feeding (though fertilized) is only 800–1000 kg/ha. To obtain these increased yield, different feeds were used, such as, certain cereal grains and oil cakes. The relative feeding co-efficient, that is, the amount of feed in kg required to yield one kg. of fish is dependent on several factors, such as, the productivity of the pond and the feed used. In most cases, however, the average feeding co-efficient was 2.5–3 for carp's upto 600 gram. Taking these data as a basis, the costs per unit area and per ton of fish can be calculated for the two methods of culture, with or without feeding on the following items:

	With Feeding		*Without Feeding*	
	Per ha.	*Per ton*	*Per ha.*	*Per ton*
Yield/ha.				
ha./ton				
Costs				
Charges for capital investment in ponds and fishing gear				

	With Feeding		Without Feeding	
	Per ha.	Per ton	Per ha.	Per ton
Water				
Fertilizers				
Maintenance				
Feed				
Labour				
Marketing costs				
Interest on working capital				
General and over head expenses				

The comparison between the costs per ton of fish cultured with feeding and that cultured on natural feed alone will indicate if commercial fish culture with or without feeding will be economical. In Israel, commercial carp culture without feeding have been found to be uneconomical.

It has been felt that the feeding of shrimp with balanced diets will result in a considerable increase in yields. This was necessary for high commercial value of shrimp in export, market and short period of culture season. From economic point of view critical examination is necessary before applying a new balanced diet commercially.

1. Considering that natural food is an important part of shrimp food, would feeding with a balanced diet cause an increase in yield or a lowering of the feeding co-efficient?
2. What should be the composition and the price of such a diet?
3. Would the feeding of this diet be economical, that is, would the gain in shrimp yield cover the added cost of feed, which would undoubtedly be higher than that of traditional feeds used?

Natural food is an important factor in the growth of shrimp in ponds. It would be a mistake, economically to neglect this food which has qualitative as well as quantitative values. The protein and other nutrients in the natural food should be taken into consideration, and the supplementary diet added in such a way as to balance the natural food.

The protein context in natural food is very high, reaching about 60 per cent of its dry matter (Schaper claus, 1961; Mann, 1961). When the natural food alone is utilized by the shrimp, part of this high quality protein is converted into energy. This source of energy could be provided even better by the carbohydrates in added feed, namely, cereal grains. The protein in the natural food could thus be released to provide for further growth of prawn to gain weight.

It is evident, therefore, that when natural food is abundant in the pond, a protein rich diet would not have much advantage on yield when shrimp are small in size and the standing crop in the pond has not reached a "critical point" of about 8 to 10 gram each.

The situation is however, different, when the shrimp grow larger (more than 12 gram each) and their standing crops exceeds the "critical point" (this critical point is different in different ponds, according to condition). The need for protein then increases, while its supply decreases. A shortage in protein develops, and it is at this point that the protein–rich diet affects yields. When shrimp are fed at this stage with a protein-rich diet their growth rate remains as before, or even increase, where as when they are fed with cereal grains there is some growth inhibition, probably due to shortage of protein.

It is possible to calculate the "break-even-price" for the protein rich feed, above which feeding with this feed is not economical, as returns in shrimp yield would not cover the additional cost for feed.

	Amounts	*Costs per ha.*	*Costs per ton*
Yield/ha			
Ha/ton			
Total costs using traditional feed			
Additional costs of labour marketing etc. due to increased yield			
Costs of traditional feed for yield above 'critical point'			
Total costs without protein rich feed			
Difference in cost per ton yield			
Protein rich feed needed per ha. when fed above the "critical point"			
Protein-rich feed needed per ton of yield			
Break-even price for protein-rich feed.			

Such an analysis is required to be made for any diet before introducing it for commercial use.

Costs and cost-effectiveness

Costs can be divided into two categories–fixed and variable Fixed costs include expenditure or allocations, pertaining to capital, such as, interest, depreciation and other expenditures arising whether the productive assets are actively utilized or not. Variable costs are expenditures incurred in the course of generating revenues and are generally presumed to be linked to the levels of gross earnings or output.

Like most industries, in culture fisheries and in capital intensive mechanised fishing, costs are exclusively borne by the entrepreneur and the worker gets a wage, which is also a cost item. In most artisanal fisheries, however, the normal system of crew remuneration includes sharing the earnings, and in some of the systems the earnings are shared after they are netted of certain costs, which results in a part-sharing or complete sharing of costs as well.

In artisanal fisheries sector, a clear polarity of interest between the owner and crew does not generally exist. With owners participating to a great extent as crew, both the apportioning of costs and earnings strictly between the factors of production–labour and capital–take a wide variety of forms and are not necessarily "rigid" in their patterns.

Costs–fixed Costs

Insurance and depreciation is included under fixed costs. While insurance is a real expenditure, depreciation is a cost allocation which does not lead to a cash outflow. Only the fishing units of the mechanised sector are insured and now this has been made compulsory. The normal insurance coverage is only for the fishing period (September to May); coverage during monsoon months requires the payment of a higher premium. Default in payment of the premium is very common among fishermen.

Depreciation is calculated on a straight line basis, using the reported initial investment and the reported average lifetime of the respective gear and craft.

The correct assessment of depreciation allocations is difficult.

Different methods can be considered are:

1. Present value of the fishing assets in actual possession;
2. Present price of new fishing assets of the same type; and
3. Initial investment or actual cost of the fishing assets at the time of purchase.

The useful service life of the various items of fishing assets can also be ascertained by questioning fishermen.

The ratio of depreciation to gross earnings is a measure of how much of the output is set apart for replacement of capital. The ratio varied from 6 per cent for the encircling net with plank canoe to 23 per cent for prawn net with Catamarans. The depreciation in the mechanised sector may be 18 per cent of gross earnings for the trawlers and 25 per cent for the gill netters.

Variable Costs–Owner's Operating Costs

Variable costs can be regarded as comprising of (i) owner's operating costs and (ii) common operating costs. From the owner's point of view the wages to the crew and to other workers on shore are cost items.

The general practice in some area is for the owner to bear only a part of the total operating costs, which may include costs of fuel and lubricating oil, repairs and maintenance of craft and gear, food for the crew and other expenses, such as, shore charges. There are instances when owners bear all the expenses of a particular trip, when the fish catch is very small and the total operating costs exceed the gross earnings. This practice is quite common when the size of the crew on a fishing unit, and the number of non-owner workers, are both large.

In the mechanised sector, the operating costs borne by the owners amount to between one fifth and one-third of the gross earnings. In the artisanal sector, the share of operating costs borne by the owner is highest for the lobster net fishing with dugout canoe (86 per cent of the gross earnings) and lowest in the case of encircling net with plank canoe, where it is less than one per cent of the gross earnings. In the mechanised sector, the main items of owner's expenses are repair and maintenance of craft and gear, and fuel costs. The former is very high because it involves purchase and fitting of manufactured mechanical parts and requires specialised skills and services, all of which are costly. Fuel costs are borne by the owner's of some gill netters; while they are generally included in the common operating costs of the trawlers.

Common Operating Costs

Operating costs, apportioned between crew and owners, and which are deducted before the earnings (from the sale of fish) are referred to as common operating costs. There are variations in what are to be treated as common operating costs incurred during the fishing trip. Food and drinks immediately before, during and after the trip, fuel during the trip, commissions for marketing of fish, traditional taxes and occasionally contribution to common funds created and utilised by a collective consensus of workers and owners after the trip, are considered as common operating costs.

In the artisanal sector, the common operating costs varies between 5 and 30 per cent of the gross earnings. They are high when the gross earnings and the number of crew members are high. The cost of food is most important and the length of the trip and the time of the day when the fishing trip is undertaken as well as the extent of physical strain of the operation, all affect this cost item. Food expenses are over 6 per cent of the gross earnings in the case of encircling net fishing, drift net fishing and hook and line fishing with plank canoes.

The next most important items of common operating costs in the artisanal sector are the traditional taxes and offerings collected for temple, charge and village fund and utilised for religious and social purposes. While it is a cost to the crew and owners it is a contribution to common good and hence part of the value added. The sales commission given to agents or auctioneers who facilitate the sale of fish is an expenditure incurred by most fishing units Inclusion of repairs and maintenance of craft and gear in the common operating costs is confined to some areas where the minor repairs to gear are borne collectively. The expenses of food and drink to crew during and after the repair work on the net are owner's operating costs.

Repayment of loans as a common operating cost is incurred generally in cases where a large number of the crew are also owner. By common agreement they set aside a fixed percentage of the gross earnings for making the repayments of the principal or the interest. This common operating cost are collectively owned.

Part or the common operating costs is classified as other expenditures, such as:

1. Cost of bait for hook and line fishing.
2. Cost for the transport of craft to a point from where they can launch safely to sea in stormy weather.
3. Mending and drying nets, recoiling wings and ropes and beaching the canoe.

These expenses are largely in the form of food and alcoholic drinks.

In the mechanised sector, major items of expenditure under common operating costs include, fuel, food, ice auction fees and materials like baskets etc.

Fuel costs are by far the largest component of common operating costs. In the case of trawlers they account for nearly half the gross earnings, one-fifth in the case of gillnetters. By including fuel as a part of the common operating costs the owner ensure that the crew economise in its use.

Expenditure on food for the crew accounts for 5-6 per cent of the gross earnings. Generally the owner advance money to the crew per trip, which is subtracted later from the gross earnings. A part of this amount is collectively used for buying rice, condiments and kerosene. Cooking is done on board, the diet is rice and fish, any amount remaining is saved or used to buy cigarettes and betel leaves.

The composition of the operating costs *i.e.*, their break down into owner's and common operating costs varies between different fishing units. In general, the operating costs borne by the owner form the smaller proportion of the whole (7 to 27 per cent), with the exception of encircling net (61 per cent). In the mechanised sector, while the owners of gill netters bear 49 per cent of the total operating costs, for the trawlers, their share is only 12 per cent. The reason for this large difference is that most of the owners of the former pay for the fuel, while it is a common operating expense in the case of trawlers. The high common operating cost of 93 per cent for encircling nets are attributable to the crew's participation in a fund to repay loans for the purchase of equipments, in general the crew is composed primarily of those with ownership stakes in the fishing units.

Remuneration to Crew and Others

Remuneration, as a cost, from the point of view of owners is between 32 per cent and 75 per cent of the gross earnings. But in the artisanal sector, where owners by and large also participate in fishing operation as crew, this remuneration partly accrues to them, to that extent the whole of this item cannot be regarded as a cost to the owners.

In the mechanised sector, the remuneration of the crew is a very low percentage of the gross earnings, 13 per cent in the case of trawlers and 26 per cent in case of gillnetters. The average income of the individual crew is, however, comparatively high.

Remuneration in kind and cash to others are costs incurred for services and for contributions to the observance of socially sanctioned customs, paid in the form of fish. They are made before the catch is displayed for sale. Some of the fish goes to owners household and is therefore not a cost for him. Further some payments made related to services that have nothing to do with fishing, such as, payment to the barber and, therefore, in a strict sense should not be considered a "cost of fishing".

Net Profit or Loss

The ratio of average operating costs to the average gross earnings varies between 12 and 36 per cent in the artisanal sector and 65 to 66 per cent in the mechanised sector.

Profits or losses are the residues after paying all costs and setting aside all allocations.

Cost-effectiveness

To assess how effectively a rupee of operating cost is used by different types of fishing, the average weight and average value of fish that can be caught by each type of fishing for a total operating cost of Rs. 100.

These parameters are important in the light of the use of scarce inputs, particularly hydrocarbon fuels, in the economy. The returns in weight of fish can be taken, for example, to be an index of cost-effectiveness in the production of animal protein. The return in value terms reflects how this is modified if the varying prices received for the fish caught by different fishing units are taken into account.

Earnings and Profitability

Revenues, or earnings from fish production/catch, and profitability can be visualised from three perspectives:

1. From that of the worker-fishermen;
2. That of the owner; who may also work; and
3. That of society as a whole.

This includes the system of sharing divisible earnings, how much of the revenues accrues to the workers and owners and finally the profitability to the owners and to the economy as a whole.

Divisible earnings are normally that part of the value realised from the fish sold which are distributed as remuneration in cash to crew and owners; the gross earnings from fishery less the common operating costs.

There is a wide range of variations in the average divisible earnings of various fisheries. Earnings are effected by various fisheries potential, the price of the fish and the amount of common costs deducted.

Sharing System

The system of sharing the divisible earnings is based on certain generally accepted notions accepted by workers and owners. Basically there are three shares–one for the crew, one for the craft and one for the gear. The crew shares are regarded as the return on labour. The share for craft and gear are the returns on capital. It is out of this that charges for interest allocations for depreciation and some classes of repairs and maitenance are made.

Though these guiding principles form the basis of the system, in actual practice there are considerable variation depending largely on:

1. The nature of fishery/fishing operation;
2. The variations of skill and methods of task sharing among the crew; and
3. The absolute size of the capital investment and the capital intensity.

In most of the simpler craft-gear combinations, the general pattern of share division is more or less the same. It includes the divisible earnings to be split into a required number of shares, which is normally one more than the number of crew on the fishing unit. The rationale is that all the working crew get on equal share and the extra share accrues to the craft and gear, appropriated to the owner. If the owner works as a crew member he gets two shares–one as worker and the other as owner. However, as the capital investment increases, the number of extra shares increase at the disposal of the owner. With catamaran as craft operating anchovy net or prawn net in Kerala, the divisible earnings are split into three shares, one for each of the two crew and the other for the owner, normally one of the crew.

Apportion into 10 shares–7 shares for seven workers and three for the own and made in fishing unit consisting of smallmesh drift net with plank canoe. On days when the operation of the unit requires the help of persons other than fishing crews, the total member of shares increase to say 13-7 for the fishing crews, 3 for the helpers on land and 3 for the owners.

The pattern of sharing of divisible earnings by the more capital intensive combinations engaged in more complex types of fishing often eatails a reward not only for labour and capital but also for management and specially skilled crew. Such considerations result in a differential share rate even among the crew. Provisions also exist for owners to appropriate one extra share which they in turn can distribute at discretion to members of the crew.

The divisible earnings of the shore scine operated with the help of dugout canoe are divided first into two and half shares, one of the owner and rest one and half for the workers. The one and half shares are further divided into seven shares more than the number of crew. Each crew member first receives one share of the extra seven shares, three are given to the person who has undertaken all the major repairs to the unit, pay the local dues and advance money to the crew members who are in need. The remaining four shares are redistributed to the best crew members and to other minor helpers. The owners may on occasions when the value of fish sold is low, forgo their shares and even bear all the operational expenses to avoid discontentment of the crews.

Sharing Pattern of Divisible Earnings

Crew Share (%)	*Owner's Share (%)*	*Type of Fishing*
30-40	60-70	Trawling and gill netting with mechanised boat.
40-50	50-60	Prawn fishing with catamarans
50-60	40-50	Shore scine, sardine fishing
60-70	30-40	Large mesh drift net, Hook and line, Encircling net, Lobster net
70-80	20-30	Boat scine
80-90	10-20	Cast net

Owner working as crew has been included in crew's share, owner's share is strictly the returns to capital, including provision for maintaining the capital assets. In most of fishing type, the crew get more than 60 per cent of the divisible earnings. The capital intensive mechanised sector provide a higher share to the vessel, ranging between 60 and 70 per cent of the divisible earnings. The prevalence and continuance of the sharing system even in the more capital-intensive fisheries of the developed countries indicates that it is rooted very much in the nature of occupation itself.

According to Turvey and Wiseman (1956), the sharing system has two effects:

1. If the investment is successful, the return to the entrepreneur would be less than marginal productivity of capital.
2. If the investment is unsuccessful, the entrepreneur would bear less of the loss.

Remuneration to the crew

The total remuneration received over a period of time is defined as the sum of all the cash payments plus the payments in kind. Payment in kind include the food (or the cash expenditure for it) and fish which is given to the crew during or after the fishing trip. The importance of this in assessing the living standards of fishermen in relation to other workers in other sectors of the economy is apparent; money wage comparisons alone may be misleading.

The per capita remuneration in different type of fishing can not be equated with the total annual income of the fishermen, because in many areas the fishermen generally do three types of fishing in a year. The average per capita remuneration to crew is highest on the mechanised units. But the annual average income in fisheries especially of those who have only their labour to offer is low, less than half the average per capita national income.

Remuneration to the owners

Considerable debate centres on the question whether investments in artisanal fisheries sector are made with the livelihood or the profit motive. The self-employment characteristic of this sector blurs the distinction between capital, management and labour. The distinction between a "wage earner" and a "fishing partner" on most of the indegenous fishing is very hazy. An owner with small resources is often no better a position than the 'pure' fishermen labour.

The participation of owners in fishing operations suggests that it is "livelihood and survival rather than 'profits' that is the prime motive for fishermen investing on craft and gear. "Profit" nevertheless seemed to be the driving force in a few type of fishing.

The owner's share of the divisible earnings less the owner's share of the operating costs is the gross profit available to the owners. From this gross profit the owners have to make allocations for capital replacement and interest charges.

At times due to extremely high operating costs, a worker's remuneration is as much as the gross profit of the owner and the mechanised fishing units are no exception to this.

Profitability-private and social

The net income generated through an economic activity is commonly referred to as net value added. The net value added shows the real addition to the national income after all physical inputs and or costs incurred in the production process have been subtracted from the output. The gross value added includes in addition one cost, namely, the sum allocated for depreciation. The net value added has three elements;

1. Income accruing to labour-crew remuneration;

2. Profits and interest accruing to capital; and
3. Revenues accruing to the state, community in the form of taxes.

In each case, income may occur not only in the form of money, but also in kind

A measure of social "efficiency" of the production process is the ratio of the value added to total gross earnings. A high ratio indicates that the material cost in producing the respective output has been very small. In indegenous fishery at least four-fifths of the gross earnings are actual income. In case of mechanised fishing more than two-thirds of the gross earnings have to be set apart for paying costs of fuel, repairs etc.

The profitability of an economic activity can be measured in different ways, according to the point of view of the different beneficiaries from the generated income. The amount of value added per unit of investment is a measure of the profitability of the production process from the point of view of the entire economy. On the other hand, a common measure of private profitability is the "return on investment" is measured by the net profit generated per unit of capital invested in the means of production. It is this private profitability, which usually decide if an owner of capital invests money in a certain economic activity or not. As the return on investment is influenced by the distribution of the value added, it is normally not in itself a completely satisfactory measure of the social profitability of the production process.

From social perspective, cast net fishing with dug out canoes and hook and line fishing with catamarans are the most rewarding, generating incomes averaging roughly three times of their initial investments.

Private profitablities are high in case of encircling net fishing, shore seine, hook and line and small meshed drift net fishing. In general, considering the high interest rates. (36 per cent prevails in the informal credit market in indegenous fishing villages), the private returns to capital are not vary lucrative in the artisanal fisheries sector and can not be regarded highly beneficial to the national economy.

The mechanised fishing units, also perform very badly from the owner's point of view as well that of the national economy.

The average gross incomes generated are only one-fifth of the initial investments in case of trawlers and one-third in case of gill netters. In both the cases the average fishing vessel is unprofitable to the owners, mainly due to fast rising fuel costs and higher wages of the crew compared to the non-mechanised units.

The collectively owned and operated encircling nets (on an average 11 of 15 crew are owner-workers) result in high private profitabilities for two reasons:

1. The technique, as such are very effective and have a high level of productivity when used at the right time and place; and
2. The collective ownership and work pattern permits the mode of allocation of returns to labour or capital to be flexible by common agreement one or the other can be raised or lowerd.

Economic Evaluation of High-tech Fisheries Project

Today modernization in the fishing industry applies to all aspects from the location of fish, through catching, primary and secondary processing to distribution to the ultimate consumers. In the culture sector, from the production or procurement of seed, feed, rearing in different stages of their life to grow out stage, harvesting before passing on to processing and distribution to the consumers.

The range of the industry is extremely wide, from the owner operated vessel or aquacultural farm to the completely integrated operation with control of catching, processing and distribution of a wide range of species and products to the consumers sales point. In the culture operation, integration of hatchery to produce seed, feed meal plant to produce feed and rearing management all are necessary before the stock is harvested, processed and distributed to consumers.

Individual enterprises may include one or more of the various elements in the productive cycle from the sale of the product as it is taken from the water to processing and presentation to the consumers in its many different forms. The total system which must be considered can become fairly complex, requiring a variety of techniques for evaluation. One must carefully examine each step in the handling of the product, from the time the raw material is produced or procured until the eventual consumption of the finished product. The more sophisticated the techniques become to produce the product, the greater the likely hood of adopting high-tech method in one area affects, or is affected by, another area. These effects must be taken into economic consideration as a part of a total evaluation.

Objectives

Before one can relate the economics, one must be quite clear and thoroughly understand the overall objectives of the enterprise. Unless this is done, putting a lot of money and time will be completely wasted and the capital expenditure committed for the project will be completely in–compatible to the over all objectives.

Assuming that high-tech production mechanisms is a means to meet main objectives, one must examine and consider the sub-objectives. These objectives may not always, in the first instance, be monetary, although in all cases they will inevitably become, or be reflected, in monetary terms. Included among the objectives will be;

1. To exploit a resource, or carry out a production process, which is impossible to achieve by any other means.
2. To fill gaps because of skilled labour shortage or to reduce dependence on manual labour
3. To improve utilization of productive, area or of unrelated pieces of equipment and materials for the benefit of the operation as a whole.
4. To provide management information for decision making.

The objectives of the investor who provides the funds for the project may be earning, appreciation of capital, social factors, and many others–all for the purpose of making a profit and survival of the industry.

Today, the fishing industry is no longer a simple operation of baiting hooks and hauling in the fish, with a buyer waiting at the pier. It has become increasingly competitive in terms of both the resource, production and the market. This competition is now forcing decisions with which the industry was not faced in the past. The future will see this decision process becoming more complicated, and the objectives must be clearly defined it the right answers, leading to economic survival, are to be made.

Criteria of Evaluation

Because of the nature of fishing industry, and in particular, the lack of control over the raw material, are such that the earnings from a particular piece of equipment or material may fluctuate widely, the choice of a method of evaluation is not an easy one, because of the multiplicity of specific goals which an investment is intended to serve. Whatever the criteria might be, the investment decision must abide by the fundamental guide line, which is that preference must be given to the undertaking which will generate larger and quicker benefits. This is particularly true in the fishing industry, where there is so little control over the basic resource and the cash flows can vary so widely.

There are a number of approaches, ranging from simple pay-back calculations to discounted benefit cost ratio. But before understanding different approaches to evaluating the return on investment, it is essential to examine the major components of economic analysis.

Components

The basic evaluation of the proposed project consists of matching the cost of economic gains, and can be carried out in many ways. The project out lay is represented by the net investment required for the propositions, while the economic benefits are represented by the operating cash flows generated by the project. Another important element is that during which the investment will be made, and benefits realized. The nature of these components, and the estimates underlying them must be examined and estimated before the various techniques can be applied.

Concept of Net Investment

Net investment in a project, the funds required will include not only for the specific fishing gear, equipment or construction of ponds, seed, feed, but also the cost of ancillary facilities, binding, changes or additions, changes in working capital requirements and financing costs etc. While the project resulting in the replacement of existing capital asset there is a temptation to include the remaining value of the old asset in determining the amount of investment. The book value of such an asset is however, of no consequence as it represents a sunk cost attributable to a previous decision. The only relevant quantity for analysis of the new project is the salvage value of the old asset. Such a cash flow from the old asset is a reduction of the investment required for the new. Like wise, any salvage value of the new project at the end of its useful life, can under certain conditions, be considered a reduction in the original investment.

The basic rationale of economic analysis is to show cash flow instead of accounting allocations. In determining the size of investment it is necessary to pose the question "what are the net cash outflows (commitments) of the various alternatives?"

Operating Cash Flow

Once the size of the investment is determined the second step is to define the relevant economic benefits normally found as recurring cash inflows provided by the project.

The second major class of expenditure includes those whose purpose is to provide profits. In this class the relevant operating cash flow will be the net cash inflow resulting from the difference between the revenue and the expenses incurred.

Finally, there are considerable number of cases which cannot be readily analyzed in their economic effects, the benefits being a matter of management judgement.

One must not look at the project itself, but also consider the effect in other areas of the operation.

Difficulties can arise when accounting allocation, such as, overheads are considered or when improvements in the labour or space requirements cannot be utilised. Accounting allocations are irrelevant unless there is a change in the cash flows. If a fraction of worker's time is released or some additional water area becomes available, which cannot be productively used, then no saving accrues. On the other hand, if the increase water area is required, there is a cost only if its use results in the loss of an opportunity to derive from it an economic benefit, such as, profits from another activity.

Thus the analysis of operating cash flows involves a step-by-step questioning of every revenue and expense category with the critical query "will it eventually have more (or less) cash from the project?" Another aspect of cash flow which must be considered is that of tax. Presumbly the projects will, in one way or another, increase the profits, or at least maintain the profitability of the enterprise. It is, therefore, necessary to calculate the taxes due because of profits. As the economic gains have been calculated in the form of operating cash flows a complication arises as taxes are paid on the basis of accounting profits. The difference between cash flows and accounting profits is found in such accounting transactions as depreciation and amortization of the assets or project expenses. These do not represent cash outlays, but are merely allocations of past expenditures.

To show the impact of the project on the accounting profits, one must consider depreciation and amortization as part of the operating picture in so far as they affect taxes.

The Estimate

The problem of dealing with the uncertainty of estimates is a complex one, particularly in the fishing industry. The concepts of net investment and operating cash flows, which have been described, are complicated by the fact that the decisions deal with future events and for all practical purposes, estimates have to be used. In so far as investment is concerned this is not usually too difficult, since quotations and bids can be solicited for a fairly close estimate of the funds to be committed. The big danger is to overlook ancillary requirements of the project.

Profits related to investments, are estimates prior to the decision and only the operation it self can prove or disprove the estimator's figures. Thus it is important to look to past experience of the similar projects of the same type. In order to avoid gross errors it is sometimes useful to establish a range of estimates for those elements of operating costs and revenues most likely to be affected, such as the most likely figures, the highest and lowest possibilities, in order to ascertain the merit of the project, should unfavourable conditions develop. It is possible to give weighted judgement to the range of possible outcomes, by applying probability factors and calculate the weighted value of the cash flow expectation.

Time Pattern

It is usually found that the economic benefits derived from a project are patterned unevenly over time. The introduction of a new product, for instance, may show slowly rising cash in flows from year to year, or even zero inflow until momentum is gained. An investment in a trawler or culture pond may produce declining benefits over the years as its effectiveness decreases, and costs to maintain its rise. In fishing industry, irregularity because of the nature of resources, weather condition, ecological changes, etc, are all factors which produce irregularity over time. It is important to identify the time dimension of the proposed project.

Economic Life

The final criterion is the importance of distinguishing the physical life of the infrastructure and equipment, usually the period over which depreciation is taken for tax or accounting purposes, and the period of economic usefulness, the economic life. It will make quite a difference whether the infrastructure or equipment will yield, economic benefits over a one or ten year period. This is of particular importance to the fishing industry. Problems are faced from rapid developments in new techniques and equipment, and also the possibility of product changes, because of the nature of fishery resources. A trawler, built today, may have an economic life of only ten years because of technological changes. Its physical life, on the other hand, given reasonable maintenance, may be fifteen or twenty years. The period during which the project is expected to yield economic benefits is the one over which the analysis should be carried out.

Relative Desirability

Having determined the net funds outlay, and what the operating cash flows will be, what pattern these flows will assume, and what the economic life of the project will be, one has to combine this data into rational measures of desirability.

Two main approaches are commonly taken:

1. Rules of thumb, or rough guides, on the one hand, and;
2. Attempts to derive more exact economic yard sticks on the other.

The rule of thumb category contains the widely used pay back period, and the simple return on investment which, at times, can be an approximation of the economic yield.

The category of economic measures uses the concepts of the time value of money *i.e.*, a rupee receive today is worth more or less than a rupee received a year from now.

Payback Period and Simple Return on Investment

Most commonly used method of obtaining a measure of project desirability, simple in calculation. The payback period simply measures the speed with which funds will be returned from the project to be used else where:

Payback

$$\frac{\text{Net Investment} = 1000}{\text{Average Annual Operating Cash Flow} = 2500} = 4 \text{ Years}$$

An outgrowth of the payback period is a simple so-called rate of return which is reciprocal of the payback period,

Rate of Return

$$\frac{\text{Average Annual Operating Cash Flow} = 2500}{\text{Net Investment} = 10000} = 25\%$$

But the factors of importance in determining investment desirability are not limited to cash flow. They include the length of time during which the cash flow is generated (the economic life) and the specific patterns of cash flow thus, the economic worth of a project is the balanced result of all of these factors while the payback devices average the cash flow, regardless of the economic life. The payback method can only be a rough guide on a screening device.

Methods of Utilizing the Time Value of Money

When investing in a project, management cannot be indifferent to the choice between projects with the same characteristics, excepting for their economic length. In the interest of maximizing the benefits from the investment common sense dictates that the longer-lived projects should be given preference. If one invest 1000 now, is worth more than 1000 at one year from now as he could have earn 6 per cent interest if kept in savings bank account. Similarly, the entrepreneur should value investment outlays and cash inflows according to the timing of each element, for investment outlays to be made in future years are less costly than ones to be made right away, and cash inflow today is more valuable than that which might be received in the future.

Requirement of Fishing Industry

Adoption of high-tech method in the industry involves investment, not only in money, but in technology and management. After meeting the costs of operations, there must be a return on the investment. Adoption of high tech method or mechanization must be conceived relative to the part of the total system within which enterprise operates and will, therefore, take different channels and forms according to the vulnerability of the particular part to the recurring economic imbalances. For an enterprise operating within a small part of the system, say catching, it would be desirable to remain flexible as to the species fished for, that is, lobster, shrimp, sardines, and ground fish, so that the fishing effort can be sufficiently diverse, enabling a variety of products to be exploited and minimizing the risk of being caught by an imbalance in a particular field. The enterprise must avoid high-tech methodology or mechanization which restricts its field of operation.

On the other hand, large integrated operation is less vulnerable to an imbalance in any particular segment, being more concerned with the results to be obtained from the total system. Thus a vessel can be automated to a higher degree; geared to maximizing efficiency over a narrow range of fish species. If the enterprise is integrated in breadth, as well as in depth, then the tendency will be towards separate specialized units, each intended for a narrow range of operation, rather than for a single unit with wide flexibility.

It is said that one must consider the total system. The system will include all or some of the following parts in marine fishing:

1. The resource itself.
2. Exploiting the resource.
3. Processing the results of exploitation.
4. Marketing and distributing the finished product.

In the area of aquaculture, the system may include all or some of the following activities:

1. Seed production or procurement.
2. Production of the resource.

3. Processing the products.
4. Marketing and distributing the products.

Fisheries Resource

The fisheries resource, with few exceptions is common properties being available on a catch-as-catch-can basis for any one, with few restrictions, who wish to exploit it. This means that the risk factor is extraordinarily high, and expenditures are not generally within the scope of industry, being mainly a matter for government, as a part of the fisheries infrastructure.

Improvement in automation, such as search missions to locate and identify new resources and their limits, information services, data bank can provide information for improving the exploitation.

In terms of this part of the system the economic value of such action would be the net cash flow generated, and this in turn becomes only one factor of the evaluation for the next part of the system, that is cash inflow, against which the cash outflow must be set.

Exploitation

The exploitation of the resource involves everything from fixed gear tended from shore, or by small boats to sophisticated fishing vessels involving huge capital costs. Expenditure in this part of the total system will generally be for one or more of the following purpose:

1. To increase the catching efficiency of vessel. This will involve devices for locating the resource, reduction in net handling time and gear technology.
2. To increase the utilisation of the vessel. This will include improving reliability, handling equipment to reduce discharge time, shore facilities to improve turn around. Freezing at sea, to enable greater time to be spent in the fishing ground.
3. To decrease the operational cost of the vessel, including such things as reduction in crew, better maintenance.

Processing

The development of successful processing unit poses problems in plant operation, particularly when dealing with a highly perishable commodity.

The material source for the most part seasonal, with wide variation between peak and valleys in the rate of input. Species normally handled have wide variations in handling and processing characteristics, such that mechanical equipment suitable for one, is not suitable for another. Except in very few instances, the economic pattern of the supply market relationship does not, at the present time, permit to take advantage of scale in any particular area. The market demand also has seasonal variations, and these do not always match the material supply pattern. High capital expenditures in processing must be matched by high utilization, and this infers much greater rigidity in approach to processing. The danger here is that of a commitment to an operation which one cannot keep employed at a realistic level.

So the need for examining the total system-cost of vessel, handling, holding, thawing, processing and market acceptability is emphasized.

Distribution

The distribution segment of the system absorbs in excess of 60 per cent of the consumer's money. This is made up of the cost of movement from the processor, inventories and storage at various points in the chain, the investment tied up, represented by goods in storage or transit; and profits taken out at the various steps in the process.

The elimination and reduction, or any of the myriad elements between the source and the consumer, can generate benefits. Quick transportation opens up new markets for the fresh product hitherto inaccessible and more acceptable to the consumer, and commands higher prices. The reduction in inventories through the distribution chain, result in lower carrying charges.

Chapter 5

Financing in Fisheries

In India financial assistance was available till 1967 from the Central and State Governments in the form of loans and grants, as well as subsidy for the construction of fishing boats, purchase of marine diesel engines and improved types of fishery equipments in the various maritime states While the loan portion for the mechanization scheme of fishing craft was financed in full by the Central Government as a part of the development plan expenditure, the amount of subsidy constituted 50 per cent share, the remaining share being the contribution from the respective states. The pattern of subsidy, however, varied from state to state depending upon the progress of the scheme. In 1967 arrangements were made to provide institutional finance for fishery schemes under which the Agricultural Refinance Corporation (ARC) helped the fishery projects through co-operatives in each state and provided loans with the gurantee of state governments. This arrangement helped the Department of Fisheries to concentrate more on developmental schemes by restricting loans to commercially viable fishing projects through institutional finance.

Of late, the commercial banks in the country have also come forward with schemes for financing of fishing boats with the assistance of the manufacturers of marine diesel engines in selected regions. Restriction has been imposed on the import of fishing vessels of less than 60 feet (18.5m.) length. The import is however, permitted subject to placement of an order for one indigenous fishing vessel of about 60 feet in length against every two imported ones. This necessitated certain amount of financial assistance for the fishing units to acquire fishing vessels on easy terms. The Indistrial Development Bank of India (IDBI) accordingly included fishing vessels under the rediscounting scheme on deferred payment basis over a period of seven years. The deferred payment arrangement however, need to be streamlined because the boat owner does not directly get the benefit of deferred payment in the manner in which facilities are extended for the supply of fishing vessels in other parts of the world.

Role of State Bank Group in Financing Fisheries

Basing on the encouraging results of experiments on the fishermen in the Ratnagiri district of Maharashtra by providing credit facilities to fishermen for the purpose of purchasing hull, marine engine and other allied accessories to the tune of 56 lakhs to 100 fishermen, fishing was included as one of the important activities eligible for financing from the State Bank of India and its subsidiaries. The progress made in this regard may be judged from the fact that the number of fisheries schemes financed by State Bank group incresed from 30 in June, 1968 to 478 at the end of october 1969, limits sanctioned increased from Rs. 20 lakhs to Rs. 293 lakhs during this period.

Basic Pre-requisites of a Fishery Proposal

The followings are some of the important requirements which must be taken into account while submitting a fishery proposal for financing by the bank:

Integrated Scheme

Integrated scheme should aim at increased production (catch) of fish.

Potentiality of Fishing Grounds

Adequate catch of easily marketable types of fish should be available within resonable distance to make the project worthwhile and viable.

Selection of Area

Should be compact and one which can be effectively supervised.

Technical Fasibility

Technical proposal like trained fishermen, and others, should be made available.

Economic Viability

The anticipated income out of the project should be such as to ensure a regular income sufficient to repay bank's laon with interest within the stipulated period.

Assured Marketing Facilities

Both in the fresh and processed form in internal and export market should be assured.

Items Eligible for Financing

1. Purchase of mechanised boats fitted with nets and other equipments.
2. Setting up of boat building yard, workshop and servicing stations.
3. Providing transport vehicles for transportaion of catch.
4. Setting up of processing plants.
5. Putting up cold storage cabinets.
6. Construction of godowns.
7. Provision of berthing facilities, such as jetties.

Terms and conditions

1. Medium term loan for meeting the capital expenditures under the project are granted for periods upto a maximum of ten yeras by the State Bank of India and seven years by the subsidiaries at 8½ per cent to 9½ per cent interest per annum.
2. Short term loan for meeting working capital requirements, such as, establishment, maintenance and other day to day operational expenses for a maximum period of one year at 8½ per cent to 9½ per cent interest per annum.

Role of Industrial Development Bank of India (IDBI) in Financing the Fishing Industry

The IDBI is willing to play a constructive role in the development of the fishing industry and its assistance is available for:

1. Setting up of shipyards for building deep sea fishing trawlers.
2. Acquisition of such trawlers.
3. Setting up of cold storage and freezing units.
4. Providing facilities for processing and canning fish.

The IDBI till 1970 provided assistance to fishing industry by way of refinance only, the amount granted being Rs. 35.76 lakhs relating to two fishing concerns. The assistance provided was for acquisition of fishing trawlers and setting up of cold storage in case of one concern and for setting up a processing and canning unit in the other case. Proposal for refinance assistance for undertaking processing and manufacture of fish products was also considered.

Role of Agriculture Refinance Corporation (ARC) in Financing Fishing Industry

Upto 1972, the corporation has sanctioned 12 schemes for development of fisheries in the co-operative sectors covering five states, namely, Andhra Pradesh, Kerala, Tamil Nadu, Mysore and Maharastra. These schemes envisages among other items, purchase of 550 mechanised fishing boats and construction of two boat building yards, by which it is expected to increase production of fish by about 22 thousand tonnes. The total outlay in these schemes is Rs. 650 lakhs.

After having created an economic climate favourable to trade through their Structural Adjustment Programmes (SAP), the World Bank and the IMF as well as other international agencies are helping industrial aquaculture expand by giving loans to both the central and state governments directly. While the World Bank loan to India is for improved fish culture, shrimp culture forms a substantial component. The expansion of this industry is justified on the grounds that it will benefit the poor by providing them with better nutrition, more employment opportunities and higher incomes, also that it makes use of land that is unfit for any other agricultural or forestry purpose. The lure of earning foreign exchange is also a key factor.

World Bank's Assistance

The World Bank is a multilateral funding agency designed to ensure liberal, capitalist world economy by enforcing rules favouring a free movement of capital internationally. To increase the volume of lending, the Bank began to encourage, developing countries to incur more debt.

The World Bank became involved with aquaculture in the seventees, when it began providing loans to governments of Asia and Latin America for development of shrimp ponds. The Bank financed development projects in Indonesia, the Philippines, Thailand and Bangladesh. By 1980, the Bank broadened its support to include China, India, Brazil, Columbia and Venezuela.

The aim of investments in prawn culture in the seventees was to set up a basis for processing and products for the market which meant an emphasis had to be given on the infrastructure in the form of roads and refrigeration, so that industrial shrimp production could expand by eightees. In 1992, the Bank invested $1.685 billion in agriculture and fisheries, of which India received $ 425 million for shrimp and fish culture.

In 1991, the World Bank began to push for semi-intensive prawn farming in India. The need for close co-ordination between production, hatcheries and feed raised the investment to a high of U.S. $ 11000/ha. Since 1992 World Bank has provided assistance to five states in India for fish and shrimp culture. They provided 80 per cent of the project cost for brackish water shrimp, 8 per cent of the project cost for inland fisheries and 12 per cent of project cost for project management.

The World Bank in its report, Dec, 1991 states that semi-intensive shrimp farming is ideal for India, and that the World Bank through its financial assistance, could help increase India's shrimp production, provide employment and help the country earn much needed foreign revenue.

Government Agencies–Marine Products Export Development Authority (MPEDA)

An autonomous body set up for assisting the shrimp culture industry and for overseeing the development of both the industry as well as its trade.

MPEDA offers the following subsidies to support the aquaculture industry.

1. Subsidy for new farm development, assistance of 25 per cent of capital investment or Rs. 30000/ha. up to a maximum of Rs. 1.5 lakh.
2. Subsidy for establishment of medium scale shrimp hatcheries of 30 million seed per year capacity and above, assistance of 25 per cent up to a maximum of Rs. 5 lakh, can be availed of by private parties/individuals.
3. Subsidy for feed and seed, assistance at 25 per cent up to Rs. 3000/ha. and Rs. 450/ha. respectively.
4. Subsidy for establishment of brood stock bank, assistance of 25 per cent of capital cost subject to a maximum of Rs. 1.5 lakh.

In addition shrimp farmers are allowed to import shrimp feed at concessional custom duty.

MPEDA has also established two hatcheries of its own, one each in Orissa and Andhra Pradesh.

Public Financial Institutions–NABARD

Financial assistance for the aquaculture industry has also been provided by several public financial institutions such as, National Bank for Agriculture and Rural Development (NABARD), Industrial Credit and Investment Corporation of India (ICICI), Shipping Credit and Investment Corporation of India (SCICI) and Industrial Development Bank of India (IDBI)

NABARD through which World Bank loans are usually routed, has been given the go ahead by the government to directly finance high-tech/innovative/export–oriented and other special projects related to storage, marketing, infrastructure development and processing by the private sector. Among the prioritised area is aquaculture, where the bank is financing integrated projects of private entrepreneurs like M/S Shakti Aquaculture in Maharashtra and M/S, Aquarious Fisheries Ltd in Goa. In 1992-93, NABARD financed the conversion of 387 ha. of land to aquaculture, from 1993-95, the conversion financed by the bank increased to 3984 ha. It also extends assistance to agencies like the Central Institute of Fisheries Education to supply large quantity of seed to prawn farms and to undertake extension activities to popularise the prawn hatchery concept.

Credit for the Unbankables

Most of the small-scale fisherfolk are not able to raise credit. To improve the lot of small-scale fisherfolk, a project has been launched by Bay of Bengal Programme in Ranong, Thailand in 1986. The project met the challenge by starting a revolving fund to be managed by those, who needed the banking most Despite several problems the results have been heartening, repayments being around 75 per cent and would seem to indicate that here is a way to offer credit to the unbankables, which should be followed through.

Fisherfolk, especially the poorer ones, do not have any assets to offer as collateral, also, the banks find it expensive to service a lot of small loans. Depending on savings of the people also do not work out, as savings are low, where they existed at all, and not sufficient for the type of enterprises planned.

The opinion is to create credit scheme. The idea, in theory, is rather simple. Villages or groups of fisherfolk would be organised. Grants in cash or kind would be extended to individuals or groups to undertake particular enterprises. The people would return the funds along with a small service charge to the groups or village communities who, in turn, would lend out money to others. In other words, a simple revolving fund based on an initial grant, which could, at least in principle, perpetuate itself in the community and help the community to get access to low-cost credit to enable it to take up enterprises which, in turn, would help increase incomes. It is hoped that such credit would not only act as an incentive to fisherfolk to take up new technologies, but would also absorb possible uncertainties and even encourage the communities to add to the funds from their savings.

The revolving fund schemes of 1989 in Thailand were managed basically in two ways, one, wherein a group accepted an outright grant and managed the revolving fund as a group activity and two, where a village committee under the direction of the village headman took the responsibility for the fund. The service charges, varied from group to group between 1 to 3 percent. The Department of Fisheries supported the activity by supervising and monitoring the activity and by helping the fisherfolk to manage the funds.

Reasons for Success of Revolving Fund

The most obvious reason, of course, is the economic viability of the enterprise they undertake.

In the first place, therefore, training and supporting fisherfolk to select viable enterprises, which can be sustained through local resources and the products of which have good and accessible markets, becomes vital. In extension terms, the readiness and the economic viability of technologies become prerequisites before extension should be contemplated. It also means extension staff have to build up their economic and enterprise managed skills in order to help and support fisherfolk.

A group that works well together, seems to earn well together and repays on time. The organisation and mobilization of groups, ensuring good and just leadership and cohesive action, does play a crucial role in the success of revolving funds.

Credit receive is sometimes used for purposes other than planned. Given the lack of resources, it is not surprising that this happens. An old unpaid loan, an illness in the family or the sudden demand from another of the family's enterprises, often diverts funds and leads to trouble. Close follow up and guidance can, and did, solve some of the problems, but on some occasions, with the groups and the leadership's support and consent, such diversions had to be regularised and the loan rescheduled to ensure its repayment. It is not enough to look at the enterprise alone but the family and all its enterprises as a whole to ensure success. Close and regular follow-up by extension staff cannot be over–emphasized. Building awareness, through training in simple accounting and record-keeping, and better management of the enterprise and of the group go a long way in building the group's ability to develop.

Lastly, is the participation of the fisherfolk in determining needs and priorities, in setting objectives, in planning the effort, organising it themselves and implementing and monitoring it. This seems to be not only a way to overcome hurdles, but also a way, of ensuring sustainability. Ultimately, the fisherfolk will determine their development.

Risk

Uncertainty and unpredictability of catches and production is the characteristic of fisheries sector; both for quantity and price dimensions. At the time of going for a fishing trip, the fishermen have no way of predicting the size of the catch. There is also uncertainty about the price that will be obtained on selling the fish due to its perishable nature.

The unpredictability of the size of the catch can be expressed numerically as a risk. The sophisticated measure would be a probability distribution of size of catches, which require more powerful data processing. A simpler form would be the number of occasions, when a fishing trip made by a fishing unit yielded no catch, expressed as a percentage of the total number of trips made. The probability of catching fish is affected by fish behaviour and fish abundance, the technique of fish catching, the time spent in fishing, the distance of fishing ground from the shore, the skill of the fishermen and other factors. There is an element of correlation between the distance of the fishing ground from the shore, and risk. Most of the inshore fishing methods seem subject to a higher risk of no catch than those fishing further offshore (hook and line fishing, large mesh drift net fishing). This may possibly be because the fishermen is prepared to accept a higher risk of failure in inshore waters, since the effort and costs that may possibly be wasted are kept within reasonable limits, but it may be no more than expression of the behaviour patterns of shoaling, pelagic species like sardines, mackerels or anchovies, they either appear inshore in large shoals yielding high catches or fail to appear.

In culture fisheries, however, the risk is little less though there is uncertainty and unpredictability in production (yield) mainly due to mortality of the stock during culture period which can be attributed to the quality of seed, availability of food, disease and the effectiveness of the culture technology adopted.

Catch Per Man-hour

The manpower is an important input expended for catching a given quantity of fish. Since the process of catching fish–or waiting for fish to be entangled in the net or hook–forms only a fraction of the total manpower expended during the trip, two different parameters can be calculated:

1. The catch per man-hour of fishing; and
2. The catch per man-hour at sea.

The former takes the account of the actual fishing time, while the latter includes, in addition, the time or effort taken to reach the fishing ground and return from it.

Both in artisanal and mechanised sector the man–hours of fishing (the product of the number of fishing hours and the number of crew in the fishing unit), and the man–hours at sea (the product of the total trip time and the number of crew) can be used for comparison between the sectors, as regards efficiency of use of manpower, but ignores the expenditure of mechanical work by mechanised vessels.

But the numerical comparisons obscure the significant differences in the nature of work of the fishermen in various type of fishing. For, example, night fishing is generally considered more ardous than fishing during day. Encircling nets require a spurt of vigourous activity, once the shoal is located. Gill net fishing on the other hand requires patience and watchful waiting. Other differences are related to the quality of the fishing equipment and skill of the crew. Catches are also influenced by the relative abundance and the degree of dispersion of the resources. The catch per man–hour at sea is in addition affected by the distance of the fishing ground from the shore.

In mechanised trawlers and gill netters, where human labour is augmented and partly replaced by engine power, the catches per man–hour of fishing and per man–hour at sea are nevertheless low. However in monetary terms, the trawlers realize the highest gross earnings per man–hour followed by next best encircling net fishing with canoes.

Capital Intensity–Productivity of Labour and Capital

Catch or gross earnings per man–hour of effort, or per trip show the efficiency or productivity of the fishing unit when it is in actual operation. But these data, however, do not indicate how productive the fishing unit over the entire year, as this is affected both by seasonal patterns of utilisation and by the intensity of utilisation during the operating season.

Two commonly used criteria of productivity are the labour and capital utilized to achieve a given amount of production. Here labour productivity is expressed as the yearly catch, or gross earnings per fisherman and capital productivity as the yearly catch, or gross earnings per unit of capital invested.

In an economy where labour is plentiful and cheap and capital goods are relatively scarce, the technology applied should make use of relatively more labour and less capital equipment. In other words, the capital intensity, or the investment per crew member in fishing assets should be low. This usually implies that labour productivity also remains low, as labour power is not augmented or replaced by machines, and technologies requiring high capital investment and offering economics of scale are not applicable. Though a low labour productivity is desirable from an employment point of view, it will set an upper limit to the personal income levels. The personal income levels on the other hand, are not exclusively influenced by the labour productivity, but are also determined to some extent by the distribution of ownership of the productivity assets.

At times a clear trend of increasing gross earnings of fisherman with increase in capital invested per crew is noticed, but this correlation is not very strong in some cases. On the positive side hooks and line fishing with catamarans, lobster fishing and cast net fishing realize high labour productivities at very low level of investment per crew member, due to intensive utilization of the fishing assets over the year.

Several fishing methods, namely, prawn nets, anchovy nets, drift nets and boat seines fail to attain average yearly gross earnings equal to their respective capital costs. This is also true in case of drift net and mechanised fishing (trawling and gill netting). These fishing methods expend very low levels of human effort relative to the corresponding investment levels in fishing assets.

In summary, the artisanal fishing units on an average make a better use of invested capital than mechanised units. An excellent performance on both accounts, labour and capital productivity is attained by encircling net operations, small mesh drift net fishing, hook and line fishing, lobster and cast net fishing.

Risks in Fish Culture

Fish culture is considered to be a high risk activity mainly because the aquaculture products are raised in the aqatic medium, generally outside the farmer's direct observation and care. Moreover, the fish culture is a comparatively new industry and is still on a learning curve to establish itself on a par with other allied industries.

Risks experienced in aquaculture can be broadly classified as those that are caused by the state of technology, technical and managerial skills of operators and uncertain financial support that is presently available to the sector. Another group of unpredictable risks are loss of stock due to diseases, accidents or poaching, losses caused by natural disasters, use of substandard quality inputs, product contaminations and loss of markets due to competition and over production.

In agriculture and animal husbandry, the farmer aims at the survival of the whole stock. On the other hand in aquaculture allowance is made for a reasonable percentage of mortality and the initial stocking density is adjusted accordingly. It may not, therefore, be justifiable to count the expected mortality as a risk factor in aquaculture business. Nature itself makes such allowances for most aquatic animals by compensating with high fecundity in their original habitats. Improved fish culture technologies strive to reduce this mortality and adjust the initial stocking density as required.

The more important risks facing aquaculture seems to show that the majority of them are susceptible to better management, if the experience gained is utilized properly and suitable measures are incorporated into the design and operation of the farm from the very beginning.

Among the risks that aquaculture is exposed to, occurrence of diseases is the one that causes greater loss of stocks and consequent financial losses. Treatment of diseases under farm conditions is difficult and expensive, and even when the diseases are controlled, there may be problems in selling the produce. The problem is compounded when there are no known methods of treatment as in the case of viral diseases. Prophylactic measures are the only means of eliminating the risk.

Risk of losses caused by natural disasters, like floods, drought, storms including typhoons and cyclones is second in importance. The effect of red tides and algal blooms causing mass mortality of fish and shell fishes are also included under natural disasters. Additional costs are involved in building disaster proof facilities and undertaking protective measures.

Poaching and theft are more or less universal and appear to be on the increased with the expansion of farming of high-value species and of cage farming.

Losses due to mechanical failure of plant and equipment can be avoided or minimized by the ready availability of engineering assistance and keeping stand-by equipment. From the above, it is evident that it is possible to reduce the extent of more important risks involved in fish culture production even if it is not possible to avoid them completely.

Under the circumstances, the most likely decision would be to explore the feasibility of covering as much of the risk as possible through insurance.

Insurance Cover

A small but active insurance industry has developed in a number of countries to offer special cover for risks relating to aquaculture production. The producers and investors have looked at insurance as a means of reducing the exposure to major risks and consequent financial losses that can occur because of unpredicted problems arising in the farming business and to serve as a surety to lending institutions and individuals that provide financial assistance to them.

Because of different types of predictable and unpredictable risks in fish farming, an all-risk insurance cover would be the most appropriate one for an aquaculture enterprise. But in the present state of aquaculture insurance industry and the weakness of farming technologies, such a comprehensive insurance would call for premiums beyond the reach of most enterprises.

Insurance of fish culture farms is not mandatory and not all farms are insured even in countries where this service is available. The aquaculture industry continues to consist of a large number of small-scale enterprises that strive to economize on operating costs. According to Morris (1992) insurers of aquaculture business all over the world have been loosing money on mortality insurance cover. As such aquaculture insurance has generally been restricted to limited coverage.

Since loss of stock due to different causes is the most important contingency that can affect the profitability and continued survival of the enterprise, many farmers limit their insurance to stock mortality risks. This type of insurance is becoming more affordable because of improvements of farm management. Producers are increasingly become aware of the need to improve their culture practices, and particularly the value of disease monitoring to reduce mortality risks. Participation in a recognized integrated health management programme that reduces the risk of disease outbreaks; location of farms in safe areas not unduly exposed to natural disasters; and easy access to expert and extension assistance, are factors that would help in reducing the magnitude of risks, and therefore improve the availability of insurance at reasonable premiums. Insurance for only "excessive losses" either of stock or of economic returns may be second option. Business risks relating to price, claims on customers may be covered through credit and trade insurance.

Credit for Small-scale Marine Fisheries

Commercial banks have so far financed mechanised fishing trawlers and marine fisherfolk under the government ATMF and IRDP schemes, which however, aims in assisting the poorest of the poor fisherfolk, who do not own any craft and gear and whose income is below Rs. 3000 per year with subsidy and margin money to acquire productive assets (boat and net). But unlike in agriculture, in marine fishing, the unproductive group without any assets, is only a very marginal one. The vast majority has an income above Rs. 3000 per annum owns craft or gear or both and is compared to the level of technology–very productive. For this productive majority of small scale marine fisherfolk no assistance available are at present deprived of the benefits of institutional finance and have to depend on the informal credit market with its unrewardingly high interest rates.

In the above background project on fisheries finance have been formulated with the co–operation of various banks and refinance and margin money by the National Bank for Agriculture and Rural Development (NABARD).

Objectives and Benefit

The finance scheme aims at spreading banking habits to marine fisherfolk with regard to both savings and credit in order to establish an adequate money supply to ensure full complement of craft and gear, which will enable fisherfolk to exploit rich marine resources to a larger extent.

The existing informal credit by middlemen and private money lenders restricts the further development of the trade by charging unrewarding and extraordinary high interest rates and establishing feudal bondages.

The question of informal finance by the banks to a certain extent depends, not so much on subsidised rates of interest or availability of subsidy (small-scale marine fishing as relatively interest inelastic activity) but more on the following criteria:

1. Ready availability of credit for individuals with a minimum involvement of institutions;
2. Simple formats and procedures;
3. Short period between application and disbursement;
4. Direct contact between bank branch and customer, mutual trust;
5. Diversified purpose for which credit is given; and
6. Flexible repayment schedules.

To achieve these goals, the technical competence of coastal bank branches with regard to small scale marine fisheries is to be strengthened and a co-operation between banks and marine extension service of the Department of Fisheries is to be established.

For the purpose, a special application/activity form for small–scale marine fisheries, formats for a village profile, respective guidelines, a detailed description of the technology, a credit strategy paper and diversified schemes for fisheries credit have to be prepared.

Application/Activity form, village profile and Diversified Credit scheme are supposed to play a major role in achieving the overall goal which is to deliver the adequate amount of finance for the appropriate purpose to the fisherman who actually requires and requests it.

Disbursement of credit may be linked with savings schemes to ensure:

1. Creating an atmosphere of trust between bank and fisher folk;
2. Increasing the credit potential of the bank;
3. Improving the moral commitment of fisher folk with regard to loan repayment;
4. Linking savings to entitlement to credit which will serve as an incentive to, accumulate savings;
5. Reducing bank's administrative costs per loan; and
6. Financial outlay and expected benefit.

The benefit of the credit scheme may be estimated in terms of quantitative achievements with regard to the following:

1. Net value added (in Rs)–total amount of wages, profits, duties etc, which have been directly generated by the investments, financed under the credit scheme;
2. Fish/prawn production (in tonnes) resulting from the financed gear and craft;
3. Benefit-net value added, net value added per Rs. of investment–total catch of fish and prawn in a year.

Other quantitative achievements are foreign exchange earning from prawn export.

Schemes and Scale of Finance

Schemes may include most of the traditional marine crafts and gears. Regarding the economics of the scheme it has to be mentioned that no systematically collected data on cost and earnings are not available. Hence, basing on the observations and assuming a good year for fishing, no major illness of the loanee and no other disturbances with regard to fishing operations, schemes are to be formulated and updated in accordance with the change in prices and materials regularly.

Under these circumstances it is assumed that in most cases the operation of one gear will generate sufficient earnings to repay the loan and to take care of consumption expenses for the family. This will, however, not be the case every year. Therefore and to ensure the optimum exploitation of the marine resources, fishermen should be equipped with different types of gears.

Credit Potential in Marine Fisheries

For any coastal state, whose rich potential fishing grounds are least exploited, the development of any economic activity needs infusion of technologies, and supportive fund flow. Adoption of new technologies by traditional marine fishermen and its assimilation by the target community could very much depend on their confidence regarding the availability of funds to buy the technology. As it happens, in this kind of developmental process, a sense of ownership develops, resulting in substantial increase in efforts for fructifying the goals. The provision of funds–right amount at right time with right kind of administrative machinery synchronised with technological innovations holds the key to the development of small scale fishery and perhaps ultimately converts today's artisanal class of marine fishermen into a vibrant entrepreneurial community.

Economic Dimensions

To provide an economic insight into the inter–related issues and for a modest estimate of funds requirements of the traditional marine community (a) to maintain, (b) to acquire and (c) to modernize the means of production, it is essential to make a quantitative analysis of the existing marine fishing industry, on the following points:

1. Length of coast line.
2. Area of the continental shelf.
3. No. of exploitor families; percentage of full-time fishermen.
4. No. of fishing villages.
5. No. of landing sites.
6. 'Man day' employment capacity.
7. Production of fish (landing tonnage).
8. Value of production, rate per kg.
9. Projection of landing tonnage in comming 2 to 3 years if condition for growth of this industry are made more congenial.
10. Projection of exports of hard currency area in comming years.

Credit and Deposit Estimates

There is a need for the commercial banks to prepare a modest credit and deposit estimates in this less exploited field

Credit Estimates

At today's technological level, one unit of the means of production, namely, a boat and one or two gears would cost in the range of Rs. 30000 (one unit of log raft and two sets of gears) to about Rs. 60,000 (to cover masula boat with a shore seine or displacement craft with gear). Taking into account the ratio between owner–cum–labourers and pure labourer as 70 per cent and 30 per cent, the number of active fishermen involved in fishing has to be calculated. Among them, over the next five years, number of fishing families who may have to acquire their own craft and gear has to be ascertained. Due to the growing demand for fish, the gap between resource and exploitation and the existing and growing labour force, the expansion of traditional craft and gear will also be feseable. Thus keeping an average requirement of Rs. 20000 to own one unit of production, the total requirement of funds are to be estimated.

A moderate infusion of technology would easily throw up a credit requirement to the order of 90 to 110 per cent of the estimated amount, which can be phased out over a period of five years.

Deposit Estimate

Male folk amongst fishermen community presents a picture of almost 80 per cent engagement during the fishing days with average earnings at the level of Rs. 60 to Rs. 75 per day during 180 to 200 fishing days. Daily they can contribute Rs. 5 towards savings, if right type of campaigning and motivation can be undertaken. Thus a fisherman has the capacity of generating roughly Rs. 1000 or so per year as deposits.

Under these assumptions during a credit plan period of five years, the banking system, by introducing special strategies and design can syphon–up about one third of the amount as deposit, which has been estimated for credit disbursals. For the non-fishing days banks should disburse consumption credit–repayment linked with production credit, thus assisting the fishermen to develop a total linkage and relationship with the banks at the end of the credit plan period. This would also enable the credit institutions to supplement in the beginning and subsequently replace the traditional lending sources of the fishermen community.

Strategy for Flow of Credit

Strategies could be enumerated as follows:

1. Approach to be followed in handling credit operations for target community.
2. Funds for increasing the number of existing craft and gear to cover in adequacies.
3. Funds for innovations/improvements/updating of technologies in displacement boats/log rafts and gears.
4. Credit for supportive activities.
5. Funds for allied activities in fishing villages.
6. Provision of innovative credit on soft terms.

While launching the credit operations initially in pilot villages, subsequently in other fishing villages two major approaches can be followed.

1. Loans for craft and gear including maintenance could be financed by banks directly to individual fishermen, upto a limit of Rs. 10000 following the scales of finance, prevailing in the area. This would mostly take care of inadequacies of equipments–craft and gear. This could be possible by initial liasoning by the local fishery officer's with banks.

For investment beyond Rs. 10000 schemes need be drawn up and cleared by NABARD for individual banks. This would be mainly required once new technologies are developed and to be introduced on a larger scale.

Considering the fact that there is a credit gap, for increasing the existing craft and gear, commercial, banks have to enter the field of financing, keeping area approach in mind. At the district level through the district co-ordination committee for bank credit, concerned official of State Fisheries Department can negotiate and allot villages or landing sites in favour of different banks operating along the coast. This would prevent over-lapping of efforts, as also, over dues. The aspirants for credit would be mainly owners of part of a unit, non-owners and labourers who desire to own either gear or boat. As regards to drawing–up of cost and earnings and repayment schedules are concerned, there shall be no difficulty as under the existing system, a share is allotted to boat and gear from the total catch.

For conveniences, a few sample repayment schedules, which have been drawn up in consultation with fisheries extension staff and fishery technicians working with the banks have been produced in following chapters. A few schemes along with the project financial outlay and physical targets have been drawn up by the credit consultant for guidance.

A working capital of Rs. 2000 per year to take care of maintenance of boat, gear and other contingencies, as well as, consumption needs (to be liquidated at the end of the year) would be most essential. The disbursals can be phased into two, namely, Rs. 1400 for meeting the maintenance of gear, repair of crafts etc. and Rs. 600 for consumption needs, to be disbursed on non-fishing days. This credit can be liquidated out of the income from the bumper catch, that is 4 to 5 times in a year.

Several government and international agencies are trying to introduce innovations/improvements in log rafts, as well as, displacement boats. In any case the cost structure would change so also the durability and return. Innovations and improvements in fishing gears would also require credit support from institutional sources.

To provide funds for supportive activities, boats financed under a credit scheme should be constructed at a central place or by private boat building yards. These yards need credit and could be covered under "small scale industries" and financed as an eligible activity. Once the quality and design of constructions are adhered, banks can place orders for construction of boats to be provided to traditional fishermen when they obtain loans.

Insulated ice boxes, packing boxes for carrying iced fish, especially designed baskets etc can be financed. Credit needs to be provided to individuals for manufacturing these items and fishermen can be financed to acquire those.

Funds for marketing fish by fisher women on head loads or at a later stage by cycles need be provided. A working capital credit of

Rs. 1000 inclusive of consumption credit should be made available in credit pilot villages to fisher women on soft terms, say 4 per cent interest.

Amongst the traditional marine fishermen two kinds of unemployment persists. Vieled unemployment and under employment in non-fishing days and under uncertain weather conditions. Unemployment amongst young girls and to some extent amongst boys owing to absence of any viable occupations. Viable bankable poultry–keeping, bee-keeping schemes, fitting into the banks system lending, may be designed befitting local conditions.

While providing finance in the "credit pilot villages" and subsequently in other fishing villages it is necessary that the total credit requirements are met and all their felt needs are provided from institutional sources. This would have a salutary effect on the fishermen community, namely, would deter them from going to money lenders and would greatly improve their quality of life thus increasing their level of efficiency and fishing effort.

Though the amounts are small, yet its effect on improving the bank–customer relationship are immense. This kind of loans can bring banks and fisherfolk closer. Banks are providing these finances elsewhere, in the instant cases they would extend it to fishermen villages. The list of such items needing finance is ever expanding and the items discussed is purely indicative, but not exhaustive.

Considering the dimensions of credit required to provide means of production to the majority of fisherfolk and for maintenance and replacement of craft and gear by the ones who already own it, and those who like to acquire it, organisational arrangements have to be thought of, deliberations and decisions arrived at.

Having scanned the need for organisational machinery and its existing level, a two-pronged approach may be initiated. Credit being a vital component of technologies has to be handled in a fashion, where understanding develops on the part of both the field level machinery.

The first approach to be adopted for initiating a homogeneous direction is to have some fisheries technicians in the bank. The second approach is to provide credit training to fisheries extension officers and have a credit expert trained in commercial banks, attached at headquarters level in the Directorate of fisheries to co-ordinate liase, negotiate credit proposals with NABARD and commercial banks on ongoing basis at least for five years period.

By adopting these two approaches, the heterogeneity of directions at field level can be reduced to a great level and the common focus on a supportive credit for fishery development can continue.

To be precise the following action plan need be initiated:

1. Developing a span of control or command for FEO's and linking them to note more than 800 to 1000 active fishermen.
2. Opening branches of banks at mutually agreed sites, keeping various economic dimensions of marine fishing in view.
3. Induction of fisheries technicians at least 2 or 3 per operating or participating banks.
4. Putting a credit-person at headquarters of fisheries administration with training in commercial banking to negotiate, liase, provide solutions to field level problems during the implementation period.

Economics of Model Schemes

Displacement Craft–Donga

Clinker built, made of sal wood of 30′ length, 7′ breadth and 6′ draft boat, operating all the year round, except when sea is too rough in May or June or July, in coastal waters upto 20 km. having a life span of 10 years.

Used in combination with various gill nets and together with other boats, in combination with encircling nets, wage labour or sharing system, where crew members contribute net pieces.

	(Rs.)
A. Capital Cost–Requirement of Material	9000.00
(a) Total wood, 75 cft at Rs. 120 per cft. size of planks range from 4" to 12", 13 planks for in each size for chine construction. Further 5 planks for each side for top construction. Total (13 × 2) + (5 × 2) = 36 planks	

	(Rs.)
(b) Nails (four face) = 60 kg at Rs. 16/kg.	960.00
(c) Cotton for caulking at Rs. 30/kg. for 6 kgs.	180.00
(d) Coal tar for preservation (1 tin contains 15 kgs) at Rs. 160 per tin for 3 tins	480.00
(e) Oil for cleaning at Rs. 12/kg–total 5 kgs.	60.00
(f) Sail cloth 18 ft × 20 ft	800.00
(g) Tarpaulin	800.00
(h) Iron for anchor, at Rs. 20/kg. 15 kg	300.00
(i) Nylon rope for anchor, 15 mm., 10 m.	220.00
Total expenditure for material	12800.00
(j) Making charge of the boat	2000.00
(k) Miscellaneous expenditure	1200.00
Total cost	16000.00
(B) <u>Annual Recurring Expenses</u>	
(a) Wages for 3 labourers at Rs. 20 per day for 210 fishing days (in addition to labour of applicant)	12600.00
(b) Repair of boat and sail	1600.00
(c) Hire charges for nets for 10 months at Rs. 500 per month	5000.00
Total expenses	19200.00
(C) <u>Annual Income and Surplus</u>	
(a) Gross earnings from sale of fish at Rs. 10 per kg., 30 kg. per fishing day and 210 fishing days	63000.00
(b) Annual recurring expenses	(-) 19200.00
Gross surplus	43,800.00
(a) Annual depreciation	(-) 1600.00
(b) Net surplus, divided into	42,200.00
(i) Return on labour of applicant (boat owner)	4,200.00
(ii) Return on capital	38,000.00
(D) <u>Economic Feasibility (Estimated)</u>	
(a) Annual rate of return on investment	238 per cent
(b) Net value added per Rs. of investment in Rs.	6.86

(Rs.)

(E) 4 Years Repayment Schedule at 18.0 per cent

Years	Principal	Interest	Principal with interest	Repayment	Balance
1st	16000	2880	18880	6500	12380
2nd	12380	2232	14612	6500	8112
3rd	8112	1458	9570	6500	3070
4th	3070	553	3623	3623	–

Displacement Craft–Salti

Carvel built boat of 32′ × 6′ × 3′, made of sal wood operate in coastal areas all the year round, except for rough days in May, June and July upto 5 km. off shore.

The boat is used in combination with various gill nets and together with other boats, in combination with encircling nets, wage labour or sharing system, where crew members contribute net pieces.

	(Rs.)
A. Capital Cost–Requirement of Material	
(a) 1" planks made of sal wood, 68 cft at Rs. 120 per cft	8160.00
(b) Arrow head 3" nails 60 kg. at Rs. 20/kg.	1200.00
(c) Putting, caulking, coaltaring materials required 5 kg, 15 kg. and 50 kg.	800.00
(d) Anchor 10 kg, anchor rope PA dia. 14 mm. 4 kg and sail cloth 18′ × 15′ and sail rope.	1100.00
(e) Labour charges for putting, caulking, coaltaring	400.00
(f) Making charges of the boat	2000.00
(g) Transportation and other charges	540.00
Total cost	14200.00
(B) Annual Recurring Expenses	
(a) Wages for 3 labourers at Rs. 20 per day for 180 fishing days (in addition to labour of loanee)	10800.00
(b) Repair	1600.00
(c) Hire charges for nets for 9 months at Rs. 500 per months	4500.00
Total expenses	16900.00

	(Rs.)
(C) Annual Income and Surplus	
(a) Gross earnings from sale of fish at Rs. 8 per kg, 25 kg per fishing day and 180 fishing days.	36000.00
(b) Annual recurring expenses	(-) 16900.00
(c) Gross surplus	19100.00
(d) Annual depreciation	(-) 1420.00
(e) Net surplus divided into	17680.00
(i) Return on labour of loanee (boat owner)	3600.00
(ii) Return on capital	14080.00
(D) Economic Feasibility	
(a) Annual rate of return on investment	99%
(b) Net value added per Rs. of investment (in Rs.)	4.02

(E) 4 Years Repayment Schedule

Years	*Principal*	*Interest*	*Principal with interest*	*Repayment*	*Balance*
1st	14200	2556	16756	6000	10756
2nd	10756	1936	12692	6000	6692
3rd	6692	1205	7897	6000	1897
4th	1897	341	2238	2238	–

Displacement Craft–Salti

Smaller boat of 27′ × 5′8″ × 2′10″, carvel built, made of sal wood has a life span of 10 years, operated all the year round, except for rough days in May, June and July upto 5 km. off shore.

The boat is used in combination with various gill nets, wage labour or sharing system, crew members contributing net pieces.

	(Rs.)
A. Capital Cost–Requirement of Material	
(a) 3/4" planks made of sal wood, 38 cft at Rs. 120 per cft	4560.00
(b) Arrow head 3" nails 40 kg. at Rs. 20/kg.	800.00
(c) Putting, caulking, coaltaring materials required	800.00
(d) Anchor 10 kg, anchor rope 14 mm.dia 4 kg and sail cloth 10′ × 15′ and sail rope.	1000.00

	(Rs.)
(e) Labour charges for caulking, putting and coaltaring	400.00
(f) Making charges of the boat	1600.00
(g) Transportation charges etc.	400.00
(h) Miscellaneous	440.00
Total cost	10,000.00
(B) Annual Recurring Expenses	
(a) Wages for 3 labourers at Rs. 20 per day for 180 fishing days (in addition to labour of loanee)	10800.00
(b) Repair	1000.00
(c) Hire charges for nets for 9 months at Rs. 400 per months	3600.00
Total expenses	15400.00
(C) Annual Income and Surplus	
(a) Gross earnings from sale of fish at Rs. 8 per kg, 20 kg per fishing day and 180 fishing days.	28800.00
(b) Annual recurring expenses	(-) 15400.00
	13400.00
(c) Depreciation	(-) 1000.00
(d) Net surplus divided into	12400.00
(i) Return on labour of loanee (boat owner)	3600.00
(ii) Return on capital	8800.00
(D) Economic Feasibility	
(a) Annual rate of return on investment	88%
(b) Net value added per Rs. of investment (in Rs.)	4.64

(E) 4 Years Repayment Schedule

Years	*Principal*	*Interest*	*Principal with interest*	*Repayment*	*Balance*
1st	10000	1800	11800	3000	8800
2nd	8800	1584	10384	4000	6384
3rd	6384	1149	7533	5000	2533
4th	2533	456	2989	2989	–

Log raft–4 log Catamaran

Made of 4 logs of wood; *Albizzia stipulata* with dimension of 8.3 × 1.5 × 0.5 m. with triangular sail of 7.5m. × 6.25 m., having a life span of 5 years are operated, all the year round upto 10 km. offshore. The raft is operated with various types of gill nets, also with boat seines, hook and line and liftnets. Operated on share basis. In the case of gill nets, 8 shares 5 shares go to the 3–4 fishermen labourer (including owner), 2 shares to the boat and one share to the net.

	(Rs.)
A. Capital Cost	
(a) Cost of wood	7000.00
(b) Making charges	1400.00
(c) Sail and accessories	1600.00
Total cost	10,000.00
(B) Recurring Expenses	
(a) Repair	600.00
(C) Annual Income and Surplus	
(a) Earnings–Out of 8 shares in which the total catch is divided, the loanee will get 2 shares for the boat. Furthermore he will get one third out of the 5/8 share for the three crew members if he works as a labourer, means in total his earnings will be 3.68/8 of the total sale proceeds	
(b) Total sales proceeds	
(i) Average catch of prawns during peak period at 5 kg/day for 45 days at Rs. 120/kg.	27000.00
(ii) Average catch of other fish for the period of 215 days at 20 kg/day at Rs. 6/kg.	25800.00
Gross earnings	52800.00
(i) Loanee's share (gross earnings) divided into	24200.00
(ii) Boat share	13200.00
(iii) Labourer's share	11000.00
Gross earnings	24200.00

	(Rs.)
(iv) Annual recurring expenses	(-) 600.00
Gross surplus	23600.00
(v) Annual depreciation	(-) 2000.00
(vi) Net surplus divided into	21600.00
(a) Return on labour of loanee	11000.00
(b) Return on investment	10600.00
(D) Economic Feasibility	
(a) Rate of return on investment	106%
(b) Net value added per unit investment in Rs.	8.72
(E) 4 Years Repayment Schedule	

Years	*Principal*	*Interest*	*Principal with interest*	*Repayment*	*Balance*
1st	10000	1800	11800	3000	8800
2nd	8800	1584	10384	4000	6384
3rd	6384	1149	7533	5000	2533
4th	2533	456	2989	2989	–

Log Raft–3 log Catamarans

Made of 3 logs of *Albizzia stipulata* wood of 4.2 × 0.76 × 0.42 m. dimension with triangular sail 4m. × 2.8 m. has 4 years life span. The raft is operated all the year round upto 5 km. offshore. Used with various types of gillnets, boat seines, hook and lines, lift net. Operated on share basis. In the case of gillnets 2 shares (out of 3) to the labourers, one share to boat and net.

	(Rs.)
A. Capital Cost	
(a) Cost of wood	4000.00
(b) Making charges	1000.00
(c) Sail and accessories	1000.00
Total cost	6000.00
(B) Annual Recurring Expenses	
(a) Annual repair	400.00

	(Rs.)
(C) <u>Annual Income and Surplus</u>	
The loanee gets 1 out of 5 shares for his labour (3 labourers) and another share for the boat = 2/5	
Total sales proceeds:	
Average catch of prawns during peak period at 5 kg/day for 25 days at Rs. 120/kg.	15000.00
Average catch of fish for the period of 230 days at 10 kg/day at Rs. 6/kg.	13800.00
Gross earnings	28800.00
	14400.00
Annual recurring costs	(-) 400.00
Gross surplus	14000.00
Annual depreciation	(-) 1500.00
	12500.00
Net surplus divided into	21600.00
(a) Return on labour of loanee	9600.00
(b) Return on investment	2900.00
(D) <u>Economic Feasibility</u>	
(a) Rate of return on investment	48%
(b) Net value added per unit investment (in Rs.)	7.4

(E) <u>4 Years Repayment Schedule</u>

Years	*Principal*	*Interest*	*Principal with interest*	*Repayment*	*Balance*
1st	6000	1080	7080	2000	5080
2nd	5080	914	5994	2000	3994
3rd	3994	719	4713	3000	1713
4th	1713	308	2021	2021	–

Log Raft–catamarans

Made of different wood, *Erythrina indica*, the raft requires 4 logs of 4.0 × 0.7 × 0.4 m. in dimesions, triangular sail 4.5 m × 3 m., with a life span of 4 years, operate all the year round upto 3 km. off shore.

Used in combination with various gill nets and hook and line. Operated on share basis. In the case of gill nets, 3 shares, 1 for boat and net, 2 for the 2 or 3 labourers.

	(Rs.)
A. Capital Cost	
(a) Cost of wood	2400.00
(b) Making charges	600.00
(c) Sail and accessories	1000.00
Total cost	4000.00
(B) Annual Recurring Expenses	
(a) Annual repair	400.00
(C) Annual Income and Surplus	
Total catch:	
Average catch of prawns at 5 kg/day for 25 days at Rs. 120/kg.	15000.00
Average catch of fish at 10 kg/day for 210 days at Rs. 6/kg.	12600.00
Gross earnings	27600.00
Gross earnings/share of loanee; 1/3 for labour and 0.5/3 for boat	13800.00
Annual recurring costs	(-) 400.00
Gross surplus of loanee	13400.00
Depreciation	(-) 1000.00
Net surplus divided into	
(a) Return on labour of loanee	9200.00
(b) Return on investment	3200.00
(D) Economic Feasibility	
(a) Rate of return on investment	80%
((b) Net value added per unit investment (in Rs.)	10.8

(E) 4 Years Repayment Schedule

Years	*Principal*	*Interest*	*Principal with interest*	*Repayment*	*Balance*
1st	4000	720	4720	1000	3720
2nd	3720	670	4390	2000	2390
3rd	2390	430	2820	1500	1320
4th	1320	238	1558	1558	-

Gill Nets

Large meshed gill nets with 15-20 cm. mesh size, operated on the surface, one set consists of 60 m. length, 20 m. depth has life span of 7 years. Operated in spring and early summer with one boat. 6 persons contribute each one set, totaling 968 m. long and go for fishing in one boat. The catch is equally divided into 7 shares (6 for 6 sets of nets and one for the boat). In other cases this net is operated by wage labourers.

		(Rs.)
A.	Capital Cost	
(a)	50 kg. nylon twine PA 210 × 12 × 3 or PA 210 × 15 × 3 at Rs. 200/kg.	1000.00
(b)	25 kg. PP rope, 12 and 14 mm. dia at Rs. 82/kg.	2050.00
(c)	Net weaving charges at Rs. 40/kg.	2000.00
(d)	20 mm.dia PVC floats, 40 nos at Rs. 8 each.	320.00
(e)	Sinkers, 40 nos at Rs. 2 each.	80.00
(f)	Charges for framing of net and miscellaneous charges	550.00
	Total cost	15000.00
(B)	Annual Recurring Expenses	
(a)	Net repair	500.00
(C)	Annual Income and Surplus	
	Expected number of fishing days–100	
	Expected catch per fishing day of entire unit–70 kg.	
	Total sales proceeds at Rs. 14/kg.	98000.00
	Loanee's share (1/7) = loanee's gross earnings.	14000.00
	Annual recuring cost	(-) 500.00
	Gross surplus of loanee	13500.00
	Depreciation	(-) 2144.00
	Net surplus divided into	11356.00
	(a) Return on labour (at Rs. 20/day for 100 days)	2000.00
	(b) Return on investment	9356.00
(D)	Economic Feasibility	
	Rate of return on investment	62%
	Net value added per unit investment (in Rs.)	1.52

(Rs.)

(E) 4 Years Repayment Schedule

Years	*Principal*	*Interest*	*Principal with interest*	*Repayment*	*Balance*
1st	15000	2700	17700	6000	11700
2nd	11700	2106	13806	6000	7806
3rd	7806	1405	9211	5000	4211
4th	4211	758	4969	4969	–

Sanla Jal

Surface driftnet, 15 cm. mesh size, 175 m. length 7m. depth is operated all the year round except for rough days from May to July with the help of displacement boats on share basis or with wage labour. The net has a life span of 3 to 5 years. Three crews are required to operate the net.

	(Rs.)
A. Capital Cost	
(a) PE twine, 60 kg. at Rs. 100/kg.	6000.00
(b) 20 kg. PP rope, 12 mm dia at Rs. 82/kg	1640.00
(c) 120 nos. of PVC floats 15 mm. dia at Rs. 6/each	720.00
(d) 80 nos. of earthen sinkers at Rs. 0.50/each	40.00
(e) Making charges at Rs. 40/kg	2000.00
(f) Framing charges and miscellaneous	600.00
Total cost	11000.00
(B) Annual Recurring Expenses	
(a) Repair	1600.00
(b) Rent for a boat at Rs. 400/month for 10 months	4000.00
(c) Wages at Rs. 20/day for two labourers for 210 fishing days	8400.00
Total expenses	14000.00
(C) Annual Income and Surplus	
(a) Expected no. of fishing days–210 days	
(b) Expected catch per fishig day–20 kg.	
(c) Gross earnings at Rs. 12/kg. 12 × 20 × 210	37800.00

(d)	Annual recurring Costs	(-) 14000.00
	Gross surplus	23800.00
	Depreciation	(-) 2750.00
	Net surplus	21050.00
	(i) Return on labour of loanee	4200.00
	(ii) Return on investment	16850.00
(D)	<u>Economic Feasibility</u>	
(a)	Rate of return on investment	153%
(b)	Net value added per unit of investment (in Rs.)	3.82

(E) <u>4 Years Repayment Schedule</u>

Years	*Principal*	*Interest*	*Principal with interest*	*Repayment*	*Balance*
1st	11000	1980	12980	4000	8980
2nd	8980	1616	10596	4000	6596
3rd	6596	1187	7783	4000	3783
4th	3783	681	4464	4464	–

Phasi Jal

Surface drifnet, of 8.5 to 11 cm. mesh size with 350 m. length and 15 m. depth is operated all the year round in displacement boat on wage labour or on share basis. The net has a life span of 4 years and operated with the help of 4 crews.

		(Rs.)
A.	<u>Capital Cost</u>	
(a)	50 kg of PE twine at Rs. 100/kg.	5000.00
(b)	25 kg. PP rope, 6 and 8 mm. dia at Rs. 82/kg.	2050.00
(c)	220 nos. of PVC floats, 8 cm. dia at Rs. 3/each.	660.00
(d)	Making charges at Rs. 40/kg.	2000.00
(e)	Framing charges	230.00
	Total cost	10000.00
(B)	<u>Annual Recurring Expenses</u>	
(a)	Rent for a boat for 10 months at Rs. 400/month	4000.00
(b)	Repair charges	1400.00
(c)	Wages for 3 labourers at Rs. 20/day for 210 days	12600.00
	Total expenses	18000.00

	(Rs.)
(C) <u>Annual Income and Surplus</u>	
(a) Expected no. of fishing days–210	
(b) Expected catch per fishig day–23 kg.	
(c) Gross earnings at Rs. 8/kg.	36800.00
(d) Annual recurring expenses	(-) 18000.00
Gross surplus	18800.00
Depreciation	(-) 2500.00
Net surplus divided into	16300.00
(i) Return on labour of loanee	4200.00
(ii) Return on investment	12100.00
(D) <u>Economic Feasibility</u>	
(a) Rate of return on investment	121%
(b) Net value added per unit of investment (in Rs.)	5.78

(E) <u>4 Years Repayment Schedule</u>

Years	*Principal*	*Interest*	*Principal with interest*	*Repayment*	*Balance*
1st	10000	1800	11800	3000	8800
2nd	8800	1584	10384	4000	6384
3rd	6384	1149	7533	5000	2533
4th	2533	456	2989	2989	–

Medium and Small Mesh Gill Nets

Surface or bottom drift net of 6 to 7 cm. mesh; 320 m. length, 8 m. depth operated all the year round with one large catamaram. The net has life span of 5 years. Out of 5 shares, 3 are for 3 labourers and 2 shares are for net and boat.

	(Rs.)
A. <u>Capital Cost</u>	
(a) 13 kg. netting	5200.00
(b) Head rope and foot rope, pp 5 mm dia	600.00
(c) PVC floats (70 ´ 35 mm) and cement sinkers (300 gm)	600.00
(d) Framing and miscellaneous	600.00
Total cost	7000.00

	(Rs.)
(B) <u>Annual Recurring Expenses</u>	
(a) Repair of net	800.00
(C) <u>Annual Income and Surplus</u>	
Sale proceeds from prawn fishing on 45 days–5 kg prawn/day at Rs. 120/kg.	27000.00
From fish 210 days fishing–15 kg/day at Rs. 6/kg.	18900.00
	45900.00
Loanee's share (1/5 for labour and 1/5 for net) gross earnings	18300.00
Annual recurring cost	(-) 800.00
Gross surplus	17560.00
Depreciation	(-) 1400.00
Net surplus divided into	16160.00
(i) Return on labour of loanee	9180.00
(ii) Return on investment	7180.00
(D) <u>Economic Feasibility</u>	
(a) Rate of return on investment	103%
(b) Net value added per unit of investment (in Rs.)	9.86

(E) <u>4 Years Repayment Schedule</u>

Years	*Principal*	*Interest*	*Principal with interest*	*Repayment*	*Balance*
1st	7000	1260	8260	2500	5760
2nd	5760	1037	6797	2500	4297
3rd	4297	773	5070	3000	2070
4th	2070	373	2443	2443	–

Bottom Drift Net

Bottom drift net of 220 m. length, 3.6m. depth and 5.5 cm. mesh is operated with one catamaran, all the year round, except for rough seas on share basis. 3 out of 5 shares are for the labourers and 2 shares for net and boat. The net has a life-span of 5 years.

A. Capital Cost

(a) 6 kg. ready made nylon netting @ Rs. 400/kg	2400.00
(b) Head and foot rope pp, 4 mm. dia.	400.00
(c) Floats and sinkers	400.00
(d) Framing charges of the net.	200.00
Total cost	3400.00

(B) Annual Recurring Expenses

(a) Repair of net	600.00

(C) Annual Income and Surplus

(a) Sale proceeds from prawn fishing on 30 days–30 kg prawn/day at Rs. 120/kg.	10800.00
(b) From fish 210 days fishing–10 kg/day at Rs. 6/kg.	12600.00
	23400.00
Loanee's share (1/5 for labour and 1/5 for net) gross earnings	9360.00
Annual recurring cost	(-) 600.00
Gross surplus	8760.00
Depreciation	(-) 680.00
Net surplus divided into	8080.00
(i) Return on labour of loanee	4680.00
(ii) Return on investment	3400.00

(D) Economic Feasibility

(a) Rate of return on investment	100%
(b) Net value added per unit of investment (in Rs.)	10.26

(E) 4 Years Repayment Schedule

Years	*Principal*	*Interest*	*Principal with interest*	*Repayment*	*Balance*
1st	3400	612	4012	1000	3012
2nd	3012	542	3554	2000	1554
3rd	1554	280	1834	1000	834
4th	834	150	984	984	–

Surface Drift Net

Drift gill net of 220 m. length, 3m. depth of 5.5 cm. mesh are used as surface net with the help of catamaran on sharing system all the year round except for rough sea. 5 of 8 shares are for two labourers and 3 shares for boat and net. The net is in this case allotted 0.5 share out of the 3 shares. The net has a life span of 5 years.

	(Rs.)
A. Capital Cost	
(a) 6 kg. ready made nylon netting at Rs. 400/kg	2400.00
(b) Head and foot rope, pp 4 mm. dia.	400.00
(c) Floats and sinkers	400.00
(d) Framing charges	200.00
Total cost	3400.00
(B) Recurring Expenses	
(a) Repair of net	600.00
(C) Annual Income and Surplus	
Sale proceeds from prawn fishing on 30 days–3 kg prawn/day at Rs. 120/kg.	10800.00
From fish 210 days fishing–10 kg/day–Rs 6/kg.	12600.00
	23400.00
Loanee's share (1/5 for labour and 1/5 for net) Gross earnings	9360.00
Annual recurring expenses	(-) 600.00
Gross surplus	8760.00
Depreciation	(-) 680.00
Net surplus divided into	8080.00
(i) Return on labour of loanee	4680.00
(ii) Return on investment	3400.00
(D) Economic Feasibility	
(a) Rate of return on investment	100%
(b) Net value added per unit of investment (in Rs.)	10.26

(Rs.)

(E) 4 Years Repayment Schedule

Years	*Principal*	*Interest*	*Principal with interest*	*Repayment*	*Balance*
1st	3400	612	4012	1000	3012
2nd	3012	542	3554	2000	1554
3rd	1554	280	1834	1000	834
4th	834	150	984	984	–

Bhasani Jal

Surface drift net of 3.3 cm. mesh having a length of 44 m. and depth 2 m. is operated all the year round from a displacement craft on wage labour or on sharing system. In sharing system 3 or 4 fishermen contribute 6 or 8 pieces of net, 50 m. length each. The net has 5 years life.

	(Rs.)
A. Capital Cost	
(a) Cost of 6 kg. nylon netting at Rs. 400/kg	2400.00
(b) Cost of 60 nos. of PVC floats at Rs. 3 each and 60 sinkers at Re 1 each	240.00
(c) Cost of 3 mm. PP rope for float and sinker line	400.00
(d) Framing and miscellaneous	360.00
Total cost	3400.00
(B) Annual Recurring Expenses	
(a) Repair of net	600.00
(b) Wages for 2 labourers at Rs. 20/day for 210 days	8400.00
(c) Rent of a boat at Rs. 400/month for 10 months	4000.00
Total expenses	13000.00
(C) Annual Income and Surplus	
Sale proceeds from 210 fishing days at 15 kg/day at Rs. 8/kg gross earnings	25200.00
Annual recurring costs	(-) 13000.00
Gross surplus	12200.00

	(Rs.)
Depreciation	(-) 680.00
Net surplus divided into	11250.00
(i) Return on labour of loanee	4200.00
(ii) Return on investment	7320.00
(D) Economic Feasibility	
(a) Rate of return on investment	215%
(b) Net value added per unit of investment (in Rs.)	11.72
(E) 4 Years Repayment Schedule	

Years	*Principal*	*Interest*	*Principal with interest*	*Repayment*	*Balance*
1st	3400	612	4012	1000	3012
2nd	3012	542	3554	2000	1554
3rd	1554	280	1834	1000	834
4th	834	150	984	984	–

Sardine Net

Surface drift net of 25 mm. to 40 mm. mesh of 125 m. length, 9 m. depth used for catching sardines from November to April. The net is operated with one small or medium catamaran on share basis. 2 out of 3 shares go to the two fishermen, one share goes to boat and net. Life of net is 5 years.

	(Rs.)
A. Capital Cost	
(a) 10 kg. nylon netting at Rs. 400/kg.	4000.00
(b) PVC floats (10 mm. dia)	500.00
(c) Lead sinkers (20 gm)	200.00
(d) Framing charges	300.00
Total cost	5000.00
(B) Annual Recurring Expenses	
(a) Repair of net	600.00

	(Rs.)
(C) Annual Income and Surplus	
Sale proceeds from 150 fishing days, 25 kg/day at Rs. 6/kg	22500.00
Gross earnings; 1/3 as a labourer, 1/6 for the net	11250.00
Annual recurring costs	(-) 600.00
Gross surplus	10650.00
Depreciation	(-) 1000.00
Net surplus divided into	9650.00
Return on labour	7500.00
Return on investment	2150.00
(D) Economic Feasibility	
Rate of return on investment	43%
Net value added per unit of investment in Rs.	6.86

(E) 4 Years Repayment Schedule

Years	*Principal*	*Interest*	*Principal with interest*	*Repayment*	*Balance*
1st	5000	900	5900	1500	4400
2nd	4400	792	5192	1500	3692
3rd	3692	665	4357	2000	2357
4th	2357	425	2782	2782	–

Anchovy net

Surface drift nets used for catching anchovies, small sardines and white bait, mesh size 12-25 mm. length 125 m, depth 9 m. The net is operated throughout the year with peak catch between November to March from a small or medium catamaran on share basis. 2 out of 3 shares accrue to the two fishermen, one share goes to boat and net. Life span of net 5 years.

		(Rs.)
A.	Capital Cost	
(a)	15 kg. nylon net	6000.00
(b)	PVC floats (10 mm. dia)	500.00
(c)	Lead sinkers (20 gm.)	200.00
(d)	Framing charges	300.00
	Total cost	7000.00

(B) Annual Recurring Expenses

Repair of net	600.00

(C) Annual Income and Surplus

Sale proceeds from 150 fishing days,25 kg/day at Rs. 6/kg	22500.00
Gross earnings; 1/3 as a labourer, 1/6 for the net	11250.00
Annual recurring costs	(-) 600.00
Gross surplus	10650.00
Depreciation	(-) 1400.00
Net surplus divided into	9250.00
Return on labour	7500.00
Return on investment	1750.00

(D) Economic Feasibility

Rate of return on investment	25%
Net value added per unit of investment in Rs.	4.78

(E) 4 Years Repayment Schedule

Years	*Principal*	*Interest*	*Principal with interest*	*Repayment*	*Balance*
1st	7000	1260	8260	2500	5760
2nd	5760	1037	6797	2500	4297
3rd	4297	773	5070	3000	2070
4th	2070	373	2443	2443	–

Inshore Seines

Encircling gill net, length 1640 m. depth 20 m. 35 fishermen operate the net between August to February in 5 boats and contribute on an average 47 m. of net each. This scheme is for 188 m. which means for 2 net shares out of total 35 shares. Longivity of net is 4 years.

	(Rs.)
A. Capital Cost	
(a) 60 kg. PER 152 tax twine at Rs. 100/kg.	6000.00
(b) Rope for float and sinker line PE 5 mm. and 8 mm.	500.00
(c) PVC floats (10 cm. dia) Rs. 6/each	1900.00
(d) Earthen sinkers at Rs. 6 for 100 sinkers	50.00
(e) Making charges at Rs. 40/kg.	1200.00
(f) Framing and Miscellaneous	350.00
Total cost	10000.00

(B) Annual Recurring Expenses

(a)	Repair of net	1000.00
(b)	1.10 charges for rent of 5 boats (Rs. 30/boat per day for 150 days)	2500.00
	Total expanses	3500.00

(C) Annual Income and Surplus

(a)	Sale proceeds from 150 fishing days at 150 kg/day at Rs. 10/kg.	2,25,000.00
(b)	Gross earnings (share of loanee is 1/10 out of gross earnings)	22500.00
(c)	Annual recurring costs	(-) 3500.00
(d)	Gross surplus	19000.00
(e)	Depreciation	(-) 2500.00
(f)	Net surplus divided into	16500.00
	(i) Return on loanee	3000.00
	(ii) Return on investment	13500.00

(D) Economic Feasibility

(a)	Rate of return on investment	135%
(b)	Net value added per unit of investment in Rs.	3.30

(E) 4 Years Repayment Schedule

Years	*Principal*	*Interest*	*Principal with interest*	*Repayment*	*Balance*
1st	10000	1800	11800	3000	8800
2nd	8800	1584	10384	4000	6384
3rd	6384	1149	7533	5000	2533
4th	2533	456	2989	2989	–

Wall Net

Tidal wall net of 3-5 km. length, 2-4 m. depth, mesh size 40-55 mm. is operated throughout the year, except for rough days from April to July. 700 m. of net is to be financed under the scheme, which is taken 1/4 of the entire length. The gear is set 3-4 days before and after full and new moon. An area is taken on lease from the local administration by a group of fishermen who contribute net pieces. The piece financed under the scheme can be operated by the loanee and a labourer. The life span of the net is 4 years.

	(Rs.)
A. Capital Cost	
(a) 50 kg. of PER 228 tax twine at Rs. 100/kg.	5000.00
(b) 30 kg. PP rope, 8 mm. dia, Rs. 60/kg.	246.00
(c) Making charges of net at Rs. 40/kg.	2000.00
(d) 300 bamboo sticks at Rs. 5/stick	1200.00
(e) Framing of net	400.00
(f) Scoope nets (1 m. dia) and vion hooks to collect the fish	240.00
Total cost	11400.00
(B) Annual Recurring Expenses	
(a) Repair	1600.00
(b) Wage of 1 labourer for 150 days at Rs. 20/day	3000.00
(c) Rent of a boat for 150 days at Rs. 20/day	3000.00
Total expanses	7600.00
(C) Annual Income and Surplus	
Sale proceeds from 150 fishing days, 120 kg/day at Rs. 8/kg for total unit	144000.00
Share accruing to loanees (¼ of the total) (gross earnings)	36000.00
Annual recurring costs	(-) 7600.00
Gross surplus	28400.00
Depreciation	(-) 2850.00
Lease to revenue department	(-) 5000.00
Net surplus divided into	20550.00
(i) Return on loanee's labour	3000.00
(ii) Return on investment	17550.00
(D) Economic Feasibility	
(a) Rate of return on investment	154%
(b) Net value added (in Rs.)	4.12

(E) 4 Years Repayment Schedule

Years	*Principal*	*Interest*	*Principal with interest*	*Repayment*	*Balance*
1st	11400	2052	13452	3500	9952
2nd	9952	1791	11743	4000	7743
3rd	7743	1394	9137	5000	4137
4th	4137	745	4882	4882	–

Set Bag Net

Set bag net, used in a river mouth, all the year round by the owner individually with the assistance of one wage labour. Life span of the net is 4 years.

	(Rs.)
A. <u>Capital Cost</u>	
(a) Twine PER tax 23 kg at Rs. 100/kg.	2300.00
(b) PP ropes of 18 mm. dia, 6 mm. dia, 20 kg at Rs. 82/kg.	1640.00
(c) Bamboos, used as floats	640.00
(d) Making and framing charges	1300.00
(e) Miscellaneous	120.00
Total cost	6000.00
(B) <u>Annual Recurring Expenses</u>	
(a) Repair and maintenance of net	1200.00
(b) Wages for 1 fisherman, 150 days at Rs. 20/day	3000.00
(c) Rent of a boat for 150 days at Rs. 20/day	3000.00
Total expanses	7200.00
(C) <u>Annual Income and Surplus</u>	
Gross earnings, sale proceeds of fish from 150 days fishing, 15 kg/day at Rs. 10/kg.	22500.00
Annual recurring cost	(-) 7200.00
Gross surplus	15300.00
Depreciation	(-) 1500.00
Net surplus divided into	13800.00
(i) Return on labour (at Rs. 20/day)	3000.00
(ii) Return on investment	10800.00
(D) <u>Economic Feasibility</u>	
Rate of return on investment	180%
Net value added (in Rs.)	5.6

(E) <u>4 Years Repayment Schedule</u>

Years	*Principal*	*Interest*	*Principal with interest*	*Repayment*	*Balance*
1st	6000	1080	7080	2000	5080
2nd	5080	914	5994	2000	3994
3rd	3994	719	4713	3000	1713
4th	1713	308	2021	2021	–

Fish Marketing–Bicycle Fish Retail Business

The cycle trader buys the fish at a landing site, transport it to the market by cycle and sells it there.

		(Rs.)
A.	Capital Cost	
(a)	Fixed capital–cost of cycle with special carrier and accessories	1200.00
(b)	Baskets for transportation of fish, gunny bag, weighting balance, knife etc.	200.00
	Working capital–For 1 day's purchase of fish, 40 kgs at Rs. 10/kg.	400.00
	Ice	20.00
	Fish market fees etc.	40.00
	Total cost	1860.00
(B)	Annual Recurring Expenses	
(a)	Purchase of fish on 210 days, per day 15 kg. at Rs. 8/kg.	25200.00
(b)	Ice for 210 days at Rs. 8/day	1680.00
(c)	Repair, replacement of baskets, gunny bags etc.	300.00
(d)	Rent for selling space Rs. 2/day	420.00
	Total expanses	27600.00
(C)	Annual Income and Surplus	
(a)	Sale proceeds of fish at Rs. 14/kg (14 × 15 × 210)	44100
(b)	Annual recurring cost	(-) 27600
(c)	Gross surplus	16500
(d)	Depreciation (life span of cycle 6 years)	(-) 200
(e)	Net surplus	16300
(D)	Economic Feasibility	
Net value added (in Rs.)		17.52
(E)	Repayment	
Within one year in monthly installments.		

Head Load Fish Retail Business

One woman buys fish at a landing site, transports it by head load and sells it in village market.

		(Rs.)
A.	Capital Cost	
	Fixed capital	
(a)	Weighing balance	60.00
(b)	Baskets	80.00
(c)	Gunny bags to cover fish	30.00
	Working capital	
(a)	Purchase of fish for the first day 25 kg/day at Rs. 10/kg.	250.00
(b)	Ice	10.00
	Total cost	430.00
(B)	Annual Recurring Expenses	
(a)	Purchase of fish on 210 days, per day 10 kg at Rs. 8/kg.	16800.00
(b)	Ice on 210 days at Rs. 6/day	1260.00
(c)	Repair and replacement of baskets, gunny bags etc.	200.00
(d)	Rent for selling space Rs. 2/day	420.00
(e)	Transport at Rs. 4/day	840.00
	Total expanses	19520.00
(C)	Annual Income and Surplus	
(a)	Sale proceeds of fish at Rs. 14/day (14 × 10 × 210)	29400.00
(b)	Annual recurring costs	(-) 19500.00
(c)	Gross surplus	9880.00
(d)	Net surplus	9880.00
(D)	Economic Feasibility	
	Net value added (in Rs.)	45.94
(E)	Repayment	
	Within 1 year in monthly installments.	

Economics of Operation of 9 Metre Dug-out Canoe Without-board

(a)	Vessel specification motor	
	Length	9.00 metres
	Beam	1.27 metres
	Draft	0.32 metres
	Engine Horse Power (out board motor)	8-10 H.P.

	Fish carrying capacity	250 kg.
	Type of fishing	Drift/gill netting
	Crew	3
(b)	Operational schedule	
	Fishing days per year	200
	Fishing trips per year	200
	Average fishing hours per day's fishing	5
	Total fishing hours per year	1000
	Total hours steaming per year	1400
(c)	production	
	Average annual catch	
	(i) Fish	14.75 tonnes
	(ii) Crustaceans	0.25 tonnes
	Total	15.00 tonnes
	Average annual landings	
	(i) Fish class I–3 tonnes at Rs. 1000	Rs. 30000.00
	(ii) Fish class II–3 tonnes at Rs. 5000	Rs. 15000.00
	(ii) Fish class III–8.75 tonnes at Rs. 600	Rs. 5250.00
	(iv) Crustaceans–0.25 tonnes at Rs. 40000	Rs. 10000.00
		Rs. 60250.00
(d)	Investment	
	Cost of hull	Rs. 36000.00
	Cost of OBM	Rs. 12000.00
	Sub total	Rs. 48000.00
	Cost of fishing gear	Rs. 16000.00
	Total capital investment	Rs 64000.00
(e)	Operating cost	
	Fuel	Rs. 4400.00
	Lubricating oil and grease	Rs. 1000.00
	Maintenance of hull	Rs. 400.00
	Maintenance of engine	Rs. 1000.00
	Insurance of vessel	Rs. 200.00

	Insurance of crew	Rs. 200.00
	Maintenance of fishing gear	Rs. 200.00
	Wages–basic wages of crew	Rs. 16800.00
	Crew allowance–Food and others	Rs. 2000.00
	Port fee and wharfage	Rs. 200.00
	Sundry items and overheads	Rs. 400.00
	Total operating cost	Rs. 26800.00
(f)	Average annual earnings	Rs. 60250.00
(g)	Net earnings before depreciation and debt service	Rs. 33450.00
(h)	Depreciation on boat at 3 per cent	Rs. 1080.00
(i)	Depreciation on engine at 33 per cent	Rs. 4000.00
(j)	Interest on total investment at 12 per cent	Rs. 7680.00
	Total (h + i + j)	Rs. 12760.00
(k)	Net earnings after depreciation and debt service	Rs. 20690.00
	Crew wages for 9 metre boat	
	(i) Master fishermen–1 × 600 × 12 — Rs 7200.00	
	(ii) Deck–hands–2 × 400 × 12 — Rs 9600.00	
	Total — Rs. 16400.00	

Economics of Operation of 15 Metre Trawler-gill Netter

(a)	Vessel specification	
	Length	15 m.
	Beam	4.12 m.
	Draft	2.44 m.
	Engine Horse Power	88 H.P.
	Fish carrying capacity	4 tonnes
	No. of crew	5
	Type of fishing	Trawling and gill netting
(b)	Operational schedule	
	Fishing days per year	200
	Fishing trips per year	200
	Average fishing hours per day's fishing	6

	Total fishing hours per year	1200
	Total hours steaming per year	2000
(c)	Production	
	Average annual catch	
	(i) Fish	90 tonnes
	(ii) Crustaceans	10 tonnes
	Total	100 tonnes
	Average annual landings	
	(i) Fish class II–15 tonnes at Rs. 5000	Rs. 75000.00
	(ii) Fish class III–15 tonnes at Rs. 2000	Rs. 30000.00
	(ii) Fish class IV–60 tonnes at Rs. 600	Rs. 36000.00
	(iv) Crustaceans–@ Rs. 12600 per tonne for 10 tonnes (Average price of 3 sizes)	Rs. 126000.00
	Total fish and crustaceans	Rs. 2,67,000.00
(d)	Investment	
	Cost of hull	Rs. 180000.00
	Cost of engine	Rs. 200000.00
	Cost of other equipments	Rs. 10000.00
	Sub total	Rs. 390000.00
	Cost of fishing gear	Rs. 20000.00
	Operating capital	Rs. 10000.00
	Total capital investment	Rs. 420000.00
(e)	Operating cost	
	Fuel	Rs. 44000.00
	Lubricating oil and grease	Rs. 4000.00
	Maintenance of hull	Rs. 5400.00
	Maintenance of engine	Rs. 6000.00
	Maintenance of fishing gear	Rs. 1000.00
	Insurance–vessel	Rs. 9500.00
	Insurance–crew	Rs. 10000.00
	Wages of crew	Rs. 42000.00
	Crew allowances, food and others	Rs. 12600.00
	Ice	Rs. 3000.00

Port dues	Rs. 2000.00
Slip-way charges	Rs. 2000.00
Sundry items and overhead	Rs. 6000.00
Total operating cost	Rs. 147500.00
(f) Average annual earnings	Rs. 2,67,000.00
(g) Net earnings before depreciation and debt serviceing	Rs. 119500.00
(h) Depreciation on boat and engine at 8 per cent	Rs. 32000.00
(j) Interest 12 per cent on total investment	Rs. 50400.00
Total	Rs. 82400.00
(j) Net earnings after depreciation and debt serviceing	Rs. 37000.00

Crew wages for 15 metre boat

(i) Skipper–900 × 12 × 1	Rs. 10800.00
(ii) Driver–800 × 12 × 1	Rs. 9600.00
(iii) Deck hands–600 × 12 × 3	Rs. 21600.00
Total	Rs. 42000.00

Chapter 6

Market

Principles of Fish Marketing

Fish marketing connect (bridge) production and consumption time and distance are required to be overcome.

Prices in general, are regarded as a good index of the degree to which the potential demand for a commodity is being met by the available supplies. Normally price should be such as to ensure for the producer a fair return for his labour and investment. The fishermen unfortunately working under the handicaps seldom receive a living wage much less a fair return, and practically all the plums of the trade are collected by the middlemen wholesalers, who constitute the sole distributing agency. The number of middlemen varies with the scope and density of fishing operations, the regularity of supplies, the nature and extent of processing operations, such as, cold storage, drying, salting canning etc., the retailing and delivery services, the transport facilities and the volume of the trade and finally the rate of turnover. As distribution constitutes only the final link between the producer and the consumer, trade margins should be subject to adjustment based on a correct evaluation of the cost of materials and charges for labour and service. The prices of various articles as also the charge for labour required for preservation and packing of fish have increased very materially within recent years, and transport cost per km. is also fairly high, but the large margins between the producer's share and the price which the consumers have to pay can hardly be justified by these increases, and such fair margins of profits which the traders must realize on the capital investment involve as also on the volume of turnover. A review of the present position leaves no doubt that while both the producers and consumers continue to suffer, the middlemen–traders make enormous profits at their cost.

In fish marketing production characteristics influence the supply, which again depend on the annual catch of fish. More than 1800 fishing villages are scattered along the Indian coast line of 6536 kms. Due to great dispersion of marine fish landings, concentrating the catches in few assembly centres poses a great handicap in organizing for large scale marketing of marine fish in inland areas.

Price structure is different for different species. Demand and consumption pattern influences the availability.

No problem is experienced in marketing prime quality fish; but marketing of large quantity of trash fish is a real problem as the consumers are not ready to accept them. A change of form (fish meal, FPC) is necessary for their disposal.

The seasonal aspect of production of marine fish also influence their marketing. At fish landing sites, the producers offer their marketable surplus catch for sale not by weight, but by measures of heaps, lots or baskets which differ from locality to locality.

In most of the fish landing centres, there is no provision of marketing sheds and fish landed in opean beach and open ground. Provision of marketing sheds at the landing centres are absolutely essential.

There should be provision of feeder roads linking fish landing sites to nearest railway station, motorable roads for transporting fresh fish in good condition to the terminal markets.

Inland fisheries both capture fisheries of rivers, lakes and reservoirs and culture fisheries also contribute to the production characteristics.

Demand and Supply

Demand and consumption patterns of fish are determined by geography, feeding habits of the locality, traditional and nutritional standards. Demand of fish may be either domestic or for export purpose. The domestic demand depend on human consumption and industrial consumption. Human consumption varies area–wise, income-wise and region-wise, while industrial consumption depend on regional variation, quantity, variety and seasonability. The cause and effect of the items in each demand segment are (a) income, (b) price, (c) family size and structure, (d) availability, (e) services and mode of marketing, (f) variety, (g) cultural and attitudinal factors, (h) promotion, (i) substitutes and complements.

The supply of highly perishable commodity (fish) is dependent on (a) regional distribution, (b) species composition, (c) utilization pattern and (d) technology, both in respect of whole sale and retailing. Infrastructural facilities consists of landing and consuming centres, where weighing, grading and peeling, packing and transportation facilities will be available. The location of markets and environmental condition of the locality also influence supply and marketing of the commodity. In the landing and processing centres, ice plants, cold storage, curing and drying yards, fish meal plant, fish oil plant affect the marketing and disposal of fish considerably. Both short term and long term financial need for effecting supply of fish against demand, includes investment on working capital and for that purpose possible source of finances are required to be located.

The economic aspects of fish marketing include:

1. Nature of business–nature of perishability of the products with consequent uncertainities.
2. Fluctuation in purchasing and selling prices of marketing intermediaries.
3. Variable costs of marketing intermediaries for various items, like transportation, storage, packing, handling, establishment, interest and commission.
4. Wastage and loss due to storage, transportation, dehydration and handling, both quantitatively and financially.

Beach Market for Mass-catch Fish in Senegal

5. Commission and profit, state wise, variety wise, region wise and season wise.

Management of Fish Marketing

An efficient marketing system developed on modern management methods is the sinequanon to bring rapid changes in the functions of production and consumption needs of a society. As the demand for product goes up and therefore, requires a higher

Retail Fish Market at Gariahat, Kolkata

output to meet the new demand created, the marketing function has to necessarily cover a wide network of economic activities.

Modern Marketing

Earlier the term marketing is used to denote simply buying and selling in a market place. The industries used to sell products according to whatever demand was existing in the market. The modern management is now not only interested in meeting the actual demand, but also tapping the potential demand in the market by converting some of the needs into wants. Marketing now, means not only selling but includes the overall activities connected with planning of production, designing and sell of products and creating favourable atmosphere for buyers in the market for purchasing the products. This naturally necessitates the utilization of financial and technical resources to gain higher profit for the business.

Modern Marketing and Fisheries Development

At present, the mechanism of fish marketing, like marketing of other products, is increasingly becoming a complex one which necessarily requires adopting better marketing management techniques and methods. With the increased production envisaged

Marketing Freshwater Fish from Reservoir, Orissa, India

during the mechanization programme in developing countries, the special marketing arrangements have to be done so that there is no spoilage and wastage of fish output. The fish production has become more that ten times during the last four decades, which calls a break–through in old methods of marketing of fish.

Traditional Marketing

In many developing countries traditional system of marketing is generally followed in which all methods and practices in trade dealings are based on some customs. They are unchanged and unimproved over decades. There is no free entry in the fish marketing since fish merchants do not allow any body. They also maintain much secrecy in their trade interests and they do not reveal how they trade, what are the marketing margins. From the field survey, it was observed that in some cases there are big margins of net profits ranging from 100 to 200 per cent over the purchase price and marketing costs. There is no fair and perfect competition in some fish markets and it is usually these markets or producing centres are faced with either monopsonistic or monopolistic methods of control over fish out put.

Fish Marketing in Developing Countries

In Singapore and Philippines, the fish marketing is completely in the hands of private merchants. In Sri Lanka, though the fishery trade is largely in the private hands, the State has given recently wide support to co-operative enterprises. There is much exploitation of fishermen by traders in the east coast of West Malayasia, Indonesia and in Philippines. Starting of fish processing in many developing nations has introduced a new type of middlemen in the distribution chain, since all the means of processing is owned by a person other than fish producers.

Besides, the auctions are held in complete secrecy and signs are used at the time of auctioning fish. The whispering auctions and clock auctions are the usual features in Manila, Philippines. In Hongkong bidding is done by forming rings amongst buyers. In India at Agra, and Kanpur fish markets the bidders show signs of their consent with fingers. In Bombay fish market, the auctions for certain species of fish are followed by the use of a vernacular word "Kodi" (Kodi means 22 pieces of fish). In Malabar area, Kerala,

Marketing of Cultured Tiger Prawn, Kantai, West Bengal

"Kalli", a local language is used (Kalli 18780 kg approximately) for disposal of sardines and mackerels. The unit of sale vary from place to place, which makes difficult to improve the distribution system in the country.

In many fish markets except in Calcutta, grading and weighing are not done for fresh fish, and sold either by heaps or in baskets. Inadequate landing facilities, sheds for auctioning are preservation inputs exist in many parts of the country.

Malpractices are the usual features in the fish trade and bargains. The middlemen always cheat the fishermen by showing wrong records, deducting unnecessary traditional charges, such as, "Hisabana" (for accounts), postage, "Bruthi" (festival money) from the gross sale of produce. At every auction they remove one or two big fishes from the lot as a customary practice. They do not also pay immediately to the fishermen after completing the auction sales.

Modern Fish Marketing

Fish marketing may be defined as all those overall economic activities performed from the point of catching fish to the point of final consumption. Handling of fish at the sea itself for its

Freshwater Fish Marketing, Bhubaneswar, Orissa, India

maintenance of quality, initial grading and processing, packing and sorting species wise from bulk catches are necessarily have to be taken care of in the modern methods of marketing of fish.

As fresh fish cannot be kept for a long time, it has to be preserved by proper means for enhancing the marketability of product. Sometimes the fish in its original form after its capture may not be demanded by consumers, but when processed into other form would find a ready market. Trash fishes of marine catch do not find a ready market, but when processed into fish meal or fish protein concentrate (FPC) are demanded for poultry or live stock feed and from those who never consumed fish before. Transporting fish in refrigerated or insulated vans from producing centres to interior hiterlands create place and time utilities for the product.

Prices at the auction market, marketing expenses required for the transportation of fish, icing and freezing costs are important determinants in marketing costs. It is also necessary to watch what is produced is matched with the demand. Increased fish output should be absorbed without waste. It means, not only meeting the present demand for fish, but influencing and creating new demand for the product. Seasonality in fish production though it is difficult to control also limits the possibilities of marketing fish through out the year unless it is frozen, dried or canned; for example, mackerel in abundance is available in Mysore coast during November to February. The consumers in many part of the developing countries is not familiar with some fish species used as food. Further development of marketing and distribution based on modern lines is, therefore, very much needed in many developing countries for bringing a rapid growth in fisheries development.

Fishery Trade on Micro and Macro Levels

At micro level, in the country, there are innumerable number of fish merchants arranging the distribution of fish from landing centres to the main consuming centres. These middlemen have acquired lot of experience from a very long time. It is difficult to eliminate them unless some other measures are adopted.

In fish transaction, there may be three or four different categories of intermediaries, namely, commission agent, whole saler at the assembling centres, interior wholesaler, and finally retailer. All of them take high margins of profit due to the high risks involved in the trade. At macro level there is no organization, which can look after the whole distribution of fish, except some unfruitful attempts of some state fisheries development corporations, which had wasted government capital in trying to distribute fish to many important markets.

Handling of Fish and Quality Control

In many areas, the quality of fish is not upto the standard due to bad handling of fish by fishermen, as well as, buyers in the market. When the fish is caught, it has to be handled carefully so that its initial freshness is maintained until it reaches consumers. It has to be chilled immediately. The small trawlers do not generally carry ice with them and do not gut and wash fish. For long fishing time, the fishermen should be insisted on this, so that from the beginning, high quality fish in the sea itself is maintained. They should be taught how to gut, wash fish and the catch stowed in ice. Methods of stowage vary but consist mainly, bulking, boxing and shelfing. Instead of throwing fish on the dirty floor for auction, they may keep the fish in basket or in aluminium or plastic trays/basket for auction.

On the shore the catches have to be handled in a hygenic way. Ice should be provided at low cost for preservation of fish so that fish is not spoiled and not used for human consumption. There should be suitable law for compulsory inspection as to the maintenance of quality, as it has been instituted in Germany in Great Britain and in India for export of prawn.

Fish Supply and Prices

Though the day to day aggregate average fish prices are determined according to law of demand and supply, during the peak seasons supply of fish is so huge that it normally fetches a distressing price for a large number of fish species. Over supply of fish may be due to two factors; sudden increases in catches or decreased demand at a particular time and place. If such changes in supply or demand occur unexpectedly lower price sometimes may not stimulate any increased demand for fish. When there are glut landings, the seasonal surpluses of fish can be processed and marketed at the time of scarcity during off season.

In European economics, in the past, there used to be gluts of herring causing the market collapses on account of big variations in prices. In Norway, a minimum price policy was introduced by the sales organizations, in consultation with the government. When fishermen in the past were virtually unprotected against low prices, caused by changes in supplies, price support measures were introduced. All fish landed in Norway cannot be sold lower than the minimum price.

So besides offering fair price, a marketing co-operative by applying the tools of economics, can bring to a great extent stabilization of fish prices.

Marketing Cost and Differential Prices

Even though the average wholesale prices are normally low at the producing centres, the retail prices, as observed, usually go to high level. This is due to the reason that the product (fish) is handled by a large number of middlemen, who are not real producers in the economy, and operate between whole sale and retail market. Not only the middlemen charge some profit, as it is often more high, but also the marketing charges are added to the original landing price. Reduction in marketing charges by removing those unnecessary costs would result in lowering the consumer prices without any reduction in the producer price. On the contrary, it may enhance the producer price in the long run and bring down the retail price.

Thus fishermen's income may be improved by changing the pattern of distribution of fish. Usually, the marketing charges do not go to a high level if the markets are located near the landing centres.

Pattern of Distribution Through Middlemen

Producer → At producing centre
↓
Wholesale purchaser → At producing centre
↓
Commission agent → Assembling centre
↓
Inland wholesaler → Assembling centre
↓
Retailer → Marketing centres
↓
Consumer

The consumer price is trebled or quadrupled over the producer price, when the market is far away from the producing centres.

Fresh Fish Prices in Assembling Centre and Consuming Centre

Type of Fish	*Assembling Centre*	*Consuming Centre*	*Distance from Producing Centre to Consuming Centre (km)*	*Percentage Difference Over the Whole Sale Price*
Seer fish	Mangalore (Mysore state)	Bangalore	320	333.3%
White pomfret	-do-	-do-	-do-	227.2%
Cat fish	-do-	-do-	-do-	352.9%
Sardines	-do-	-do-	-do-	535.7%
Mackerel	-do-	-do-	-do-	227.3%
–do-	-do-	Belthangady	55	172.7%
–do-	-do-	Coimbatore	320	227.3%
–do-	Karwar (Mysore State)	Bombay	1280	437.5%

Marketing expenses incurred by private merchants in sending seer fishes to markets at a distances of 320 km.

Items of Expenditure	*Percentage to Total Charges*
Price of fish at producing centres	
Transport charges per basket of fish	38.4
Three blocks of ice per basket	14.4
Packing labour charges	12.0
Empty basket charges	14.4
Municipal charges	1.2
Miscellaneous charges	2.4
Commission charges on the price of fish	17.2
Total marketing expenses over producer's price–	
Margin of profit over the purchase price and–	
Marketing expenses	
Percentage of profit–	158.4%

Auctions in Fish Market and Price Determination

There are two types of auctions; (1) bidding and over bidding and (2) "dutch" auction. In the former method, auction start with low price and it ascends to a higher price until a buyer quotes and accepts the bid. In Bombay Crawford fish market, the former method is followed. But "dutch" auction, which slightly varied in principle, is followed in Netharlands, Portugal, Spain, Belgium and to some extent in Japan. In india it is limitedly followed at Kakinada coast in Andhra Pradesh and also in some Calcutta markets of West Bengal. In this system, instead of starting with a lower price and going up, a higher price is started and is lowered until a buyer is found.

Auction is a better system of marketing as if operates on the principle of demand and supply, specially perishable commodity like fish. This system of selling is beneficial one at least for a few important varieties giving a higher price if there is good demand. But it is not certainly helpful for those species, such as trash fishes, which have relatively no demand at a higher price in the market.

Mode of Auction

Simple method of auctioning displays a limited quantity of fish, either one kilogram or one basket of single species. The bidders may see the fish basket and bid a higher price for the fish displayed and after the bidding is over, he takes all that particular species from the vessel. This system is advangeous in reducing the time required for auctioing. This method of sampling auction for prawns is followed in Mangalore market and in some of the wholesale markets in Calcutta.

Auction is useful when there are large number of buyers in a market Auction do not work well whenever rings are formed in the market (a number of buyers bidding collectively instead of competing individually); and also may not work due to:

1. Cash discounts privately arranged after the bid has been completed;
2. Sometimes fishes are being removed before it is auctioned.

If auctions are not necessary, fish sales can be arranged by direct delivery system, price support measures and fixed price systems.

Delivery System

Direct delivery to the wholesalers from the vessels may maintain better quality of fish and auctions may not be required for frozen fish. In Germany, auctions are waning and being replaced by delivery contracts. Delivery contracts are valid for one fishing season and it is delivered to firms, fish meal industry etc., after fixing prices. Direct deliveries to consumers or whole salers by fishermen without auctions in respect of certain classes of fishes is also in vogue in India.

Minimum Price System

Due to over abundance of supply of fish, there is greatest danger of falling prices and to avert this, the minimum price system has been introduced in many European countries, particularly Norway, Sweden, Iceland and to some extent in Germany. The usual type of minimum price system, is to see that the prices do not fall below the market price. In certain cases a gurantee prices is given to the producer over and above the market price which he obtains.

In India and in South Asian Countries there is legally no minimum price system so far; though in Sri Lanka, fishing corporation purchases fish from co-operatives at prices fixed seasonally, but it is not working satisfactorily due to the fact that it receives gluts of supplies of fish when free market prices are low and very little when prices are above the fixed level.

Marketing of Processed Fish

Marketing Decisions and Financial Decision

Important function of management is to see that good amount of profit and return on capital investment are earned. This naturally depends upon the proper decision of marketing manager takes with regard to the nature of cost of production and the correct price to be charged for the product and the knowledge he possesses about financial matters of the company. Thus every marketing decision is a financial decision.

Nature of Costs

Though it is difficult to fix fish prices in auctions, prices can be fixed for all those fish which is processed and manufactured in a factory. Before fixing the selling price for processed fish goods, the nature of costs have to be taken into consideration. These costs are known as fixed and variable costs. Variable costs are of those vary with the production, whereas the fixed costs do not alter for some period for certain estimated capacity of production. For instance, fishing labour, fuel, food for the crew are variable costs and depreciation on boat and engine, salaries and establishments are fixed costs. The total costs come by adding the variable and fixed costs.

Break-even Analysis

The company starts earning profit when certain volume of sales takes place at a point, where the costs and revenue are equal, If small change in volume of sales makes a big change in profit, the company is in a position of volume sensitive situation. Similarly, in small change in costs would make a significant change to the profit than a small change in volume, the company is to be in cost sensitive situation.

Selling Price and Volume of Sales

The selling price which gives greatest amount of profit depends upon the relationship between the quantity that would be sold and the selling price for any period of time. Normally low price, the greatest volume of sales and vice-versa. Bigger volume of sales may bring maximum profits and not a higher profit per unit sold. The concept of elasticity of demand is useful in determining the maximum revenue. For example, when there is in elastic demand for fish, the small reduction in price does not bring increased maximum revenue. Small turnover of business at a fixed price at certain level would bring bigger margins. Thus selling price which produces maximum profit is quite different from the selling price which produces maximum revenue.

Factors Affecting the Selling Price

To produce the maximum profit the relationship between selling price and sales volume for that particular period has to be taken. Fixed costs do not come into picture. At any selling price which produces the maximum profits, the variable cost is taken into account.

Fixation of Selling Price

One method is that is done by calculating the cost of unit of product and certain percentage of profit is added to the cost for deriving the selling price. But to derive the exact cost of production is a difficult thing since costs always change from time to time. As such, selling price has to be determined every now and then.

For assessing the price which produces maximum profit is to apply trial and error method, either by reducing or increasing price and see how much maximum profit is earned at each level of volume of business. A very higher price may reduce very much the volume of business and very low price may increase the volume of sale but earning a lower profit per unit. Therefore, selling price has to be determined in such a way that sales may not fall off completely as there are other competitive products in the market. A suitable selling price has to be fixed between the maximum and minimum prices. Thus the most profitable selling price depends upon the relationship between price and volume of the product. Arbitrary price based on costs should not be taken.

Selling Costs and Profits

If the expenditure on production materials is doubled, it would also double the output. But doubling the selling costs would not increase the sales volume of a company as it is not directly related on the amount of expenditure increased on advertising and promotion or the number of salesmen employed. Internal efficiency of the organization determines the profit gained. At a certain point the given number of people employed to get certain sales volume with profits is the optimum point. Beyond this point, even if the number of salesmen is increased, it would not increase the sales. But there is an exception to this. If the salesmen are employed on commission basis without any salary, and to meet their own expenses then the selling costs would be directly proportional to the business turned out. This is also true with the advertisement expenditure.

Volume of Sales and Capital Investment

At certain level of capital investment in the business, the returns on investment is directly proportional to the profit, and it would also bring an increase return on investment. While planning a change in marketing it should be taken into account what effect it would be on the capital investment. In certain cases it is not economic to think of expanding the capital when there is much to recover as realisations, treated as capital for the period, from a large number of buyers, which means net return on investment.

Market Research

For successful marketing a correct and intimate knowledge of the market-who and where the potential consumers and their motivations to buy a product–is very much required. This should be based on facts and figures collected on a scientific basis from a market, but not on guesses or personal opinions.

The desk study and investigations can be conducted on the material available which might have already been gathered either by governmental bodies or public organisations. If the material required from the published material is not available, then, the field investigation has to be conducted, when the material is directly collected from consumers or industrial users. The informations can be sought by observation or by interrogation by preparing suitable questionnaires. For getting the accurate results constitution of the sample should fully represent the constitution of the whole population.

With the limited resources available, it is not possible to conduct intensive research for developing new products. But the new products developed by Central Institutes of Fishery Technology are likely to have good markets. For instance, canned sardines is preserved in its own juice, which means no additional cost for oil or any other preservatives. Prawn waste shell is being developed for industrial uses. Shell waste converted into paste form and is used as a flavouring agent in the food preparations made in South East Asian Countries. Sardine oil is used in printing ink, rubber industry and paints. Fisheries college, Mangalore had processed some food products with trash fish; fish jam, cake, bread fish paste, vegetable fish sausage etc. Some of the new products can be taken up for further marketability in the country.

Market Forecast

Forecasting means, the predicting the future trends of market under given set of conditions. Market forecasting is different from sales forecasting. Market forecasting gives the idea of total demand in a particular market and that of sales forecast gives company's own sales.

The art of predicting the future trends is not easy in all cases. A change in the general economic climate can be forecasted when there is a particular change in taxation on a product going to effect the total demand for the product.

Field Surveys

On the basis of field surveys of buyer's opinions, the forecasting is possible. A field survey has to be carried out to find out the buyers intentions. But this has some limitations. Even if correct intentions are obtained, it is not possible to rely completely on that data since a buyer intends to do a thing may be different from when he does not actually buy and when actually makes a buying decision. While giving the information, they might have given the information without taking the actual position of their financial position.

Projections Trend

On the basis of past trends future trends can be predicted. The simple method is to draw a graph showing the sales of last few years and extending the line forward for the future years. But this has some limitations, since sales fluctuate month to month. On the basis of mean sales trend over past years it can be forecasted. The sales of a few years are taken to establish the mean sales trend. Taking moving annual totals, which are the total sales for the preceeding twelve month, mean for one month or for a year can be predicted. Short term forecasting is easy with this method, but not long term projections.

Extension

In many countries extension agencies are not playing a significant role with regard to the development of fisheries, though there is considerable extension work done in agriculture and industry. Because of lack of proper extension agencies in many developing countries, the fishermen are completely ignorant of modern techniques of fish production. At present there is a big gap between the scientific knowledge in fisheries and the fishing methods followed by fishermen. The extension, through, co-operatives can play an active role disseminating knowledge in the following field of fisheries:

1. Extending knowledge in fish processing, modern as well as indigenous preservation and marketing
2. To train the fishermen in using the modern mechanized boats and in new fishing methods.
3. Educating the fishermen in principles of co-operation and method of keeping accounts.
4. Educating the fishermen on the advantages in fish handling and quality control.
5. Giving the marketing information and intelligence regarding the prices of fish and their demand in markets.
6. To make them know where processing facilities are available and with whom they are available and how to make use of them.
7. Educating them in proper exploitation of fishery resources.
8. Bringing out fishery bulletins and news.
9. Arranging films, exhibitions and discussion meetings.

Guidelines for Developing Fish Marketing

1. Retail distribution of fish should be developed as it is profitable. New markets have to be investigated for increasing the role of fish and fish products.
2. Whenever there are facilities for processing, the co-operative should purchase surplus fish during the peak season, store it and release the product when there are scarcity conditions for realising a higher price. Trash fish which can not be sold should be converted into fish meal.
3. Prices for processed products should be fixed in such a way to give maximum profit.
4. New fish products have to be developed.
5. Prices given to fishermen should not be lower than the market prices.
6. Supply of fishery requisites and loans should be properly linked with sale realisation of fish proceeds.

7. All facilities like jetty, fuel depots, workshop facilities for repairing boats have to be created.
8. Normally all the marketing activities, such as, auctions should be centralized at one point and the control of this should be in the hands of co-operative. No private person should have any control over that.
9. Only trained personnel should manage the fish marketing and in the affairs of society business interests should precede political interests.

National Policy for Fish Marketing

Since many marketing co-operatives are not working satisfactorily in this country and elsewhere, it is essential to draw up a national policy and plan for the development of fish marketing. Government should prepare a national Plan for the development of fish marketing in the country. Different fishing zones may be drawn up and accordingly it should work out how many marketing co-operatives are required for different zones and areas. There should not be overcrowding of co-operatives, so that every marketing co-operative gets an opportunity to cover a wider area for distribution of fish. It is necessary to bring out suitable legislation to curb the activities of the middlemen and as far as possible the activities of the government should support the co-operatives. Tax holiday and certain other fee concessions have to be given to co-operatives until they develop their own resources. Encouragement should be given by the government in purchasing the fish products manufactured by them.

In Hongkong, Singapore and Malyasia to some extent the governments have introduced some legislation to curb the activities of the middlemen. In Singapore a system of licensing controls and regulations of the traders is in vogue. The wholesale trade in Hongkong is directed by the Fish marketing Organization In Malaysia fish landings have to be done only in a few points. In Bangkok and Kaulalumpur centralized fish markets were established where government supervision is there.

Marketing Committees

To give better status and improvement in fish marketing, the governments have to create marketing committees for directing there co-operatives. As they are investing huge capital, the matters connected with finance and important decisions should be taken by this committee, which should be a top level body consisting of marketing and technical experts. When such direction from top level is comming, it is difficult for the vested interests to play an active role in the affairs of the co-operatives. To exercise control over them and for realising the loans, from the fishermen, this committee can have proper supervision. In India, there can be one committee for every state.

Technical Assistance

Besides giving the education in co-operative principle to the fishermen, it is essential for the governments to give free technical help to the co-operatives. Plans for construction of jetties at the landing sites, processing plants, storages and factory establishments have to be provided by the government technical experts. Fishing harbours, feeder roads, other infrastructure inputs, which require heavy capital outlay, should be taken up by the government for maintaining a high productivity in fisheries.

If in any area there are many marketing co-operatives, they should combine and co-ordinate their activities so as to derive economics of scale and better bargaining strength to fight private interests.

To sum up, it is emphasized that organisation of marketing co-operatives would aim not only to improve the economic conditions of fishermen, but also mobilizes the resources and increases investment in the fishing industry. For fishermen, who are predominantly of small means and who had great difficulty in marketing and processing, the organisation of co-operative would render much needed service and gradually uplift the fishermen from the economic clutches and bondages of middlemen. Co-operative is by far the indispensable organisation for the overall development of fishing industry.

New Products in the Domestic and Export Market

There are bright prospects for new products developed out of fish both in domestic as well as export markets.

Surveying the domestic marketing environment for fish, the following facts have been noticed.

People want fish. As a deviation from the tradition, more and more people become non-vegetarians and fish eaters. The fish usage rates tend to shoot up, but the availability acts as a barrier to a prominent and fast increase in per capita consumption. People have become more selective to choose those items of food which will not consume much of their time. They prefer to buys prepared food from hotels, restaurants etc. But the demand can not be satisfactorily met because of short supply.

Even at the face of such inadequacy of fish supply, the following facts exists;

During period of bulk landings of fish prices fall drastically. The result is that the increase in harvest which theoretically should favour the producers considerably fails to do so, *i.e.* there is no proportionate enhancement of income to the producers.

Presumbly due to the lack of a price to support the costs and efforts to store and land, the industrial fishing sector discards approximately 90 per cent of the by catches, which in ideal conditions would have angmented the income of the fishermen and boosted up the economy of fishing.

The above examples of under utilization and non–utilization of the available harvests have been diagnosed as a severe economic problem. The solutions of the above problems must be thoroughly explored and identified. The marketing environment must be probed into indepth and stress should be given to dispose the marketing forces favourably.

An agreeable proposal for efficient and effective utilization of fish will be to develop new products and develop the market. Product diversification to suit the consumer's habits, life styles etc. must be introduced.

The new product development out of fish requires new technology, sufficient infrastructure, sufficient trained manpower. In order to establish all these, there should be appropriate and adequate institutional support. Entrepreneurship mush be effectively attracted to the field. Since such ventures require considerable amount of investment, there must be measures to retain attraction to the field.

The most important and difficult part of new product development is market penetration and market development. This is especially so in the absence of adequate market study and flow of information. It is in this context that indepth market study is relevant and essential. The service of an expert professional marketing research team is inevitable in this regard.

However, there are indications to assume that strong market can be developed for diversified fish products in the country. Some of such indicators are, the changing life styles of the people, the rising purchasing power of the consumers, the changing eating habits of the people. That the new products cannot have market penetration in coastal states is felt to be an outdated concept. The fast transition to urbanisation and increasing population concentration in urban areas point to the prospects of developing markets for more sophisticated products. the spread of education and growing awareness of quality and hygiene also point to the possibility of market development. It is a discernible factor in this wake that the Indian consumers is ready to buy a quality product at higher price rather than resorting to an inferior product at lower prices. The frequency of repeated purchase may be a little less depending on the buying capacity of the individual consumer. The fast growing markets for baked food, fast food even at the face of rapidly increasing prices indicate the trend.

Moreover the population in the country is such strong that even a very small percentage ready to buy any products constitutes a very strong market for that product. Therefore, it is beyond doubt that for a quite meaningful time (possibly decades) there will be no lack of market for any product in India provided the product is of good quality and attributes. But there are strong hurdles to take products to the consumers, if the products are frozen. Fish being a highly perishable material and subject to fast deterioration, requires great care on storage, transportation and distribution. The need for developing a cold chain and refrigerated transportation becomes relevent in this context.

A strong back–up of marketing is essential element in the market introduction and growth of diversified fish products. A well planned marketing strategy has to be formulated. The marketing mix has to be very efficiently deployed to catch up market. Institutional support has to be extended to the entrepreneurs in this respect also. A proper combination of promotion mix, namely, advertisement, sales promotion, publicity and personal selling, with planed distribution system, the establishment of cold chain, and very desirable product attributes and hygiene will definitely succeed in bringing in market growth. Ideal market segmentations and targeting the segments with the most fitting product messages are sure to bring a tremendous achievement. It is simple logic to think that one kilogram of product per person per year for one per cent of the Indian population will facilitate utilisation of a minimum quantity of 17000 tonnes of fish.

We have a good resource of a number of species of fish which normally are caught in pelagic and demersal fishing and enjoy only very low prices. These fishes can be utilized for drying for human consumption, fish mince preparation, filleting slicing etc. The main species are scianids and croakers, lizard fish, pink perch, cat fish, ribbon fish etc. The fishes with white or light meat can be used for mincing and development of battered and breaded products and also for "Surimi" Battered and breaded products can be introduced to the domestic market and "Surimi" is a material of great export potential. This is a basic material for development of several fish analogues. Fish pickles can be developed from most of the fish varieties which can be introduced to domestic market and exported as well. Fish sausage can be developed from ribbon fish.

Export Markets for Diversified Products

There has been spectacular growth in the world export of fish and fishery products over the years due to substantial increase in consumption of sea food world wide mainly because of consumers confidence in its nutritional benefits and wholesomeness. It is expected that the growth in consumption of seafood will continue and remain for many years. In the international market, fish and fishery products are mainly exported in fresh/chilled/frozen/dried/smoked/salted and canned forms. Improved economic status, changing life styles and consumption pattern and various other socio-economic factors have triggered off major structural changes in the world seafood markets for diversified products, particularly value added/ready to cook products in convenient packs. Considering the current trend, consumption of sea foods in diversified forms in the sophisticated overseas market is expected to increase substantially. Suppliers in developing countries will be able to expand their business in this line of diversified products, typical products of which are categorised below;

Shrimp

1. Breaded butterfly shrimp (clean tail)
2. Breaded round shrimp (clean tail)
3. Shrimp patties/shrimp burgers
4. Shrimp sticks/mini shrimp sticks
5. Shrimp scrips
6. Shrimp salad royale
7. Shrimp dried cooked (P and D)
8. Shrimp dried cooked (HL)
9. Shrimp in Spanish tomato sauce

Squid

1. Shrotfin squid tubes
2. Shortfin squid fins/arms
3. Shortfin squid rings
4. Breaded squid
5. Fried squid
6. Breaded squid rings

Cuttlefish

1. Cuttlefish salad imperial
2. Cuttlefish in lemon butter sauce

Fish

1. Fillets
2. Burgers
3. Sticks
4. Seafood mix
5. Scallops
6. Fish pickles
7. Fish kababs.

Principal Markets for Diversified Products

Japan, USA and W. Europe are the principal markets for shrimp, fish and cephalopods. Among these products, shrimps are considered the most valuable products of the sea and consumed all over the world, particularly in the developed nations where the taste for them in various forms and from various species is still growing. Though these three principal commodities are already having good market, acceptance as raw materials in the principal world markets, of these commodities are really converted into successful diversified products in value added form. There is potential for growth of consumption of seafood in diversified forms in these markets because of change in eating patterns that are taking place there.

The developing country like India has necessary advantages to alter these habit as we have key benefits like plentiful and reliable labour at low cost as well as consistent sources of high quality raw material for production of diversified seafood products for export.

Market for Frozen Fish Fillets

The world demand for fish in fillet form (frozen) has been on the increase from 551846 tonnes to 793847 tonnes over the last few years. USA is the principal importer. Europe, UK, Germany, France and Italy are the major countries importing substantial quantity of fish in fillet forms. There is substantial demand for fish sticks/fish portions/processed fishery products and breaded shrimps particularly in USA.

There is therefore, good scope in India to utilise low value fish for conversion into these diversified products for export.

Coated Fishery Products

With the new exciting technology, as well as, increasingly sophisticated consumers, who demand more differentiation in terms of taste, as well as structure and flavour, coatings have become more diversified. The European seafood market is rapidly changing from the traditional fish finger to the microwavable crumbed products which form convenience and taste. The tendency in USA is also towards preparing coated products in line with diet conscious trends. Therefore the market for coated seafood products in USA and Europe are expected to go further.

Surimi/Analogue Products

Surimi has become one of the most dynamic commodities in the international seafood trade because of its utilisation. Surimi is an intermediate product made from minced fish meat that has been washed, refined and treated with cryoprotectants. Surimi is used to manufacture various types of fish cakes (kamaboko)/imitation crab meat (Kanikama). It is also used to make fish hams, and sausages. Japan plays an important role and is the largest producer as well as importer of Surimi. Though Japanese annual surimi production has been to the order of 3.5 lakh tones, she has been importing substantial quantity of surimi as below.

(In tonnes)

Country	*1987*	*1988*	*1989*
USA	5500	30500	90000
S. Korea	8000	8000	10000
Thailand	21000	25000	30000
Total	34500	63500	130000

Japan has been exporting substantial quantity of imitation crab meat to USA, Canada, Europe and Australia over the years as below.

(1000 tonnes)

Year	*USA*	*Canada*	*Europe*	*Australia*	*Others*	*Total*
1982	6.7	–	0.5	1.8	0.3	9.3
1983	13.8	0.2	2.8	1.6	0.4	18.8
1984	26.8	0.8	2.3	1.6	1.0	32.5
1985	30.9	1.4	3.8	1.9	1.0	39.0
1986	25.3	1.9	5.8	1.6	1.1	35.7
1987	17.6	1.8	7.1	1.1	0.8	28.4
1988	8.7	1.7	7.5	1.3	0.6	21.1

A study team of experts from the Japan External Trade Organization has indicated that the feasibility of commercial production of Surimi in India. The team identified a few varieties of low value white meat fishes that are presently landed in India in larger quantities (eg croakers/ribbon fish etc.) that could be utilised for Surimi production on a commercial scale. There is, therefore, good scope to produce Surimi in India for export.

Fish Trade–in Chilka Lagoon

The development of fresh fish trade of the lagoon has been initiated due to the enterprise of outside merchants from Bengal, who adopted the system of taking the lease of lagoon fisheries from the owners and sub-leasing the same to the local fishermen at higher rates reserving monopoly rights, receiving the entire catch from their respective fisheries since 1920. The export figure from the lagoon, before second world war was on an average 50,000 maunds annually, most of it having been sent to Calcutta.

With the establishment of an ice factory at Kaluparaghat, the export of fresh fish from the lagoon developed rapidly making the Kaluparaghat, the largest export centre of the lagoon.

In the beginning the merchants utilised the services of some fishermen to collect catches from the fishing grounds and bring them to the assembly points near the rail heads. These fishermen, in course of time, gave up fishing and took the profession of commission agents. They weigh and took fish from the fishermen in the fishing grounds keeping sufficient margin for any shortage due to future loss in weight and weigh out the same to the merchants depots at the assembly points.

As many as 300 heavy country boats, using bamboo mats as sails, were engaged to transport fish some time covers a distance of 32 kms. or more. If the wind is favourable the fish could be booked to Calcutta on the same day, or else it has to be left over for the night, for which fishermen were paid lesser price for fish brought late.

Usually fish were packed in horizontal layers with fish and ice alternately in the ratio of 1:1 as fish and ice in summer and 3:2 in winter, leading more than 40 per cent spoilage on arrival in Calcutta. The fish was auctioned publicly at Howrah and the *adatdars* (the principal merchants) used to charge 3 to 12 per cent if no money was advanced, double otherwise from the auction value of fish sent.

Out of Rs. 11 lakhs as pre-war gross income of the lagoon from fisheries, the fishermen for all their troubles used to receive Rs. 1 to 2 lakhs per year. During the war, the income from the lagoon fisheries went up to nearly Rs. 20 laksh per annum. By introducing a system of control of prices both at the source as well as at the retail market for a portion of the supplies, it was possible to apportion 40 per cent of this amount to fishermen.

This fact, that the fishermen to get due reward for his labour, the Central Fishermen Co-operative Marketing Society Ltd was formed at Balugoan, in one of the fish assembly point of the lagoon to successfully market its catches in Calcutta.

But individual merchants operating in the lagoon always maintain their upper hand than that of co-operative society and the percentage of fish exported by the merchants were also higher than that of the co-operative societies. Between 266 to 281 individual merchants and 166 to 231 co-operative societies were operating between 1944 to 1954 handling 55.3 to 64.4 and 35.6 to 44.7 per cent of fish export trade of the lagoon during the period. Fresh fish of the lagoon was then exported only from rail heads, like, Kaluparaghat, Kuhuri, Gangadharpur, Balugaon, Khallikote and Rambha, of which Balugaon and Kaluparaghat were the main centres handling 60 to 65 per cent of fish traffic.

Inspite of the creation of the co-operative system in 1959 (CFCMS Ltd.), the main part of the fish is traded through the private market. It seems to be difficult to change this as the private market is the traditional way for selling fish. The private ownership over parts of the lagoon and the merchants and traders monopoly over the fish market, which grow to prominence during the British rule, created this dependence. Though in bye–laws of the CRCMS it was stated that all the fish caught should be sold through the Apex, yet this is not done in most cases, not even through primaries, and there is no provision in the bye–laws to punish the offenders. Some of the prosperous societies though, impose a fine if their members do not sell a stated part of their catch through the primary's marketing channel. The percentage of fish sold through the co-operative system has varied over years and has only been about 10 per cent on an average. During 1968-69, fishes and prawn from the lagoon were exported by the co-operative societies and individual merchants leaving nearly 30 per cent for local consumption inside the state. A total of 2423.95 tonnes (68.51 per cent of the total fish catch) from the lagoon was exported to Calcutta during 1968-69.

Marketing of lagoon fish through co-operative societies further dropped down to 4.74 per cent (322 tonnes) of total catch (6862 tonnes) in 1987-88 through 16 co-operative societies out of 68 primaries.

The export of prawn from the lagoon outside the country was first started in 1970-71 and with the increasing export potentialities of the lagoon prawn, eight prawn processing plants were established along the periphery of the lagoon for processing and export of prawns. In 1990-91, the total fish and prawn catch of 4273.3 tonnes from the lagoon were marketed in the following manner:

1. Export of fish to outside state — 2134 tonnes
2. Export of prawn to foreign markets — 180 tonnes
3. Consumption of fish and prawn inside the state — 1745 tonnes
4. Consumption of fish and prawn by local in habitants in and around the lagoon — 214 tonnes

Price Structure

The producer (fishermen) of the lagoon used to get only 10 to 40 per cent of the sale price of prawn and fish. The bulk share was going to the wholesaler (adatdar) commission agents.

The average pre-war price of Chilka fish and prawn per maund (37.5 kg.) were

	Pre-war Price	*Price in Rs. per Maund (37.5 kg) Price During 1944–46*
Mullet	5.50–6.50	13.00–15.00
Persia	4.00–6.00	12.00–12.50
Thread fins	4.00–6.00	15.00–30.00
Hilsa	5.50–7.50	13.00–15.00
Bhekti		
(a) big size	16.00–20.00	40.00–60.00
(b) small size	12.50–15.00	12.50–15.00
Jew fish		
(a) big size	6.50–8.00	25.00–32.50
(b) small size	3.00–3.50	12.50–15.00
Prawn		
(a) big size	7.50–10.00	25.00–30.00
(b) small size	3.50–5.00	7.00–10.00

The fishermen of the lagoon used to get three fold higher price than that of pre-war period during 1944–46.

The merchants used to incur Rs. 22.12 for packing 5 maunds of fish with ice including loading them into railway wagon during 1940 on the following items.

Six blocks of ice	13.50	
Carriage of ice	0.50	
Cost of 5 baskets of one–third rate as each basket being used thrice	0.75	For Packing
Three blocks of ice	6.75	
Carriage of ice	0.25	
Carriage, weighment and loading of fish	0.50	For supplying to collecting agents

The price of Chilka fish gone up gradually due to increased export outside the state.

With the export of prawn to the foreign market and increasing demand of Chilka fish in domestic market outside the state, the price structure exhibited sudden increase in all varieties of prawn and fish of the lagoon and the average price during 1991 is indicated below:

Varieties	*Price in Rs. per kg.*
Mullet	20.00
Penaeid prawn	100.00 (40 count)
Medium sized prawn	40.00 (100 count)
Small sized prawn	20.00 (160 count)
Hilsa	20.00
Clupeids	12.00
Perches	15.00
Threadfins	12.00
Cat fishes	10.00
Beloneforms	8.0
Etroplus	10.00
Misc.	5.00

Catch Disposal and Fish Price

Since fish is very perishable, fishing communities even in an artisanal fisheries come to engage in trading at an early stage in their development.

Fish catch is disposed in two ways; (i) fish taken for consumption and (ii) fish that is marketed.

Fish Taken for Consumption

Not all fish caught is sold. When an artisanal fishing unit lands, the first allocation of fish is for the consumption needs of the house holds of the crew, the non–crew owners of the fishing unit, the people whose services are generally paid in kind and the old and physically handicapped of the village. But the price of prawn and cuttle fish are so high that no fisherman will give it away as gratis; even fish for consumption for his own house is bought from the market.

Quantitatively, fish taken for consumption varies from region to region and on the type of fishing. It is a function of the crew size and the quantity as well as of the species landed. On an average 5 per cent of the catch is disposed of in this way.

Marketing Structure and Prices of Fish

To a large extent the marketing structure determines the shore price which in turn affects the gross earnings.

The marketing structure at any landing place and its subsequent forward linkage are determined by several factors. These include the main groups of species landed, the accessibility of the sea shore, the number and type of buyers the mode of selling, the socio–religions custom, the proximity of consumption centres or the ultimate market, and these factors are mutually inter–acting. The pattern of marketing structure and linkages is first of all related to the structure of the fishing fleets and their pattern of operation.

The inter–relationship is best seen in the scales of operation, of production and marketing in the different regions. This may be reflected by small–scale distribution system, composed of men carrying 25–50 kg of fish on cycles and of women carrying 5–15 kg on their heads and may be of large-scale carrying 1 to 2 tonne of fish in each lorries.

Another important factor that influences the web of relationships that make up the marketing structure is the mode of selling or more precisely, the organisation of transactions between producers and buyers. Negotiations between producers and buyers are generally conducted through an intermediary, who is either a commission agent or an auctioneer. The function of this intermediary is to facilitate the process of exchange of fish and money at first sales. This is done by mediating in a bargain or by conducting an auction which disposes of the produce at a value mutually acceptable to the fishermen and the buyers. In general the responsibility of the intermediary ceases once the value of the fish placed before him is settled. His services are paid for by the fishermen and form part of their common operating costs.

Buyers of fish on the seashore are generally the women or men who distribute fish by head load, the men who use cycles, the wholesale merchants who use lorries, a limited number of other hawkers and purchasing agents of exporters.

The type and the number of these buyers and the mode of disposal of fish both have a bearing on price and hence the earning of fishing unit.

Bargaining is generally practised where the buyers are fewer in number and exercise an element of monopsony power. Auctioning works when there are numerous buyers and supplies fluctuate, as they do in most fisheries.

The small scale fish distributors are generally in business primarily in the interest of survival and livelihood. They buy small quantities of fish, transport it over short distances, serve a more or less regular clientele and make small profits. Wholesale merchants handle large quantities, move them over long distances indulge in a considerable amount of speculation and attain high profit.

The nature of end market influences the pattern of organisation of the marketing activity and in turn has its bearing on prices at first sale and earnings of the fishing units. Exportable species of fish and shell fish which are of high unit value require, and can support a more capital–intensive and hierarchical marketing structure. In the neighbouring rural hinterlands of fishing villages, purchasing power is low and fish that is to be sold to them necessarily must go through less costly distribution networks and be bought at a lower price at first sale.

The higher average fish prices in city markets are due to:

1. Closeness to capital city, megacity with many high and middle income earners may result in a higher effective demand;
2. High quality fish species in large quantities;
3. The existance of several efficient and genuine fishermen's co-operatives, which have improved fishermen's economic power vis-a-vis intermediaries like money lenders and middlemen; and
4. The less monopolistic and more atomistic marketing structure may result in higher competition and higher fish prices.

The developing nations, as a means to improve their way of life and secure animal protein sufficient to feed their rapidly increasing populations are striving to effectively utilize marine resources and, as a first step, are exerting efforts to promote coastal and offshore fishing converting conventional fishing boats into powered vessels.

In consideration of the enormous amount of time and the huge investment required by large-scale fishing projects, this course of development is regarded more sound from the stand point of "Feasibility".

Mission and Mechanism of Fish Markets

Unlike general commodities, perishables, like fish, meat, fruits, vegetables carry with them the following drawbacks:

1. Quick loss of freshness and deterioration of quality,
2. Difficulty of long storage,
3. Wide variation in amount of supply and sharp fluctuation in price.

This is especially true with fresh fish where, with the exception of cultured products, it is almost impossible to carry out planned production and supply. Merchandising characteristics, such as, these have, rendered the existence of fish markets extremely meaningful. The mission of fish markets is:

1. To protect consumers from unreasonable inconvenience due to price instability;
2. To provide equitable quality evaluation and pricing for the producers;
3. To uphold standards of public health by improving facilities for ensuring product freshness.

In Japan there are three kinds of fish markets:

1. Markets managed by fishermen's co-operatives (producers associations) in local villages;
2. Markets operated by corporations founded jointly by wholesalers and/or producers in large fishing ports;
3. General wholesale fish markets established in both producing and consuming areas of large cities in compliance with the Central Wholesale Market Law. In this case, the local self-government prefectural or metropolitan community is the market founder.

Producers of marine products are generally small and scattered along coastal or inland water areas, of the country. Consumers as well are widely scattered and their demand per unit is small. Thus the three fold function of the central market is:

1. To stabilize supply by collecting fish hauls from the markets of the fishermen's co-operatives in various fishing villages or from fish wholesalers in large fishing ports;
2. To sort fish according to type, shape, freshness and then to set a fair price by auction or some similar method;
3. To expedite prompt and smooth transport of the purchased products.

With respect to the method of establishing the price of fish consignees/wholesalers and middlemen set the prices standing side by side, in every market in both the producing and consuming areas.

Tokyo Wholesale Market (Tsukiji–56 acre Fish Mega Market)

Tsukiji is a fish market though fairly substantial quantities of meat, mushrooms, maple syrup, pickles, potatos, peaches and other foods move through this market every day. Tsukiji ranks at the top

Auction Hall for Blue Fin Tuna, Sukiji Market, Japan

in every measurable category. It handles more than 400 different types of sea food, from peny-per piece sardines to golden brown dried sea slug, caviar a bargain at 473 dollars a pound. It imports from 60 countries on six continents, eel from Taiwan, sea urchin from Oregon, octopus from Athens, crabs from Cartagena, salmon from Santiago, tuna from Tasmania. Tsukiji moves about five million pounds of sea food every day, seven times as much as Paris's Rungis, the world's second largest wholesale market and eleven times the volume of New York City's Fulton Fish Market, the largest fish market in North America. In dollar terms, that comes to 28 million dollars worth of fish per day.

Handling that incoming ocean of sea food is the work of some 60000 people and a fleet of 32000 vehicles that seems to operate in Tsukiji's work day (3 Am to 8 Am) without any fishy aroma. The fish and fish products sold at Tsukiji moves through the market so fast that it is long gone before it starts to smell. The place does its job so well, like a community where everyday work together toward the common goal of moving fish as fast as possible from the sea to the sushi bar or the super market.

The market has a clear hierarchy. At the top of the pecking order are the employees of the seven major first tier wholesalers, who buy up fish around the world and get them to Tokyo. The big seven in turn auction off the daily catch to more than a thousand middle wholesalers who cut, package and deliver the goods, sometimes to yet another tier of distributors, sometimes directly to stores or sushi bars. Associated with the fish market, there is a

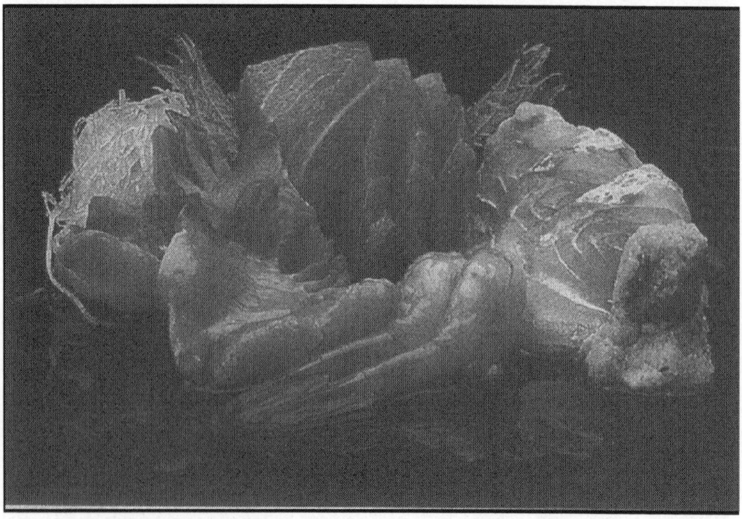

Highly Valued Raw Blue Fin, Sukiji Market, Japan

separate world of small businesses, knife sharpeners, box makers, boot sellers and three dozen restaurants on the site to serve fish sellers.

A torrent of transactions from the auctioneer, who sells 200 tuna in half an hour or about one in every nine seconds. The auctioneer has to recognize the highest bidder instantly. Silent buyers signal bids for numbered tuna with hand gestures. By fingering slivers of flesh beneath a flash light, bidders discern subtle distinctions in fat content and colour, key selling points that sway the price of a premium tuna from 6500 to 11000 dollars and up.

Longer than a man and weighing from 200 to 1000 pounds each, hundreds of tuna arrive in Japan by cargo jet every day. The first–tier wholesalers contract with agents on the charter dock in Miami to buy those big tunas as soon as they reach shore. From the airport the tuna are trucked to Tsukiji and bounced out on the floor of the big tuna shed. They are lined up in long rows, while workers weigh them and label them with bright red characters.

In the crowded market the frozen fish quickly begin defrosting (flash–frozen to 76° below zero) and a cold eerie mist rises from the long lines of tuna. Around 4 A.M. buyers from several hundred second-level tuna wholesalers, who cut a morsel of dark red meat from each tuna, feel it, smell it, check its colour and oil content, constantly making notes on scraps of paper.

There are dozens of auctioneers working for big seven wholesalers and each one has his own chant, his own rhythm. Each auctioneer has to work fast, to sell 200 of them in about half an hour. The auctioneer constantly scans the arcane hand signals of the buyers circled around him, stepping up the pace and his own rate of bounce as the bidding goes higher, till the fish is sold. A bluefin from the cold, rough seas around Tasmania will have different meat than a bluefin from tropical water as so the difference in price.

In fact, virtually all the tuna and more than half of all the seafood Tsukiji sells each day will be eaten raw either sliced into small rectangles as sashimi or placed as the topping on a cube of sushi rice. And it will all be expensive.

Japan is famous for outrageous price and the country's inefficient distribution system is a key reason. The tuna which was caught by an American fishing boat, sold to a Japanese trading company, shipped via air and truck to a first–tier wholesaler at

Tsukiji, sold at auction there to a smaller Tsukiji wholesalers, cut packaged and transferred to various distributors. Delivered to the restaurants throughout central Japan. By the time fish finally got to the end consumers tuna has passed through at least seven intermediary companies, each one taking a profit along the way.

Almost every developed nation is running a trade deficit with Japan and companies around the world face problems getting many goods and services into this rich country. But when it comes to food, Japan is the biggest importer on earth. Tsukiji, of course, is the biggest importer of sea food and people working in the market seem proud that their daily labour helps offset Japan's big trade surplus. Fish has long been the protein staple of sea grit Japan, which consumes more than a tenth of the world catch.

Tsukiji puts heavy emphasis on education to pass along essential skills to the next generation. On any given day there will be classes at the market on topics like auction protocol, knife handling or time tasted techniques for making a spicy kamaboko or fish sausage.

Officials from Tsukiji's Puffer Fish Harmonious Association are teaching the proper way to carve a fillet of puffer fish. Puffer fish is an expensive and cherished delicacy in Japan. Unfortunately, it can also be lethal. Enzymes in organs of fish are fatal to humans. A national license is required for every puffer fish chef. In the class, young candidates are being prepared for the rigid licensing examination. There is a written examination that lasts two hours. In the practical, they hand the candidate a puffer fish, a knife and two pans. In 20 minutes the candidate have to put every poisonous part of the fish in one pan and all the edible parts in the other.

Modernization of Fish Sale: Introduction of Bidding System

The main forces which support the promotion of coastal fisheries are the productive spirit of small–scale and artisanal fishermen and their drive to achieve better position in the society. Because coastal fisheries using small boats have small catches, very different from fisheries involving large boats, which are based upon mass catch and mass sales. The former should seek to increase their profits by catching medium to prime fishes and by selling them at high prices. In fishing villages, their main task is to perfect the selling

Bidding for Blue Fin in Sukiji Market, Japan

system for their catches and the transportation system to big cities, with fisheries co-operative associations as their bases of activities.

However, in areas where the local market in the producing area is still not systematized, or where merchant capital still remains a strong controlling power, efforts for favourable control of fish prices and the maintenance of freshness of catch can not be very successful.

In many fish producing areas, the fishermen are selling the catches to the local middlemen and the middlemen transport the fish to the markets in big cities. In many places the fish price is not determined through bidding. The middlemen in consultation with Aratdar were setting low fish prices which were only benefiting themselves and ignoring the difference in value of each daily catch.

The middlemen for their own advantage, were said to be always setting prices lower than other places, because they fixed the price by agreement without confirming the daily quantitative fluctuations of the catch, the fish size or the quality of the freshness.

For this fishermen's main interest became just to catch more in order to raise their sales, which resulted in their negligence in sorting the catch by size or in trying to maintain their freshness.

The productive spirit of fishermen thus declined and they easily abstained from going fishing on days with bad weather.

The introduction of bidding system has brought about the following results:

1. Increase of sales profit for the producing fishermen as a result of the average fish price increase;
2. Breaking away from the old custom tradition of the fishing village;
3. Increased income for the fisheries co-operatives through sales commission fees, and
4. Active offering of loans for facilities and operation funds to fishermen of the co-operatives.

Finally, the biggest achievement was (e) the improvement in the will to produce of the fishermen. By breaking out the old tendency to catch at random, fishermen are now seeking to maintain the freshness of their catches using fish preserving equipment for live fish, and are placing their efforts to transport high quality and high priced commodities by sorting even the general catches according to type and size.

Furthermore, the middleman side has also improved. With the start of free competition through the bidding system, they have tried to find new market routes. The catches are not only sent to nearby market, but a part of prime fish catches are sent to distant large fish markets.

Distribution System of Marine Products

Japan catches a larger quantity of marine products per capita than any other country of the world, and these products are supplied to the people in the form of fresh fish, frozen products or processed foods, through a complex and sophisticated distribution system. Japanese fish markets are divided between those in the producing areas and those in the consuming areas.

In Japan almost every coastal city, town, or village has some fishing port facilities as well as an established local fishery co-operative association. Every day small coastal fishing boats, belonging to the fishery co-operative set sail from these base ports, returning in the evening or early morning. Usually the fish catches are landed at the fish market conducted by the fishery co-operative.

Fish catches consist of fish kept live in the fishing boat's corf and also dead fish. The live fish are transferred to the market's water tank, while the dead fishes are packed in fish boxes with ice before landing. The price of catches is set by means of competitive bidding by middle men in the presence of staff of the fishery co-operative.

The middlemen who have made the highest bids will once more fill the fish boxes with ice and send the fish off to a large central market in the producing area or else send them directly to fish markets in the consuming areas of large cities by common truck, refrigerated car, live fish car or a special goods train.

To meet the needs of the fishermen and aid in the distribution system, the regional fishery co-operative operates an ice making plant and an ice storage house along with the fish market. The co-operative also has a fish refrigerator so that it has the option to with hold the transport of fishes by storing them temporarily when low prices are expected at the destination markets. Some co-operatives also have fish preserving ponds and processing factories, again to help allow them to make the most profitable use of their products and sell when the price is good.

In general the distribution pattern for marine products in coastal villages is described above. However in villages where the fishery co-operative is weak, individual fishermen will often transport catches directly to the central markets without first passing through their co-operatives market system. Special products, such as, live fish, cultured fish, processed products are in most cases transported through distribution routes established by specialized dealers.

There are local wholesale markets in the consuming areas, in main cities around the country and "nation wide wholesale markets" in the large metropolices. These markets handle large quantities of products gathered from the producing areas through the efforts of middlemen, who sell them in turn to wholesalers by means of modern auction system at these markets, after which the products are sold to retailers. The large majority of marine products are consumed in cities having such wholesale markets, and the distributors have the option to transfer their goods from the nation wide markets to the local markets or vice versa, depending on the price fluctuations.

Fishery markets of developed countries have undergone a great change. The economic function of the fish market has been expanded from a regional economic block to a market of nation wide scale. At present approximately 80 per cent of Japan's annual catch is handled by the nation's fish markets, about one-fifth of which comprises the nation wide distribution network branching out from the central wholesale markets in the six largest cities.

Improved Product Commodity Attracts Large Market Economy

Horse mackerel in Japan is eaten as opened and dried horse mackerel slightly salted.

The method of salting and drying fish begins by slicing open either the abdomen or the back of the fish to dry it without actually cutting it in two. Fish weighing 70 to 120 gram is considered suitable for opening and drying. For shipment of opened and dried horse mackerel to remote markets the fish are refrozen to prevent loss of freshness in transit, and subsequently thawed by the consumer. Heat proof styro–foam containers have made this possible with the result being a vastly enlarged market. Although these are comparatively simply processed marine products, their commodity value is greatly influenced by such delicate factors as how clean the opening is, colour and totality of the meat, the selection of fish of uniform size packed individually in cases etc. These factor consistently spur competition among the producers.

Recently in Japan, the demand of canned fish has shown a sharp increase, particularly mackerel. One of the main causes of this increase in popularity of canned fish is that since the material costs of canned crab, tuna and salmon, which have traditionally been used as canned fish, have risen, causing the processing industry to strive to increase canned fish sales by revising their course of production items like "mackerel meat preserved in salad oil" have appeared on retail markets. As a result, mackerel can now be prepared and eaten in many different ways, boiled, steamed, broiled, fried, stir fried, in salad, in sandwiches what so ever and all to the convenience of the consumers.

Another major application of mackerel is dry processing. The most important thing in fish processing is to reduce body fat as much as possible other wise, the excessive fat in the processed fish will easily oxidize and quicken the deterioration or decay of the fish meat. Japan succeeded in developing good quality products by putting its new technology to practical use.

Generally dried mackerel undergoes a process where in the fish is first boiled down in a special oven, then broiled and dried. Alternately the fish is first boiled down and dried in a hot–air drying machine and then is steamed and boiled down by pressure in a retort. This new process, then, has made possible the manufacture of good quality dried mackerel from fatty mackerel. At the same time, the method has produced a number of secondary merits which surpass the old conventional way.

1. The new method reduces loss of flavour to a minimum.
2. Environmental pollution is eliminated because no fish stock is drained out.
3. The skin of the fish comes off clearly. In the old method it is impossible to completely peel of the skin.
4. The largest quantity of subcutaneous fat can be easily seperated.

Eel Business: A Pride for its Long and Unique Tradition within the Fishing Industry in Japan

In recent years, the consumption of eel by the Japanese has increased greatly. The annual consumption of eel has increased by one to two eels per person on average. The Japanese people of long ago ate eel more as a nourishing meal or medicine than as delicacy. In 15th to 16th century, however, a dish called "kabayaki" (eel broiled with sauce) appeared and established in "kabayaki restaurant", specializing in eel dishes in the cities. "Kabayaki" is an expensive, elaborate dish whose taste depends highly upon the specially skilled cook's techniques. The eel has to be alive until the moment it is cooked and for this purpose, a special channel passing the live eel from producer (eel cultivator) $\rightarrow$ wholesaler in the city $\rightarrow$ specialized restaurant exists. The eel business takes a pride in its long and unique tradition within the fishing industry in Japan.

In Japan most eels are artificially raised. A very tiny elver (eel fry) weighing 0.2 gram grows into an adult eel of about 200 grams.

With present eel culture techniques, the survival rate of the eel during the culture period from elver to adult eel is 50-80 per cent. The marketing size of the eel is 180-250 gram in weight.

If the physical productivity of eel culture is calculated using the standard value of (growth rate ´ survival rate), an average culture farmer would obtain 400-500 kg of adult eel by inputting 1 kg of elver and a competent culture farmer would get about 600 kg of adult eel with the same input.

The economic activity relating to eel production centre in capturing eel fry (elver) and raising marketable eel through culture process.

The problem in culture fishery which has to be solved in the very beginning is how to capture the eel fry which will become the seedlings for culture.

The fishermen normally capture the fry when they approach the coast and are just entering the rivers in the bay or estuary areas. Fry approach the coast after sunset and start ascending the river. Their movement, having a close relationship with the ebb and flow of tide, become active with the flowing tide, and gradually decreases with the ebbing tide. Towards the dawn, the fry stop ascending the river and bury themselves in the sandy mud. This pattern of activity proceeds in accordance with the cyclic movement of the tide.

The size of the catch vary enormously. Because the eel culture solely depends on the supply of natural seedlings, the amount of captured elver totally affects the marketing condition of adult eels one year later.

The marketing of eels in Japan seems complicated; however, basically two methods are in existence, the traditional and the modern, with both forces trying to find out their future directions for development.

As the recent cultured eel production has spread widely throughout the country and the local production has stimulated the local demand, the consumption by big cities in recent years has dropped to as much as 50-60 per cent, of the whole country. Yet with the live eels, the wholesale dealers in the big cities are still in charge of approximately 90 per cent of the total marketed eel, so only after passing through their hands, the eels will be delivered to the local secondary wholesalers.

Along with the eel wholesaler in the major consuming areas, what characterizes the eel marketing is the function of middlemen in the production areas.

The fact that eel culture is performed in two stages (nursing from elver to fry and rearing from fry to adult). Here the middlemen in the producing area play an important intermediary role:

1. Purchase and selling of elver
2. Purchase and selling of young eel and
3. Purchase and selling of adult eel.

In the above process, the middlemen in the producing area maintain a close relationship with the culture farmers. In other words, these middlemen not only support the management of eel culture farmers by their financial power but also take the initiative in price determination at transaction of each stage of 1–3.

Lately along with the increase of eel culture farmers who carry out integrated production from elver up to the adult eel, modernization toward joint purchasing and joint transporting has come to take place by the activities of eel culture co-operative unions. Thus the middlemen's control in the market has become less powerful.

Eel Product in the Market Economy

Four phenomena of increased domestic production, increased import, sales of processed goods and participation of supermarket has altered the situation of market economy with respects of eel and its product.

Demand

Until 30 years ago eels were seasonal goods and 50 per cent of the annual sales was transported during the three months period from June to August, as the Japanese people preferred to eat eels in summer. This eating habit was in accordance with the management methods of speciality restaurants who sell eel dishes in spring, summer and autumn. With the introduction of the age of mass consumption the demand has been spreading through out the year and the sale in summer dropped down to less than 30 per cent.

Problems of Domestic Eel Culture Farmers

Since 1977 the producers price has remained unchanged. The reason is firstly, there is increased production due to lower costs in foreign eel culture fisheries and secondly there is the movement away from fish by consumers opposing the rises in fish prices. Due to increases in fuel, feed and labour costs some farmers are experiencing difficulties.

Problems of Super Markets

Super markets have monopolized the popularization of eel. Since eel preparation (Kabayaki) is a food depending on people's preference, there is a high price flexibility. It is a very suitable strategic sales product. Taking advantage of the difference in production cost, the import of eel from certain countries have been promoted. In order to increase the volume of sales in the future, the maintenance of a stable supply will become the biggest task.

Marketing of Aquaculture Products

Basically the aquaculture industry is divided into three components, like support, production and marketing system.

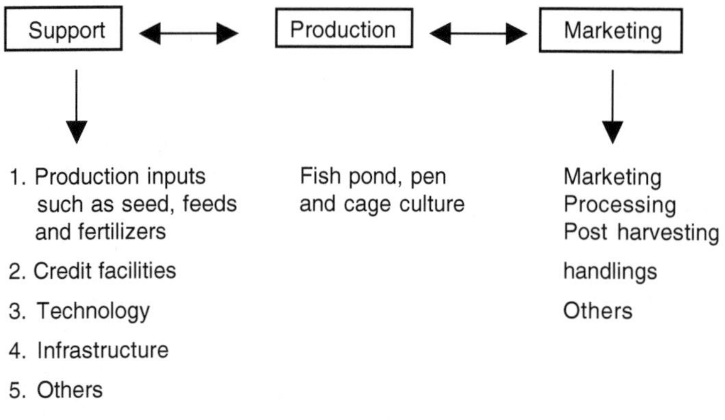

The core of aquaculture is its production, the success of which depend on the availability and efficiency of the support systems such as technology production inputs, credit etc. There has been recorded increase in aquaculture production, extensively and intensively, as a result of the improvement of management of resources available for the use of the industry.

The support systems are provided by the business and government sectors engaged in various enterprises and activities whose outputs serve as inputs to aquaculture production. The business sector provides inputs such as seeds, feeds, fertilizers, technology and credit facilities. The government sector provides infrastructures, credit facilities technology, credit assistance as well as laws and regulations needed to protect the interest of aquafarms.

With the improved productivity in various aquaculture production units, such as, fishponds, fish pen and fish cages, an efficient marketing system need be adopted to measure smooth flow of aquaculture products from the production unit to the ultimate consumer. Marketing therefore, plays a very crucial role in attaining the primary objective of aquaculture of providing the food needed by the population.

Marketing System

Aquaculture product marketing is saddled with a lot of problems starting from the nature of the product itself. Aquaculture products are highly perishable. The duration of time between the actual harvest from the pond to actual consumption affects the quality of the fish and ultimately the price.

The chart below shows a simple marketing system of an aquaculture business enterprise.

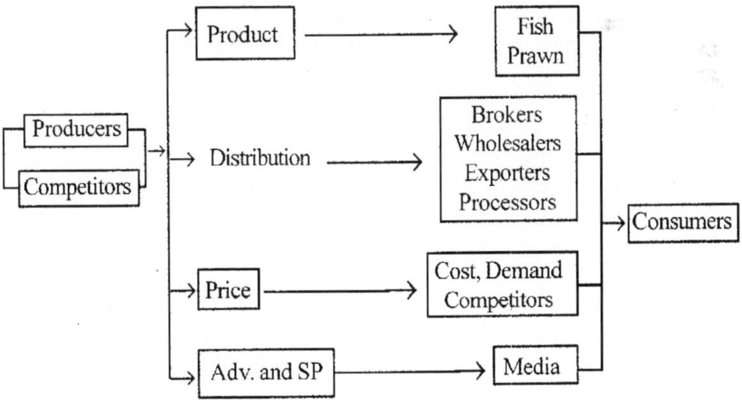

Product Strategy

Unlike commercially manufactured products, where in product quality can be distinguished from one brand to another, aquaculture products do not have the same degree of differences except may be in size and species. However it is claimed that there has been general distinctions in taste of fishes depending upon the source. In retail market, hilsa of Ganges and Padma rivers enjoy premium price over the others because of better taste. Tiger prawn from wild and sea demand better price than the cultured ones. The better taste "distinction of certain species only shows that there is some degree of differentiation in aquaculture product. 50 to 80 per cent higher price is faced by live major and minor carps than the fishes coming from Andhra and other states in Kolkata fish market.

The same is true with prawns. The selling price to exporters depends upon the size and shell hardness. The bigger the sizes, the higher the prices.

Distribution Strategy

From a macro point of view, the success of fish marketing, specially in the area of distribution depends so much on the availability of physical facilities, such as, cold storage and icing, transport facilities, packing, port facilities etc. In agriculture, especially rice it has been noticed that the improvement of rice marketing resulting to price stabilization and better physical distribution of rice on a nation wide scale. The civil supply was able to improve the national rice distribution and price stabilization through a procurement scheme supported by a network of warehousing facilities all over the country.

In aquaculture, the same can be achieved if the government can improve the present facilities for the physical distribution of fish and fish products. This is one area of marketing wherein the producer will depend on the support of the government.

From a micro point of view, the aquafarmer should analyze several criteria in determining the efficiency of a distribution channel to reach a specific target.

Economic Criteria

In choosing a distribution outlet, whether a broker, a wholesaler or an exporter, profit optimization should be a primary consideration. Will the wholesaler give a high price at the least cost of selling? This condition is not always obtained. There are always trade-offs in any trading day depending upon the supply–demand situation in any fish market.

Control Criteria

How much control the producer have over distribution outlet? In the aquaculture business, there is no one producer that can have a high degree of control over a distribution outlet, such as, broker or a wholesaler. The degree of control of a producer (supplier) over a distributor will depend upon the consistency of the former in supplying volume and quality products and probably credit terms extended to the distributor.

Market Coverage

The scope of market coverage in the distribution of aquaculture products will depend upon the network of retailers supplied by the broker/wholesaler. This is critical in the marketing system in view of the perishability of the product. The more retailers there are in a given geographical market, the faster the products will be sold during the day.

Aquaculture products are perishable and it requires more direct marketing to reduce spoilage. The following channels are being predominantly used by commercial fish producers.

MARKETING CHANNELS

Broker → Wholesaler → Retail

Producer → → Consumer

Wholesaler → Retailer

Generally, there are three intermediaries between the producer and the ultimate consumer. Each intermediary gets a share of the final price to the consumer. The more intermediaries there are, the higher the ultimate price and the longer it takes the product to reach the end-consumer.

Pricing Strategy

There are three criteria to be considered in settling pricing of fish products. They are cost, demand and competition.

Cost Criteria

The cost criteria of the product shall include production and marketing cost in general including a reasonable allocation of over head expenses. The cost shall be divided into fixed and variable expenses in order for management to determine the controllable and uncontrollable expenses.

Fixed expenses shall include amortization of fixed assets such as ponds, tanks and major equipments such as pumps, aeators etc. The period of amortization shall depend on the estimated economic life.

Variable expenses are divided as follows

Production Cost

1. Cost of seed stock, fertilizers, feeds and other supplies and materials.
2. Cost of energy.
3. Direct labour.
4. Others.

Marketing Cost

1. Packing and icing materials.
2. Transport and handling.
3. Salaries of salesmen.
4. Commission.

Other Cost, such as, Interest Charges, Overhead Expenses

The company may either adopt full costing or variable costing depending upon the competition. In full costing, the company computes the total cost and adds a reasonable markup. In variable costing, the company sets the price covering only the variable cost plus a reasonable mark-up to cover a portion of fixed cost.

Competition Criteria

Prices must be competitive. The price of similar species of the same quality should be within the same range.

Demand Criteria

In the aquaculture industry, demand–oriented approach to pricing is generally followed. The price of carps varies on the intensity of the demand in any given trading day in the wholesale fish market. In the case of prawns for export, the exporters and processors dictate the price acceptance in the export market.

Some Internationally Traded Fish Commodities

CAVIAR

The commercial exploitation of aquatic species provides food and income in most areas of the world and especially in developing countries and transition economies.

The sturgeon industry is a commercial one, with traditionally high capture and export patterns. Caviar, the unfertilized sturgeon roe is the most important and expensive product from sturgeon fisheries. Three traditional varieties of caviar are beluga taken from the sturgeon, the osetra taken from Russian and Persian sturgeon and the sevruga taken from starry or stellate sturgeon. Some species of sturgeon kept in captivity (white sturgeon and Siberian sturgeon) provide appreciated caviar varieties.

The main processed product from the sturgeon fishery is the unfertilized sturgeon roe, caviar, a gourmet delicacy. The other important processed product is its meat, which may be sold smoked, frozen or marinated. Small quantities of caviar are used as ingredients in cosmetic products such as facial creams. The swim bladders are dried and then used to produce isinglass which is used to clarify wine and beer. Finally, live specimens, namely, of sterlet *Acipenser ruthenus* are used for ornamental purposes and some companies have started marketing handicraft made of sturgeon leather.

The main producers of caviar on a world-wide level are four states bordering Caspian Sea; Azerbaijan, Iran, Kazakhstan and Russia. The fifth Caspian State, Turkmenistan is not a producer country, however Russia and Kazakhstan allocate to it a portion of their yearly sturgeon catch and export quotas.

Caviar is prepared by removing the egg masses from the freshly caught fish and passing them through a fine–mesh to seperate the eggs and remove lumps of tissue and fat. 4 to 6 per cent salt is added to preserve the eggs and bring out the flavour. The high quality caviar with low quantity of salt is known by the trade name *Malossol* (little salt). Salted caviar is packed in cans, glass or porcelain containers. In some cases it is pasteurised to obtain long term storage.

There are three types of caviar, Beluga (from *Huso huso*), Osetra (from *Acipenser gueldenstaedtii* and *A. persicus*) and Sevruga (from *A. stellatus*) They are all graded according to the size of the eggs and the processing method.

1. Grade–I–Firm, large grained, delicate intact, of fine colour and flavour.
2. Grade–II–Fresh caviar with normal grain size of very good colour and flavour.
3. Payusnaya–Pressed caviar.

In the pressed caviar, external factors have caused the fracture of more than 35 per cent of the roe skins before being removed from the fish. The product consists of blend of roes from osetra and sevruga, which is heated to 38°c in a saline solution and stirred until it has absorbed salt and regained its natural colour. It is then put into fabric pouches in which it is pressed to remove excess salt and oil. The resulting pressed caviar appears as a dry, spreadable black paste. It contains four times more roe than fresh caviar of the same weight, as it takes four pounds of fresh caviar to prepare one pound of Payusnaya.

A cheaper caviar product is the *Jastichnaja* obtained from roe that has not been properly seperated from the connective tissues (unripe). It is more salty in flavour and irregular in egg size than other caviar.

Roe comming from a fish other than Acipenseriformes is not caviar but "caviar substitute. Appreciated fish roes include those of salmon, trout, carp, pike, tuna, mullet, cod and other white fish, lump fish and flying fish. Cod roes are marketed fresh or smoked. Smoked cod roes are used to prepare *tarama*, which is a mixture of roes, oil and other ingredients like bread, garlic, lemon juice, pepper, etc. and kaviar *e.g.*, cod roe in tubes. Lump fish roe makes a cheap caviar substitute. Salmon and trout roes may also be considered as relatively up market caviar substitutes. Mullet and tuna roes are processed into dried–salted paste, called battarga in Italy and poutargue in France, a gourmet delicacy. Herring roes are considered as a delicacy in Japan. Fresh and frozen sea–urchin gonads are used to prepare a sauce or may be added to speciality recipes.

In 1976 global trade of caviar substitutes generated some 11.4 million dollar, while trade in caviar generated some 16.5 million dollar. Over the seventies and eighties the value of the trade in caviar, caviar substitutes and other roes followed an upward trend. In 1991 revenues from caviar exports reached 96.7 million dollars, while revenues from exports of caviar substitute and other roes amounted to 85.9 million dollars.

The Soviet Union used to produce and export 90 per cent of caviar entering international trade. Following the break up of Soviet Union in late 1991 the value of the world exports of caviar fell from 96.7 million dollar in 1991 to less than 63.8 million dollar in 1992 and continued to follow a decline until 1998 when export value was less than 21.5 million dollar. The depletion of the resource in its main production basin, the Caspian Sea, was largely responsible for this. However, trade in caviar showed some recovery in 1999, when the export increased to 29.4 million dollar and continued increasing to 48.5 million dollar in 2000.

At the same time trade in caviar substitutes and other fish roes expanded to reach the record value of 160.8 million dollar in 1996. Exports some how declined in value in 1997 (123.4 million dollar) and 1998 (121.4 million dollar) to rise again in 1999 to a value of 144.9 million and on to 157.1 million dollar in 2000. The scarceness of caviar and its rocketing prices led consumers to explore caviar substitutes and other roes, which are sometimes considered a delicacy in themselves.

In Russia caviar industry has been a state monopoly. Subsequently the Soviet State took control of the sturgeon fishery. Prior to the breakup of the Soviet Union, only two states, the Soviet Union and Iran controlled the sturgeon fisheries in the Caspian basin and therefore more than 90 per cent of world caviar production.

The production and export of caviar represented an important source of foreign exchange for both Soviet Union and Iran. From 1976 to 1991, exports of caviar generated average revenues of 24 million dollars for Iran and 19 million dollars for the Soviet Union. Both countries used to channel their exports through a small group of foreign partners.

Four new states, after the dissolution of the Soviet Union, Azerbaijan, Kazakhstan, Russia and Turkmenistan bordering the

Caspian Sea, have been struggling to implement an effective management of sturgeon resources and to fight against caviar poaching and smuggling. According to Tayler, 2001, poachers collectively take some 8000 MT of sturgeon per year, ten times the legally allowed catch quota for the Caspian basin in 2002. The establishment of a large structured parallel industry has been facilitated by:

1. Economic hardship in the region, which turned illegal activities relating to caviar into extremely lucrative options;
2. Difficulties in adapting to CITES enforcement legislations; and
3. Assistance from structured criminal organizations.

In 2000, the main importers of caviar (excluding caviar substitutes) were:

1. The United States, with 90 MT valued to 22.1 million dollars.
2. France with 36 MT valued to 15.9 million dollars.
3. Germany with 34 MT valued to 15.7 million dollars.

Despite its scarcity and high price total imports of caviar increased from 243 MT (valued to 13.8 million dollars) in 1976 to 488 MT (valued to 80.95 million dollars) in 2000.

Retail Price of Caspian Caviar Adapted from Beluga Caviar Pricing Guide 2002 Price for 1.1 lb (0.5 kg) in U.S. dollar

Update	*Supplier*	*Beluga 000*	*Beluga 00*	*Golden Osetra*	*Osetra*	*Sevruga*
1.11.2002	Markys Caviar	781	693	781	507	445
1.11.2002	Paramount Caviar	NA	1188	NA	792	660
1.11.2002	Seattle Caviar company	NA	1496	NA	880	704
1.11.2002	Dean and Deluca	NA	1584	NA	1112	1050
1.11.2002	TsarNicouli Caviar	NA	1254	NA	850	724
1.11.2002	Caviateria	1950	1463	1850	850	812
1.11.2002	Petrossian Caviar	1670	1500	1500	1300	990

Sturgeon stocks are seriously depleted, especially in traditional producing basins, such as, the Caspian Sea. The main factors behind the depletion of Caspian stocks include illegal fishing and habitat degradation. This has generated an increase in official prices for caviar, which in turn has led to a rise in demand for less expensive caviar from poaching and smuggling. However, in some countries, such as, Iran, the strict implementation of the CITES (Convention on International Trade in Endangered Species of Wild Fauna and Flora) regime and tightening of controls on poaching and smuggling have had a beneficial impact against illegal activities.

At the same time, aquaculture is gaining momentum as an alternative to sturgeon capture fisheries. Environmental groups advocate increased production of caviar and sturgeon meat from aquaculture to reduce pressures on wild stocks. However a decline in the demand for Caspian caviar may also entail negative consequences for employment in the Caspian sturgeon industry.

Economic and Social Aspects of the Sturgeon and Caviar Industry in the Caspian Sea During 2000

Country	*Estimated Employment Fishermen*	*Landings MT*	*Aquaculture production and Value (MT)*	*Export Value US $*
Azerbaijan	60	71	NA	403000 in 1997 (caviar) 1309000 in 1999 (caviar and caviar substitute
Iran	150	1000	NA	13785000 in 1996 (caviar) 37413000 in 2000 (caviar and caviar substitutes)
Kazakhstan	170	270	NA	5020000 (caviar)
Russia	130	648	2050 MT 20500000 dollar	17094000 (caviar) 4577000 (caviar and caviar substitutes)

Shark Fin

Sharks have been exploited by humans for various purposes, from food to medicine. But they often have been considered as low-value fish, mainly landed as by catch of other more profitable species.

Shark meat has been consumed and traded since IV century B.C. Despite the high value of some shark products, such as fins, shark have always been considered as low-value fish.

According to FAO Fish statistics, 828364 MT of chondrichthyans were landed, excluding quantity of sharks and skates taken as by catch which were estimated at the end of 1980s at 260000 to 300000 tonnes.

The main processed products from sharks include:

1. Meat, whether fresh, frozen, salted or in brine and smoked;
2. Fins, to prepare shark-fin soup;
3. Liver oil, for cosmetics and pharmaceuticals;
4. Skin, to prepare shark-skin soup, for leather and sand paper;
5. Cartilage, ground to powder and used to produce a supposed anti-cancer cure;
6. Teeth and jaws, in jewellery and sold as curios.

The main commercially-exploited shark species are:

1. The silky shark, *Carcharhinus falciformis*
2. The sandbar shark, *Carcharhinus plumbeus*
3. The basking shark, *Cetorhinus maximus*
4. The tope shark, *Galeorhinus galeus*
5. The shortfin mako shark, *Isurus oxyrinchus*
6. The probeagle, *Lamna nasus*
7. Smooth-hounds, *Mustelus* spp.
8. The blue shark, *Prionace glauca*
9. The whale shark, *Rhincodon typus*
10. The small-spotted catshark, *Scyliorhinus canicula*
11. The piked dogfish, *Squalus acanthias*.

The world's top producers of sharks are Indonesia, Spain, India and Pakistan. Indonesia's catch increased from 1000 MT in 1950 to 11973 MT in 2000. Spain's catch fluctuated 99320 MT in 1997 to 77269 MT in 2000. India's catch reached the peak of 132160 MT in 1996, to decrease to an average of some 74000 MT thereafter. Pakistani shark catch increased from 18243 MT in 1983 to 54958 MT in 1999 and 51170 MT in 2000.

The main catch areas are the Western Central Pacific (144603 MT) the East Indian Ocean (117562 MT), Western Indian Ocean

(114126 MT) and the North East Atlantic (103192 MT). The North West Pacific, once the most productive area, declined from 121700 MT in 1950 to 46494 MT in 1990 recovering slightly to 57103 MT in 2000.

The value of the world chondrichthyan production was estimated at 719 million dollars in 1994, 747 million dollars in 1995, 754 million dollars in 1996, 755 million dollars in 1997, 710 million dollars in 1998, 746 million dollars in 1999 and 742 million dollars in 2000.

Chondrichthyans constitute an extremely important fishery resource for developing countries. Over past 20 years they have gradually increased their role in terms of food and income generation, due to the increase in demand for shark fins. While developed countries catch remained relatively stable around an average of some 220000 MT per year over the entire 1950-2000 period, developing countries catches increased from 76000 MT in 1980 to 575031 MT in 2000 for a value of 515 million dollars.

Shark fines are the most valuable of all shark products and therefore, the main source of income for developing countries. The commercial exploitation of sharks started after the First World War. The belly flaps of piked dogfish started to be marketed in smoked form in Germany and shark meat was introduced into the "fish and chips" trade in the United Kingdom. Despite its nutritional content and appreciable taste, shark meat was considered a poor person's food and sharks were mainly caught, in the fifties, for their vitamin A-rich liver oil. However, the waste of upto 75-80 per cent of raw materials led businesses and countries to improve fishing and processing technologies and marketing and distribution strategies inorder to generate a wider acceptance of shark meat. In the late fifties a wider acceptance has been achieved due to better handling, the use of ice and freezing, the awareness of widespread malnutrition and thus the need to fully utilize all available protein for human nutrition, the contemporary shortage of bony fish in some areas and the marketing efforts to promote shark meat (Vannuccini,1999).

Shortfin mako shark is considered the World's best quality sharkmeat. It is marketed in the United States and Europe. Other largely appreciated species are thresher (Alopias spp) and porbeagle. The meat of smaller species like dogfish is also appreciated as it contains smaller amount of urea and mercury than other species and is also easier to process. The backs of these sharks are marketed in Europe and Australia as fillets, steaks, portions and used in the "fish and chips" trade. The fresh whole carcasses are marketed in South America. Other important sharks for the production of meat are requiem sharks.

Non-food uses of sharks include shark liver oil products, cartilage, skin and teeth. The shark's liver is saturated with oil to maintain its buoyancy in water. Shark liver oil has been traditionally used as a lubricant in the tanning and textile industry, in cosmetics, skin healing and other health products, as a preservative against marine fouling of wooden boats, as fuel for street lamps and to produce vitamin A, before synthetic vitamin A was discovered. Currently, demand is mainly for squalene a highly unsaturated alipuatic hydrocarbon, present in certain shark liver oils (Fam-*Squalidae*). Squalene is used as a bactericide, organic colouring matter, rubber, chemical, aromatics, in the textile industry, as an additive in pharmaceutical preparations, cosmetics and health food. A related compound of squalene is squalene, a saturated hydrocarbon obtained by hydrogenation of squalene. Squalene is used in skin care products as it is a natural emollient.

Shark cartilage, processed into powder and tablets, is used as a health supplement and alternative cure for several disease and beneficial in inhibiting the growth of tumours by impeding the vascularization of malignant tissues (angio genesis). Cartilage from blue sharks is believed to be of the best quality as it is richer in chondroitin than other species. Chondroitin is an acid mucopolysaccharide used for various health problems.

Shark skin is used to produce leather. There was a good market until a few years ago, when leather from shark was used to produce handbags, shoes, wallets, cigeratte cases, watch straps, coin and key fobs. With the increase in the market for shark meat, shark skin lost its commercial importance. In fact, shark carcasses are sold with the skin intact in order to protect the meat and prevent oxidation. Furthermore, sharks have to be bled, dressed and iced immediately after catch to prevent urea from contaminating the meat, but exposure to fresh water and ice damages the skin. Therefore now a days the market for shark leather is limited.

Other non-food uses of shark include the sale of teeth and jaws in jewellery and as curios, the use of certain shark parts in traditional medicine, aquarium trade, production of fish meal and glue.

World trade in sharks and dogfishes increased from 19908 MT in 1976 for a value of 34.7 million dollars to 78652 MT valued to 269.6 million dollars. Main exported products are shark fins in terms of value (88.45 million dollars in 2000) and frozen shark in terms of quantity (37259 MT in 2000). Main exporters are Spain in terms of quantity (16539 MT exported in 2000) and China in terms of value with 55 million dollars worth shark exports.

Total imports of shark and dogfish commodities increased from 24228 MT worth 47.6 million dollars to 76253 MT valued to 182 million dollars in 2000. The main importers in 2000 were Spain (13913 MT), Italy (13708 MT) and China (8599 MT)

Shark fins are among the most expensive fish products in the world, with prices quoted from 45 to 88 dollars/kg in the Singapore market. They are processed and marketed in the following way:

1. Wet (fresh, chilled and unprocessed);
2. Dried, (complete with denticles and cartilaginous platelets);
3. Semi-prepared (with the skin being removed but the fibres still intact)
4. Fully prepared, frozen, in brine, and as fine nets, *i.e.* with cartilaginous fin needles being boiled, seperated, re-dried and packaged in loose groupings.

Shark fins are classified as "black or white". There is no distinction between black and white. Some traders say that it is a description of colour of the fin, other that it depends on the depths in which the sharks live. But the fins of white groups give a higher percentage of fin needles and have a better flavour.

World trade in shark fins increased from 2666 MT in 1976 worth 13 million dollars to 5181 MT in 2000 valued to 116.6 million dollars. Exports peaked in terms of quantity in 1996 (6396 MT) and in value terms in 1997 (140.8 million dollars)

The price for headed and gutted chilled shark is 1.01 to 2.75 dollar per kg at the Honolulu market in the Hawaii.

Shark fins are sold in Singapore market in the following rates:

1. A full set of oceanic whitetip (*Carcharhinus longimanus*) fins, half moon cut, from the South Pacific–57 dollars/kg–whole sale;
2. A full set of blue shark fintip, half moon cut, from South Pacific–47 dollars/kg–wholesale;
3. A full set of mako shark fins tip, half moon cut from South Pacific–45 dollar/kg–wholesale;
4. A whole set of white shark fins from Austratia–88 dollar/kg wholesale.

Economic and Social Aspects of Shark Fishery in 2000

Country	*Estimated Employment (Fishermen)*	*Landings (MT)*	*Export Quantity (MT)*	*Export Value (US $)*
China	1500	10 to 15000	2237	55020000
Costa Rica	NA	5453	3987	13125000
Indonesia	NA	111973	1313	13280000
Japan	NA	33072	3818	19890000
Senegal	NA	10757	37	4331000
Spain	NA	77269	16539	42675000
Yemen	NA	5100	371	13855000

Chapter 7

The Co-operatives

Movement of Co-operation: Its Origin and History

"Economics" defines Prof. Marshall "is a study of mankind in the ordinary business of life" and co-operation says CR Fay "is one way of conducting certain parts of this business". The earliest concept of co-operation was probably given by Plato, when he said "In seeking the good of others, we find our own". The concept of money, property and their ownership gave rise to the inequality in the distribution of incomes amongst the people. Though all started the race together, some secured advantage over the others, which resulted into an economic system known as "Capitalism", Capitalism established itself with age, become too advanced and showed symptoms of decay. A variety of malpractices, maladjustments and exploitations began to have a free play in the system. Wealth concentrated in a few hands and gave them social strength and political power; as a result of which the mass became poor and poor, practically slave to the previlaged class. Hence an alternative economic system had to be thought, which gave rise the idea of socialism and communism at a later stage.

Socialism is a scheme of social organization which places the means of production and distribution in the hands of community and replaces competition by association. It acts in a way which is conducive to the people in general. A socialistic state is, therefore, analogous to a welfare state and it aims in the even distribution of wealth and opportunities.

Communism is a step further on socialism. It is a thought introduced during the second half of the 19th century by Karl Marx. He felt that capitalism possessed the seed of its own destruction and that the class struggle which is the natural outcome of the capitalism would increase to such an extent that the system itself would perish in its own fury. Marx precluded that surplus value in any industry was created by the labourers and not by the capitalist and hence a major portion of the profit should be given to the prolatariat class. He further claimed that political power should be vested on the working class and private property should be abolished.

Among these three divergent economic systems, co-operation acts as a balancing factor. It is said to be a "double-edged axe which strikes at the same time, at the dead obstructions of the socialistic state and at the sterility of individualism".

Co-operative Movement–Defination

Co-operative movement was born out of adversity. The circumstances which gave rise to adversity were different at different places. The movement, therefore, assumed diverse shapes, commensurate with the environments which gave birth to it. It is, therefore, difficult to search out a defination of co-operation. Some co-operators, have, however, tried to define it in the context of circumstances in which the movement was studied by them.

Mr. C. R. Fay defines, a co-operative society "as an association of persons varying in number, who are grapplings with some economic difficulties and who, by joining together on a basis of equal rights and obligations endeavour to solve their own difficulties mainly by conducting at their own risk. It is a joint undertaking to which they have transferred certain economic functions corresponding to their common need and by utilizing this undertaking for their common mutual and moral benifit"

Mr. V. L. Verma, a veteran co-operator of India states "co-operation is only one aspect of vast movement which promotes voluntary association of individuals having common needs, who combine towards the achievement of common economic ends".

Sir M. L. Darling describes "co-operation as something more than a system. It is a spirit which appeals to the heart and mind. It is a religion applied to business. It is a gospel of self–sufficiency and service".

Dr. K. N. Katju in the following defination tries to explain co-operation in the context of Indian conditions, where credit societies claim a major share in the movement.

"Co-operation is self–help as well as mutual help. It is a joint enterprise of whose, who are not financially strong and cannot stand on their own legs, and are therefore, come together not with a view to get profits, but to overcome disability arising out of want of adequate financial resources and thus better their economic conditions."

The Maclegan committee described co-operation thus, "The theory of co-operation is, very briefly, that an isolated and powerless man, can, by association with others and by moral development and mutual support obtain, in his degree, the material advantage available to the wealthy and powerful persons, and thereby develop himself to the fullest extent of his natural articles. By the union of forces, material advancement is secured, and by united action self-reliance is fostered, and it is from the interaction of these influences that is hoped to attain the effective realisation of the higher and more prosperous standard of life which has been characterised as better business, better farming and better living".

Mr. M. T. Herrick defines co-operation as "the act of persons voluntarily united, for utilizing reciprocally their own forces, resources or both under their mutual agreement to their common profit or loss.

Mr. H. Calvert, an illustrious Registrar of co-operative societies of Punjab defined co-operation as "a form of *organization,* wherein persons *voluntarily* associate together as *humanbeings* on a basis of equality for promotion of their economic interests and themselves."

Summerizing all these, it may be stated that the movement of co-operation is guided by the following principles:

1. Voluntary association.
2. Democratic management (1 member 1 vote).
3. Open door system of admission.

4. Self–help through mutual help.
5. Limitation on utilization of capital.
6. Political and religious neutrality.
7. No profit motive.

It is interesting to note that in co-operation one can find an amalgamation of conflicting interest. The employer is his own employee, the purchaser is his supplier, the borrower is his own lender and producer is his own consumer. It is the absence of diverging interests, which accounts for the motive of exploitation in it.

In the International co-operative congress held in Vienna in September, 1966, the following principles were adopted as the guiding principles for co-operative societies:

1. Membership of a co-operative society should be voluntary and available without arteficial restriction of any social, political, racial or religious discrimination to all persons who can make use of its services and are willing to accept the responsibilities of membership.
2. Co-operative societies are democratic organisation. Their affairs should be administered by persons elected or appointed in a manner agreed by the members and accountable to them. Members of primary societies should enjoy equal rights of voting (one member one vote) and participation on decisions affecting their societies. In other than primary societies, the administration should be conducted on a democratic basis in a suitable form.
3. Share capital should only receive a strictly limited rate of interest, if any.
4. Surplus of savings if any, arising out of the operations of a society belong to the members of the society and should be distributed in such a manner as would avoid one member gaining at the expense of others. This may be done by decision of the member as described below:

   (*i*) By provision for development of the business of the co-operative;

   (*ii*) By provision of common service, or;

(*iii*) By distributing among the members in proportion to their transactions with the society.

5. All co-operative societies should make provision for the education of their members, officers, employees and of general public, in the principles and techniques of co-operation, both economic and democratic.
6. All the co-operative organizations in order to best serve the interests of their members and their communities, should actively co-operate in every practical way with other co-operatives at local, national and international levels.

Birth of Co-operative Movement

Co-operative movement in England, which has shown light to the whole world, was the brain–child of an exploited class of people during the "Industrial Revolution".

In the first half of eighteenth century, conditions in agriculture and industry in England could be favourably considered to those in any under–developed country. Agriculture was carried on for mere subsistance by primitive methods and industry was confined to small artisans, who produce goods in their cottage unaided by any machinery.

But during the middle of 19th century, far reaching changes took place. The small fields changed ownership from small farmers to big capitalist cultivators, mechanized farming was introduced. New inventions in industrial field were followed by big factories, the quality and cheapness of whose products ousted the cottage industry.

As a result, working class was thrown at the mercy of factory owners. Their conditions were extremely miserable, their working hours were long and wages small. By and by their conditions came down to such a point that the working class was forced to react. Riots, massacres and revolts directed against the mill owners, became the order of the day.

Amidst this, turmoil, emerged a number of reformists who tried various means to improve employer–employee relationship. The name of Robert Owen is outstanding among them. He made a series of experiments for improving the condition of working class, like labour colonies, equitable labour exchange etc.; but could not, unfortunately, meet with any tangible success. He gave the idea of "commune" for the first time to the world by way of starting a coloney, consisting of 30,000 acres of land and 900 members. But the faith of Robert Owen in the goodness of every individual did not prove to be true in practice and the selfishness of the members led towards the self liquidation of the coloney. Although Owen's life has been said to be a life of "poor performance", it created a pool of experience, made people wiser and set them thinking about new ideas.

Inspite of the efforts made here and there, the conditions of workers remained nearly unchanged. They formed trade unions to settle terms with employers and various methods including strikes were tried but could not meet with desired success.

Later, one afternoon in 1843, twenty eight weavers met in the Charlist Hall, London. One of them Mr. Howrath Charles suggested that it was difficult to get wages increased, but they could certainly decrease their expenditure by making collective purchases. It was decided that they should start a store wherein all sales would be made on cash at market rates and only such members would be taken who may pledge their loyalty to the store. These 28 persons saved one pound each in full one year and in 1844 started. "The Rochdale Equitable Pioneer's Society", registered under Friendly Societies Act and rented a shop in Toad Lane at an yearly rent of ten pound. With the efforts and loyalty of members, the store made a speedy progress. In the very first year, the membership rose from 28 to 74. By 1851, the store had a hole time manager and sufficient funds. In 1852, the store started its own shoe making and tailoring departments. In 1888, the store had 1123 members and £3.4 lakhs as share capital. At the end of 1959, it had 93 branch stores, 40,000 member consumers and distributed about £2.5 lakhs as dividend every year. The Society, at present, has three magnificent central buildings and owns a bakery, so big as to turn out 1200 loves of bread per hour with a turn over of £1 million. The original store exists even to day symbolizing monumental success in the field of co-operation.

Almost simultaneously with the movement in England, the movement of co-operation started in Germany and put a land mark in the field of its progress. This has been followed by Italy, Canada, Denmark, Norway, Japan, China, France and many other countries of the World. It has widened in area, enlarged the field of activities and as it stands now, it has turned to be an international movement revolutionising the present day economic system. A new expression "Co-operative commonwealth" has come to epitomise and symbolize an ideal to which has attracted the alligiance of several thinkers and political parties all over the world.

Movement of Co-operation of India

The concept of co-operation is very nicely depicted in " Joint Family System", wherein an excellent example of practical co-operation has been found. Members of such a family earn and spend jointly. They have common ties and are joint in food, worship and estate.

In its modern form co-operation was introduced in India by the Government, in order to save the people from money lenders and their malpractices. A big majority of Indian people had since ages been living in villages and developing for their living on agriculture– a profession in which unforeseen expenses could not be ruled out. Failure of crop, natural calamities, cattle mortality and other such cases made it necessary for the cultivators to borrow. During all these ages, little attention was paid towards education of masses and being illiterate they invariably remained ignorant of developments in agriculture and hence involved in poverty and debt.

During pre-British period, however, borrowing remained at a very low level because spare money for lending purposes was very little. There was not much legal protection to the lenders and machinery for recovering loans did not go beyond local sanctions. The borrowing power of the people was also negligible as they had little repaying capacity. The value of the land was very little and they had no other property to offer as security. Naturally the amount of debt remained very small. During British rule, security of property increased and legal protection to the money lenders was assured. The amount of debt, therefore, rose higher and higher and malpractices crept into the profession. A major defect of this indebtedness was that most of the loan was given and used for unproductive purpose.

Soon things began to take a serious turn. In 1875, a large number of residents in Poona and Ahmedabad areas rose in open hostality against the money lenders and broke out in violence, which could only be quelled with the help of army. The problem began to be viewed with seriousness after the Finance Commission in 1901, reported that, " Sahukars has from being a help to the agriculture become in some places an inculus on it. The high rate of interest that they charge and the unfair advantages that they take of the cultivators necessities and ignorance have, over large areas, placed a burden of indebtedness on the cultivator which he can not bear. Passed on from father to son and continually swollen in the process of compound interest, the burden of indebtedness has become hereditary and retain the cultivating classes in poverty from which there is no escape that we can perceive". Mean while the Government of Madras, considering co-operation as a possible remidy deputed Mr. Frederic Nicholson to give his report on the possibility of introducing co-operative movement in India, as a measure to remove indebtedness. Mr. Nicholson accordingly studied the theory and practice of agricultural banks in Europe, particularly in Germany and in his report submitted in 1877, suggested ways and means by which the co-operative movement could be popularised in India.

In the mean time, some experiments had started on local initiative, a few of which are as follows:

In 1850, Mutual Aid Fund was established to give persons, with a fixed income, a chance to borrow money in the time of need at equated rate of interest. The fund thus started continued upto 1857. When accounts were made after 7 years, it was found that every member who had contributed Rs. 84 had been repaid Rs. 102=50. The next step was taken to make this type of society a permanent features. In 1901, there were 200 such societies called "Nidhis" with a membership of 36000 and a working capital of 2 crores. This "Nidhis" which were founded mostly in Uttar Pradesh and Bengal suggested the possibility of introducing " Co-operation in India."

Another excellent example of early co-operation may be found in a society, called "Panjawar Society", started in 1892, in the village Panjawar of the Hoshiarpur district of Punjab. The society carried on very well till 1920. During the life of the society, the members received a profit of Rs. 51000 and about Rs. 12000 was spent on improving the condition of the village.

Mention may be made of Sir William Wedderburn, a magistrate of Poona who started the first Agricultural Bank in India, in the year 1882. In Mysore, the landlords formed an association in 1894 to give loan to members for agricultural development of land. Similar banks were started by Lord Macdonald in Uttar Pradesh. Mr. Dupernex, I.C.S on being deputed by the government of Uttar Pradesh to enquire into the possibilities of establishing Agricultural Banks, published his report in 1900. The report is known as "Peoples bank for Northern India".

With all these backgrounds, the government of India realising the necessity of introducing the co-operative movement appointed a special committee under the chairmanship of Sir Edward Law to make suitable proposals for enacting a special law for this purpose. The result was the passing of the co-operative credit Society Act of 1904. This Act provided for the registration of credit societies of primary nature which were classified into two groups, namely, Urban and Rural. Thus began the era of co-operation in the Indian economy.

Fishermen's Co-operatives

Fishermen, all over the world, suffer from short commings like, limited education, weaker organising capacity and lesser capital forming ability; and the only way to get over these disabilities is to derive strength through establishment of co-operative organisations of fishermen. Most countries have both marine, as well as, fresh water fisheries co-operatives. Their problems are different, so also are their solutions. Similarly these societies can be divided into (*a*) credit or thrift societies and (*b*) multipurpose or service societies. These societies, if managed efficiently, will serve the cause of the fishermen in different ways in many countries.

The history of Fishermen's co-operatives in India is traceable to the report of the Royal Commission on Agriculture (1927) which recommended that fisheries should be placed on a sound footing. By 1944, there were 200 Fishermen's Co-operative Societies in the country. As a result of the recommendations of the "Fish sub-committee" of the Agricultural Policy Committee (1944) and the Saraiya Committee (1946), set up by the Ministry of Agriculture, the organisation of Fishermen's Co-operatives received a further impetus.

Control and Structure of Fishermen's Co-operatives

In, India, jurisdiction over co-operatives including Fishermen's Co-operatives rests with the State Governments and Union

Territories. Responsibility for the development of inland and marine fisheries within territorial limits also vests with them. Accordingly the work relating to the organization and implementation of co-operative fisheries programmes is in the hands of state governments and union territories. The Union Ministry of Agriculture performs the function of co-ordinating and promoting the activities from a national angle. Primary Fishermen's co-operative Societies consisting of individual fishermen as members constitute the base of the organizational structure of Fishermen's Co-operatives. The Primary societies are affiliated to District or Central Co-operative Fishermen's Federations (in several states) There are now 56 such federations in the country. These federations are again affiliated to the state level Apex Societies as in the states of West Bengal, Bihar, Gujrat, Maharashtra and Andhra Pradesh. The main function of the Federation is to assist the Primaries in securing credit and providing necessary supplies and arranging technical advice. Like wise, the Apex Bodies help the Federations and sometimes the primaries directly. Apex Societies and Federations also undertake processing and marketing activities. Federations and Apex bodies also implement Co-operative Fisheries Projects through primaries or individual members directly enrolled by it. The Director Fisheries function as the Registrar of co-operative societies, so far as fishermen's societies are concerned in states like Andhra Pradesh, Kerala, Orissa and Tamilnadu. In other states fishermen's societies are controlled by the Registrar of co-operative Societies, but in active consultation of Fisheries Department. In order to ensure a single line of responsibility, the government of India advised state governments to transfer the work relating to the Fishermen co-operatives to the Fisheries Department from co-operative Department.

In addition to the process of setting up of co-ordination cells at state level for the promotion of activities of Fishermen's co-operatives, which is already set in motion, it is felt that an arrangement at national level, to pool up and analyse the problems faced in various states in the implementation of co-operative fisheries programmes, to formulate remedial measures and advise the states suitably will be useful.

Progress of Fishermen's Co-operatives in the Country

During 1940-61 the fishermen's co-operative societies were organised with the main objective of rendering credit facilities to the fishermen to tide over lean fishing seasons, and to free them from the clutches of fish merchants and middlemen. Thereafter, during the Third Five Year Plan (1961-66) emphasis shifted to the linking of co-operative credit to the various activities connected with fish production and assisting this depressed and disorganised sector of fishermen, (who depend heavily on merchants and middlemen) financially, as well as, organisationally. The government of India granted financial assistance to the State Governments during the period to enable them to provide subsidies to Fishermen co-operatives towards the share capital, working capital, managerial costs etc. During the three subsequent annual plans (1966-69) and the Fourth Five Year Plan emphasis has been laid on assisting viable Fishermen's co-operatives in the formulation and implementation of Integrated marine fisheries projects with financial assistance from the Agricultural Refinance Corporation. In regard to inland fisheries, most of the state governments have adopted the policy of granting long–term leases (5-15 years) at concessional rents to fishermen's co-operatives. Ending 1970-71, there were 4721 Primary Fishermen Co-operatives with a membership of 4,43283 and having a total share capital of Rs. 13.436 million, and a working capital of Rs. 59.332 million. The value of fish catches landed was Rs. 26.463 million and the turnover inclusive of fish sales and sales of fishery requisites to members etc. was Rs. 44.607 million. Of the 4721 societies, 1246 earned profits (Rs. 2.235 million) and 2046 societies sustained losses (Rs. 4.786 million). The State Governments contributed Rs. 4.692 million towards share capital and gave Rs. 2.289 million as loans and subsidies.

Kerala state has the largest number of societies (948), followed by Andhra Pradesh (598), West Bengal (556) Tamil Nadu (442), Maharashtra (381), Assam (343), Bihar (287), Madhya Pradesh (162), Orissa (154), Mysore (121), Uttar Pradesh (97) and Gujrat (59). The remaining states and Union Territories have societies between 39 in Manipur and 2 in Delhi. 80 per cent of the Fishermen's co-operative are situated in coastal states, totalling 3259 out of 4271. With regard to membership, Kerala top the list with 112007 members, followed by Tamilnadu (65,592), Maharashtra (57137), Andhra Pradesh (55561), West Bengal (32841), Mysore (23780), Bihar (21636), Orissa (19732), Assam (18753) and Gujrat (11891). In the remaining states and Union Territories the membership varied from 6313 in Madhya Pradesh to 28 in Hariyana.

Formation of Society

Formation and registration of a society depends on the co-operative law obtainable or prevailing in a particular region. In India, for example, only 10 members is the minimum requirement for formation of a co-operative society. They have to collect at least Rs. 500 as initial share capital and deposit that amount in the nearest co-operative bank in the name of one or two promotors out of the first 10 members. These 10 members are treated as promotors and they apply to the Registrar of Co-operative Societies or his representative in the area concerned for registration. The promotors usually form the first executive committee. The other requirements and by laws for the formation of the co-operative society depend on the standard by-laws prevailing in the area. These by-laws are given by the Registrar's office and the members have to sign them in acceptance. These details vary from state to state and from country to country.

Objectives of the society have to be exhaustive, clear and indicated right from the beginning and they have to be approved by the Registrar's office.

Formation of Working Capital

Requirement of capital for the working of the society depends on the projects to be persued in fulfilment of the objectives of the society. Consequently the members have to draw up a statement of the requirement of capital and approach co-operative bank with the Registrar's recommendation for short term loans. Borrowing limit for fisheries co-operatives is ten times the share capital and can be extended to 15 times with special permission of the Registrar. If fund for heavy capital investments are required, other sources like government departments or Agriculture Refinance Corporation have to be approached.

Management

Management of the co-operative society is usually in the hands of a small body of persons, styled as 'Board of Directors or as Executive Committee, intended to supervise the working of the society. The committee is assisted by a manager, who is the head of the staff of the society, working under him. It is actually the manager, who is in charge of day to day affairs and conduct the business in consultation with the chairman of the Executive Committee. In smaller societies, it is the secretary of the society who looks after all these functions. The strength of the staff depends on the nature and extent of activities undertaken.

Ownership and Profits

As regards ownership, it can be stated that the assets of the society are the joint property of the shareholders. Here, if a division of profits has to be made, it should be in the proportion of number of shares a member holds individually in the joint ownership. The deficits also are divided similarly. But the losses or liabilities are limited to the shares of the individuals and are not a burden on the properties.

Though profits are divided in relation to number of shares there are certain statutory liabilities which have first to be fulfilled based on the provisions of the co-operative Act under which a particular society is registered and its by-laws before declaration of dividents. These liabilities are:

1. Contribution for co-operative education;
2. Reserve fund for the society;
3. Dividend equalisation fund for future purposes;
4. Building fund, etc.

As many be determined by the by-laws. Further, percentage of profits in co-operative venture is limited to 9 per cent only, extra profits, if any have to be reserved for future purposes. Thus being a co-operative venture, the profits have their relation to the objectives–mainly the well being of the members, improvement of their social and economic conditions and provisions for future progressive business; the actual financial returns or the net profit being a secondary consideration.

Member Education

For successful operation and management of a co-operative society, the members constituting the body should have proper co-operative attitude or co-operative frame of mind. In developing countries, especially in fishermen's co-operative societies, the fishermen members may not be literate. The promotors, executive committee members and other intellectuals supporting the organisation have to convince other members in this respect by good precepts and practices. Though this calls for very special efforts in fishermen communities, it is not very difficult if community leaders are convinced first.

Problem of Fishermen's Societies

Marine fishermen's co-operative societies have to face problem of their own, especially because the member fishermen have to spend large part of their time out on the sea for the purpose of fishing. They are on the land only for a short time. Hence if his society is undertaking marketing and that too in an efficient manner, it is a matter of great help to him. Secondly, the fishermen has to depend on some organisation for his fishing requirements, such as, accessories required for his fishing vocation. This comes in handy if there is a co-operative society which procures these on bulk basis and supply them to the members. If these are supplied at rates cheaper than the local merchants can supply, it certainly works as incentives to him. It happens that on account of certain statutory requirements, advances required for member fishermen from banks etc. are not immediately available to him and he feels annoyed, because on the other side, the fish merchants are prepare to advance him money without much difficulty or delay. His formalities are few and he can observe them only when required depending on the exigencies of his business. But the fishermen should not forget that for this easily obtainable advance he has to pledge his entire season's catch to the merchant and sometimes at a predetermined low price. In such cases the fishermen member has to be wise and use his judgement. The office bearers of the society have also to exert their influence to reduce the hardship of the members in obtaining advances also to convince him of the advantages of marketing through co-operative channels.

Incentives to Fishermen Members

One of the major problems in the co-operative sector is lack of proper and effective working. Such incentives are all the more essential in the case of fishermen's co-operatives, but their nature will depend on the working of the individual society, and the nature of business transactions. However, the following incentives for fishermen's societies needs worth consideration:

1. Rebate on quantities of fish sent for marketing, special gifts and concessions.

2. Cash advances for non-fishing periods.
3. Cash advances for fish-scarcity periods.
4. Supply of fishery requisites on credit.
5. Fishery requisites to be at cheaper rates than in the market.
6. Care and diligence in handling marketable fish, thus reducing losses
7. Combined shore operations without payment in cash for beaching of boats, their launching, painting etc. In inland fisheries, raising and repairing bunds; out-lets etc as in production of fish seed.
8. Transport of catch on the sea by groups of members, in turn, so that others can continue to fish.
9. Special rates for quality fish seed produced by members.
10. Leases of fishing rights of waters areas taken by the society and members charged nominally for fishing permits.
11. Welfare schemes for widows and disabled members.

Credit Facilities

Credit facilities for fishermen's co-operative society depend on the existence of financial institutions in a particular area. In some places there are separate fisheries banks, which provide credit to fishermen's co-operative societies and the fishing trade, where as, in other places there are State Co-operative Banks as well as Refinance Corporations, which extend these facilities. In less developed areas, Government extend credit facilities to co-operative societies for investment, such as establishment of ice and cold shortage plants, boat building yards, mechanisation of fishing crafts etc.

Education and Extension

Proper management of fishermen's co-operative societies which will lead to socio-economic emancipation of the member fishermen requires proper understanding of the procedures involved in functioning of the societies and the observance of rules and regulations by the members of the office bearers. For this purpose, Managers, Secretaries and other office bearers must be properly trained in the methods of co-operative functioning. Similarly, there must be inflow of new members to the executive committee, so that more and more members will be acquinted with the procedure and as well as the nature of the business of the society.

Marine Fishermen Co-operative Societies–Case Study

Gujrat Fisheries Central Co-operative Association Ltd.

The Gujrat Fisheries Central Co-operative Association Ltd. (G.F.C.C.A) is a State level Apex Co-operatieve Fisheries organisation having its jurisdiction all over Gujrat and branches for processing and marketing at Bombay, Delhi, Verabal, Porebandar, Palanpur, Okha, and Jafrabad. The membership of the association includes 57 fishermen producers co-operative societies of Gujrat state, 908 individual members, besides the Government of Gujrat, who are controlling a major percentage of shares.

The authorised share capital of the Association is Rs. 15 lakhs, of which Rs. 9,80,960 has been paid up. Out of the paid up share capital, Rs. 8 lakhs are that of share purchased and owned by Government of Gujrat and the rest are owned by member co-operative societies and individual members.

The Association was registered during the year 1956 under the then Saurashtra Co-operative Act, which subsequent to the formation of the State of Gujrat, during the year 1960 was renamed as Messrs. Gujrat Fisheries Central Co-operative Association Ltd.

The management of the Association is vested on a Board of Directors consisting of 21 members, out of which seven are nominated members as detailed below:

1. Commissioner of fisheries, Ahmedabad (Chairman)
2. Commissioner of Industries, Ahmedabad.
3. Registar of Co-operatve Societies, Ahmedbad.
4. Deputy Secretary and Financial Advisor, Department of Agriculture and Co-operation.
5. Managing Director, Gujrat State Financing Corporation.
6. Managing Director, Gujrat State Co-operative Bank.
7. General Manager of the Association–Ex-officio member.

Rest forteen members are elected from amidst the individual and co-operative societies.

The operating organization structure of Gujrat Fisheries Central Co-operative Association Limited are shown in Figure on next page.

- General Manager Ahmedabad
  - Senior Manager Verabal
    - (a) Manager Porebandar (Fresh and Dry fish)
    - (b) Manager Jafrabad (Dry fish)
    - (c) Manager Verabal (Fresh and Dry fish)
    - (d) Asst. Manager, Okha (Fresh and Dry fish)
    - (e) Manager Verabal, Trading (Nylon twine and net sales)
  - Senior Manager Production Manager Asst. Production Manager Ahmedabad, nylon net plant
  - Senior Manager Ahmedabad Asst.. Branch Manager
    - (a) Manager Palanpur, Fresh water fish production distribution to Bombay, Delhi, and Ahmedabad)
    - (b) Manager, Delhi (sales)
  - Senior Manager Bombay
    - (a) Freezing
    - (b) Fresh fish
    - (c) Dry fish

The Association has taken up trading activities, industrial complexes and service institutions within the State with a view to develop marine fishing industry of the state to stabilize price of fish to the producers and to uplift socio-economic conditions of fishermen community at large.

To achieve this purpose the Association is owning a few survey and exploratory fishing trawlers at Verabal, a few transport trucks and carrier launches, ice factories, cold storages, a boat building yard and a fish cannery at Verabal. It also controls a diesel outlet at Verabal towards its service to fishermen community. Advances are given to fishermen and societies to modernise the equipment and to change over to the newer and advanced techniques of fishing. Catches are handled; processed and marketed through processing units.

At Bombay the Association is owning freezing plants and cold frozen storage, which is mainly export oriented, the value of which is about Rs. 16 lakhs. At Ahmedabad the Association owns a nylon twine-cum-net making plant, established during 1963-64 at an investment of Rs. 8.5 lakhs.

In order to stabilize the prices of fish to the producers, the organization started giving loans to the fishermen on hypothication of their catches and conducted trading activities in the initial stages which, has by and large protected the price line for the producers and kept them free from bondages of vested interest, by providing an effective alternative for marketing their catches.

The Association has been conducting a dry fish trade ever since its inception and has contributed to the amelioration of the dry fish zone, fishermen in a great way. The dry fish trade by this Association is continued as the fishermen need a protection. This is because of the fact that the risk to the Association is limited in this trade and the prices are not subjected to huge variations.

The contribution played by this Association towards stabilizing fish prices and thereby, supporting the socio-economic standards of the fishermen is very great and outstanding towering over such services rendered by any organization at least as far as India is concerned. The leadership, the institution has given towards the new changing ideas to the fishermen has also lead to the State towards better fishing techniques and towards the trawling operations, which contribute towards the production of exportable varieties of marine products.

The Association has also established a modern service station to extend service facilities to fishermen for servicing of marine engines.

The Association had started a production oriented scheme to built fishing boats to operate them through prospective fishermen, who will be made ultimate owners, once the capital cost of the boats are collected from their own profits. This will lead to an additional production of 20 thousand tonnes of fish.

To establish a shark–liver oil factory and to install fish meal plant and freezing units and frozen storage by the Association for export of fishery products is a novel idea for uplifting the socio–economic condition of the fishermen.

The activities of Gujrat Fisheries Central Co-operative Association Limited showing locations are given below:

1. Boat building yard–Verabal.
2. Engine servicing and spare parts–Verabal.
3. Diesel outlet and petrol pump–Verabal.
4. Ice plant–Verabal and Mangrol.
5. Fresh fish (marine) production–Verabal, Porbandar and Okha.
6. Fresh water fish production–Palanpur.
7. Fresh fish sales centres–Bombay, Delhi and Ahmedabal.
8. Dry fish production–Jafrabad, Verabal, Okha and Porbandar.
9. Dry fish sales centre–Bombay.
10. Freezing plant–Bombay.
11. Nylon twine and net making plant–Ahmedabad.
12. Nylon twine and net sale centre–Mainly from Verabal.

Nylon Net Plant, Ahmedabad

The plant owned by G.F.C.C.A. was commissioned in the year 1964 with an investment of Rs. 8.5 lakhs, as detailed below:

1. Land–Rs. 46000

2. Buildings–Rs. 218000
3. Machineries–Rs. 5,32,000
4. Electric Installations–Rs. 30,000
5. Furnitures–Rs. 16000

The annual turnover from the plant are 57000 kgs of twine and 27,000 kgs. of fishing net

Boat Building Yard, Verabal

Fishing boats in the G.F.C.C.A. boat building yard are constructed on indegenous design with minor alterations for motorization to reduce cost and these indegenous boats, powered with marine diesel engines have been proved to be perfectly seaworthy and is accepted by the local fishermen. The boats are 25 to 48 feet in overall length with displacement tonnage from 10 to 28 tonnes.

The period of construction for 35-38 feet boat takes two and half months, while for 42 feet boat it take 4 months to complete. The hull is made up teak and sajad (*Tarmenilia* sp.). Iron fastening is used and found satisfactory in places of copper tack and no copper seathing is made to reduce the cost. A 48 feet hull without engine costs Rs. 75000 plus sales tax during 1971. The hull was provided with 88 HP. Ashok Leyland water cooled marine engine at a cost of Rs. 68000 during the year. 4:1 reduction gear with 39 inch propeller with non-magnetic stainless steel propeller shaft was provided.

The capacity of the yard is 30 boats at a time.

Jafrabad Marine Fishermen Co-operative Society

In 1953 Jafrabad Fishermen Co-operative Society started with eleven members, of which two were promotors and nine fishermen. Rs. 18000, as loan was received from Government, by mortgaging the properties of two promotors. The loan was distributed to nine fishermen for purchasing fishing boats and equipments. Of these seven fishermen repaid the loan and two fishermen failed to repay the loan. During 1971, the members of the society was 600 with issued share capital of Rs. 50 000. The subscribed share capital, collected during the year, was Rs. 13000 at the rate of Rs.10 for each share. The society has undertaken the responsibility of procuring fishery requisites, like, fishing twine, anchor stone, preservatives to the fishermen. The loan to member fishermen was limited to Rs. 500 each.

The executive committee of the society consists of eleven members, of which two are Government nominees, namely, Asst. Director of Fisheries, Rajkote and Suptd of Fisheries, Jafrabad. But the problem the society is facing that the middlemen and fish merchant pay interest-free loan to fishermen and bind them to hand over their entire catch to the middlemen and the society has no control over the members of the society in handing over their catch to the society for marketing. On the contrary they give their catch to the marchent at a lesser price to get the interest-free loan, during lean and non fishing season.

Inland Fishermen Co-operative Societies: Case Study

Central Fishermen Co-operative Marketing Society Limited

Central Fishermen Co-operative Marketing Society Ltd (CFCMS Ltd.) an Apex Society, came to existence in July, 1959, basing on the recommendations of Mr. Leid Law. The basic objective of the co-operative organisation, which was set up for the Chilka fishermen, in 1959 was that the primary societies would receive the whole catch from individual member fishermen and finally supply the same to the Apex Society (CFCMS Ltd) for marketing.

After the formation of Central Fishermen Co-operative Marketing Society Ltd. (CFCMS Ltd), Fishermen Co-operative Societies have been organised in Chilka Lagoon areas from time to time as below:

Year	*No. of Primary Societies Formed*	*No. of Members enrolled in the Societies*	*Nature of Fisheries Assigned to the Societies*
1960-61	21	5795	Jano and Prawn fisheries
	13	4972	Net fisheries
1990-91	68	16117	Jano, Prawn and Net fisheries

Every year the Revenue Department, Government of Orrisa present a list, with all the fisheries sources of Chilka lagoon with an annual demand for lease payment to the Apex Society. The lease value is paid in four installments annually. Apex Society sub-leases these fisheries sources to the primary fishermen co-operative societies (PFCS) in the name of head fishermen.

But the aim of marketing the catch of the members of the PFCS by the Apex Society has not been achieved. The primary societies are marketing their catch at their own convenience, and as a result, of which, the individual fishermen faces a great loss, being exploited by the middlemen. Most of the PFCS have also not been able to give their members a better economic situation. The members still go to the private merchants and the merchants and middlemen continue to take advantage of this situation and it weakens the already weak co-operatives. As a result even though there are 68 primary societies (PFCS) in the Chilka lagoon area with 16117 members, some primaries have no activities of their own and exist only as a name.

OF the 51 active PFCS, the three biggest are:

1. Soren Chilka Societies with 1094 members of which 1060 are active.
2. Uttar Chilka with 1066 members, of which 964 are active.
3. Rambha with 784 members of which 729 are active.

The smallest two societies are:

1. Mamu Bhanja Society with 11 members of which 10 are active.
2. Baseli with 22 members of which 20 are active.

Also the primary societies have little interest in the Apex Society. Many of them exist only as a name, so that they will be able to take lease of the fishery resources. As much when a primary (PFCS) cannot pay or would not like to take lease of certain fishery area, that are being leased out to another primary (PFCS) on a negotiation basis. If no other primary is interested, the Apex can decide to put it on public auction to be utilised till the end of the three year period. But there are exceptions like Satpada, Soren, Durga PFCS, which have a working structure and their own marketing The members of these primaries are more active in their societies.

Primaries are sometimes unable to operate the fishery source of the lagoon on taking lease from CFCMS Ltd, due to:

1. Fisheries areas are disputed, as they are not demarcated properly in the lagoon and not surveyed by the Revenue

Department. The maps that exist are not understood by all fishermen.

2. Un-authorised fishing activities in the area by other fishermen and non-fishermen.
3. Natural calamities and changes of ecology are at times making the areas as in-operative and un-productive even though the lease value is increased 10 per cent after each lease period.
4. Conflicting interests in utilizing the resources. Because of the very high price for prawns every one wants to catch more prawns. This led primaries or individuals for unauthorised conversion of a system to other than stated in the leasing system.

A primary may not want to take lease the fishery source with such conflicts. As a result the number of leased out fisheries decreased to only 163 in 1987–88 from 265 sources leased in 1986–87.

Some Keuta dominated villages have been economically strengthened, even if the economic situation for most fishermen has not improved since the Apex took its form. In those villages, there exist strong primary societies with their own marketing structures, and the fishermen do not have to rely so much on the middlemen any more. Even though the income of many of the primaries have increased manifold yet, they are not paying the lease values to government through the Apex with some pretext or other. They neither look to the welfare of their members, nor strengthen the primaries, through which they are getting lease of fishery sources. These primaries are catching substantial quantities of fish and prawn and their catches are mostly sold through private traders instead of societies, though they enjoy fishing rights and lease of fisheries under co-operative fold and thereby weakening the primaries and ultimately the Apex Society.

After 31.3.91 till 31.3.94, eleven new PFCS have been organised and registered in Chilka area. Of these 11 PFCS, one PFCS is organised by Fisheries women to take up dry fish trade, 4 PFCS meant for fishermen by profession and rest 6 PFCS are organised by the fishermen by caste on capture fisheries of Chilka lagoon, thus totalling 75 PFCS till 31.3.94.

Eight PFCS have been put under liquidation on the ground of violating the Rules and provision of Orissa Co-operative Societies Act and Rules, specifically relating to violation of co-operative Election Rules, 1992.

Co-operation Among Fishermen: An Answer to Fishery Possibilities Around Islands

Normaly a small fraction fish catch is consumed on the island. For economic gain of the island fishermen, the surplus catch has to be sent in mass consumption markets. Because all of their products have to bear the burden of shipping costs to the large markets the fishermen of the island have to concentrate their production around high priced fish and "speciality" products.

Co-operative Fishing System by Independent Fishermen

Drive-in type bottom lift net fishery in some of the islands (Kozushima Island) in Japan is a type of fishery managed by an autonomous co-operative group called a "net group". The distinctive characteristics of this management group are:

1. Independent fishermen become a part of the co-operative by supplying either labour or their privately owned fishing boats for the fishing operations.
2. The fishing season for drive-in type bottom lift net is limited to summer and autumn, and when the season ends the individual fishermen use their privately owned boats to engage in other type of fisheries, thus supporting themselves as year round fishermen.
3. The gross income from the co-operative fishery activities is divided among the members in proportion to the size of their contribution in labour and equipment (fishing boats) according to a prearranged system of distribution.

In recent years, the introduction of simple diving equipment and net haulers brought about a sharp increase in catching activities.

The steady increase in fish prices throughout 1970's came to an end in the 1980's and thus no further growth in fishery production based on increases in fish price has to be taken into account.

There has been an increase in boat size by some of the participating fishermen from 3-5 ton class to 5-10 ton class boats, which has led to an increase in the scale of fishing activity possible and the size of the usable fishing grounds. This, in turn, has caused a recent trend for individual fishermen to leave the co-operative fishery in favour of their own private fishery management.

In this way there are a growing number of new social tensions surrounding the fishery of the island. Until recently, however, the socio-economic condition which supported this type of co-operative fishery can be summerized in the following way:

1. Island's fishermen were able to maintain a successful fishing industry by making use of a few specialized fishing techniques, even if the scale is reduced, as is common among all isolated island communities.
2. They were able to concentrate their catch in high-priced marine products, and thus overcome the handicap of being an island isolated from the large markets.
3. They were able to establish a successful system of co-operative fishery operation by independent fishermen.

In one of the village of the island a Fisheries Co-operative Union has been established. The Fisheries Co-operative consists of several affiliated groups, including a net group, small boat fishing group, pleasure anglers transporting group and other voluntary group, each of which conducts its own independent activities. However, it is the net group that has the long tradition in fishery of the island and that holds the authorized rights to engage in drive–in type bottom lift net fishery.

Net Group

In the island from the beginning of the nineteenth century, there existed a fishery group (about 80 members who voluntarily affiliated to the "net group") consisting of (i) boat owner fishermen contributing boats and (ii) labourer fishermen (mostly under 30 years of age), engaged in pole-and–line fishing for skip jack and primitive stick–held dip net fishery. Both the group (i) and (ii) were heriditary roles. During the off-season of pole and line skip jack fishing and stick–held dip net fishing, each member (ii) engaged individually in activities, such as sea weed gathering, angling, gill net fishing from their own small boats.

The "net group" has four officials, who manage the group by council. These include a chief sponsor, two associate sponsors and one sponsor in charge of net repair.

During the lift net fishing season, the members of the net group must supply the group with their own privately owned boats and themselves as labourer on these boats. Younger members who do not yet own a boat of their own supply only labour. The nets and other fishing materials are bought communally by the net group and become their communal property.

The members of the net group receive their pay from the gross sales margin, according to a pre-determined pay scale based on the contribution of the individual member in labour and boat use (called the "shiro") after such expenses as the:

1. Fisheries Co-operative's sales commission–7 per cent.
2. Fishing boat fuel costs.
3. Fishing gear depreciation.
4. Repair costs for boats, gear and machinery.
5. Other expenses, have been deducted.

Women's Division of Fishery Co-operatives

In a number of fishing villages on the coastal districts of Japan, house wives belonging to the women's division of fishery co-operatives work together in an effort to put miscellaneous catches to more effective use. Their efforts are well rewarded.

These housewives are engaged in joint processing of shrimps caught by small scale trawl net and dredge net fishing.

A recent drastic rise in fuel and material prices became a heavy burden on their household economy and they had to safe guard their family finances by increasing earnings with their own labour.

At first each housewife started shrimp processing as her homework. Dry–processed shrimps sent to the market of the fishery co-operative, were more favourbly received by the consumers than had been anticipated.

Encouraged by this success, all housewives concerned agreed to develop shrimp processing into a joint work, so that operation efficiency could be increased. Improvements included the purchase of an oil dryer which allowed them to continue dry processing work even in rainy weather, while sun drying was the only means of home processing. The use of the dryer has greatly improved the quality of the processed products and sales have continued to increase. Dry processing has increased the commodity value of shrimps.

The excellent results of this joint work are summarized as follows:

1. Improved individual household economy.
2. Contributing to the more effective use of miscellaneous catches.
3. Fostering better human relations within a community.

The success of these housewives have made with their new project is due to the following pre-environmental and technical conditions:

1. Their joint work started based on desirable human relations which had previously been fostered through various circle activities such as movement for the improvement of living conditions within the village community.
2. Their community had been in good relations with neighbouring farming villages
3. Quite a simple processing method served the purpose. Processed products were sold as local specialities or exhibited for spot sale on various occasions.
4. All necessary arrangements were smoothly completed, including drawing up of an annual production plan.

Chapter 8

Socio-Economic Condition of Fishermen

Nutritional Status of Artisanal Fishermen Community

Nutritional status is an indicator of socio-economic well being of a community. Survey of artisanal fishermen community in Cochin revealed that 57 per cent of the households are engaged in marine fishing as labourers, 38.5 per cent using own fishing equipment and the rest are engaged in other fishery related activities like trading and post harvest operations. The daily household income ranged from Rs. 25-50 in the case of fishing labourers, Rs. 50-125 in the case of fishermen owning craft or gear and Rs. 125-130 in the case of traders. The average family size being 5.6, the percentage of men, women and children was 37.5, 36.0 and 26.5 respectively.

From the food consumption pattern of fishermen households, it is seen that their diets consist mainly of rice and fish which meet 75 per cent of the calories and 50 per cent of the protein needs. The quantity and frequency of consumption of the protein foods, such as, egg, meat and milk are considerably less. The foods like vegetables which form the major source of vitamin and iron are rarely included in the diet. The average nutrient intake of consumption unit for calories has been worked out at the rate of 2400 for adult, though it is felt that an active sea-going fishermen requires the allowance of 3900 calories, the type of activity being heavy work with considerable environmental stress.

Food Habits and Consumption Pattern

A balanced meal is rare in most of the households of fishermen excepting during peak season in fishery. Breakfast is seldom prepared and food prepared in the previous night, if left over, is used in the morning. Difficulty in providing breakfast was identified as a reason for children dropping out from schools.

In households which own fishing crafts and where men leave for fishing in the early hours, breakfast is prepared usually out of rice flour or wheat flour, plantain and jaggery and black tea. In such households it is customary to serve breakfast for other crew men who accompany the craft owners. Of late sea shore tea shops have come up in increasing numbers and fishermen consume breakfast from these shops before or after fishing trips. Cake made of rice flour and coconut, Bengal gram curry, pakora made of bread and Bengal gram flour, rice flakes and banana fry are the items of breakfast generally served in these shops. A Rs. 10 worth breakfast would provide 150 gram rice cake, 100 gram Bengal gram curry and one cup tea with milk and sugar. The advantage for fishermen with these tea shops is that they can eat food on credit and pay when they get money. Fishermen prefer to carry plain cooked rice or in the form of gruel and chutny on their fishing trips as they think that such diet prevents them from getting thirsty and reduces the demand, on water which is to be carried with them. During longer fishing trips food is often cooked on board.

The lunch mainly consists of rice, fish curry and fish fry made of oil sardines and other fish. Vegetables and butter milk find a place in the diet rarely. Milk is not included in the diet excepting in the case of pre-school children. Meat is cooked once a month or during festivals. Men make it a point to bring a portion of the daily, catch of fish for the family. The first cereal to be introduced in the diets of infants is usually ragi. By the time the child completes first year it is introduced to all foods consumed by the family.

Food is scarce during lean fishing season. Women bear the consequence of food scarcity which occurs four to five times a year. Even during such hard times neighbours extend mutual help and exchange small quantities of rice so that some gruel can be prepared. The assets including ration cards are pledged during lean season to buy food and medicine. As difficulty faced by them of buying weakly rations on ready cash payment, there is a tendency to buy food at higher price from open market on credit. Firewood and coconut husk and shell are used for cooking which have to be bought on price.

The houses, generally small huts have some space set apart for preparation of food. Earthen and aluminium wares are used for cooking. A few stainless steel and glass wares are kept aside for the use of guests. The women are in the habit of listening women's programmes over the radio and try to understand the importance of nutrition, but it is difficult for them to practice the knowledge because of low income. It is difficult to grow vegetables in the coastal areas especially during summer due to high soil salinity and scarcity of fresh water. Coconut and seasonal paddy are the major crops. Among vegetables cowpea is the most commonly cultivated one along the bunds of prawn field. Corporation taps on the main road are the major source of drinking water and women have to spend a good portion of their time in fetching potable water. Water in small pools stagnated in the homesteads is used for washing during monsoons.

Allergic bronchitis, asthma due to coastal climate, worm infestation, anaemia and dental caries are the health disorders commonly found among fishermen children. The coastal villages fall victim to gastro–enterites during monsoon due to lack of adequate sanitation.

Socio-economic Characteristics of Traditional Fishing Communities

The socio–economic characteristics of small-scale fishing communities are as below:

Living Conditions

These traditional fishing communities live in villages in the coastal areas, generally at the very edge of the landmass, where land is least productive and subject to sea erosion. Exposure to floods, fires and storms is high. The community is overcrowded, so are individual houses, which are semi-permanent structures, generally mud huts with thatched roofs. Supplies of potable water at close proximity are rare while basic sanitation facilities are inadequate, almost non-existent.

Social status with low income and educational levels, fishing communities in the region generally have a low status in society. They are regarded as socially inferior to those engaged in most other occupations.

Social Services

The fishing communities are poorly served in the areas of health, education and community welfare. The education of youngsters in fishing communities does not continue to a desired level–their families cannot afford the cost of education or the manpower loss that education entails for their fishing operations. Further, roads, electricity, postal and telecommunication facilities are woefully inadequate.

Income and Expenditure

The income of fishing communities are generally below the poverty line. Within a fishing community, incomes are uneven, and desparities are frequently quite high. Due to seasonality in fishing, incomes are not evenly spaced throughout the year. The uneven pattern of earnings, together with spending on non-essentials, inhibits savings and infact leads to indebtedness.

Ownership

Patterns of ownership and settlement vary widely for immovable property, such as, land and houses. Some fishermen are migrants, some are temporary occupants of land, some hold short-term leases, some are tenants, a few are owners. A common feature in most of these arrangements is that the land is small in area, usually unproductive.

Likewise, there are many variations in the ownership character of boats and fishing gear. There is individual ownership and co-operative ownership. Only a small portion of the fishermen own their boats. Many of the larger, powered craft and gear are owned by individuals not actively engaged in fishing. Many members of the fishing community are hired employees–they possess no assets themselves, and engage in fishing for wages or for a share of the catch.

Power Structure

The power structure of fishing communities is linked to the ownership of such assets as land, houses, boats, fishing gears etc. Outsiders who are in a position to influence the prices of fish catch or control the supply of credit distribution and marketing, also yield power.

Fishing communities have not been able to produce a sufficient number of leaders who can represent or further their interests. However, in communities where political consciousness is high, a few leaders have emerged.

Religion

Fishing communities are, by and large, greatly influenced by religion. They celebrate festivals, visit places of worship and perform religious rites. The religious elite of the communities often yield great authority.

Probably due to the risks and hazards of their occupation, the fishing communities happen to be highly superstitious, and fear the super natural.

Participation in Organization

On paper, a number of agencies (Co-operatives, associations etc.) exist, covering a wide spectrum of activity concerning, fishermen–supplies of inputs, fishing operations fish processing and marketing. But the functioning of these agencies leaves much to be desired, and the fishing community is poorly represented in them, particularly, at the level of decision–making. In government organisation too, the fishing community is very inadequately represented, and this tells on the services offered to the fishing communities by these organizations.

Women's Role in the Community

Women are not active partners in actual fishing operations, but in the coastal areas, they sometimes are engaged in the collection of crustaceans or shell fish. However, the women do play an active sometimes dominant role in fish marketing and processing. They also engage in cottage manufacture of fishing gear. But mechanization and modernization of the fishing industry have tended to push woman out of jobs, and reduced rather than expanded their employment potential. Fishing gear is an example–mechanization has hit the prospects of cottage manufacturers.

Socio-economic Conditions of Marine Fishermen: Gujrat, West Coast

The coastal fishermen engaged in marine fisheries belong to Kharwa, Wagher, Miyana, Machhi, Machhiyara, Kahar, Mangela, Tandel, Mitna, Dubla, Nayak, Koli, Sidi communities. Their population, during 1991 census was 1,24,595 of which 37,738 are actively engaged in fishery industry. Of these Waghers, Miyanas, and Machhiyara are Muslims and the rest are Hindus. Fishermen belonging to both regions have customary social organisation of their own, with a chief of Patel in each village. The Patels are the heads of the respective communities in each village having a great deal of authority over the socio-economic life of the fishermen and exercising the powers of Panchayat chief. In addition to the fishermen Patels the Pramukhs (Presidents) of the co-operatives also have a great influence on the socio-economic life of the fishermen.

The families of fishermen are often involved in ancillary fishery activities, such as, curing, drying processing and marketing. A few fishermen families are also engaged in agriculture.

A major portion of the income of fishermen is spent on food and household expenses. Detailed survey of house hold expenses reveal the following average annual domestic expenditure of a fishermen household.

Expenditure Pattern of a Fishing Family of Junagadh District, Gujrat

Description of Expenditure	*Average Expenditure per family (Rs.)*	*Percentage*
Food	5080	56.21
Clothing and footwear	1230	13.61
Housing	341	3.77
Social ceremonies of functions	363	4.02
Services (Barber, Laundry etc)	319	3.53
Miscellaneous (Tobacco, liquor and other unaccounted expenditures)	1162	12.86
Lighting and fuel	542	6.00
Total	**9037**	**100**

The incomes of fishermen are variable from place to place depending on the type of craft operated and the type of fishing engaged in. The income of the average fishery family does not exceed Rs. 3500 per annum.

The majority of fishermen are indebted to fish merchants, as a result of which they are bound to surrender their catch to the later. This factor has been mitigated to some extent by the intervention of co-operatives and government agencies and the availability of industrial finance for operational investments.

The Department of Fisheries runs schools for the children of fishermen. Specialised training is also given to fishermen in training centres organised by the Fisheries Department.

Most fishing villages have medical facilities. Hygienic sanitary facilities, however, are lacking in most of them.

Among the socio-economic needs of the fishermen the followings are considered to be the most important:

1. Berthing and landing facilities in fish landing centres.
2. Ice for fish preservation in landing centres.
3. Provision of duty free diesel at landing centres.
4. Liberalization of procedures and conditions for attaining credit facilities from the banks.
5. Approach roads to landing centres.
6. Arrangements to off-set the high cost of wood or provision of a cheaper substitute for boat building.
7. Better facilities for marketing and transport.

Accounting and Management for Fishermen

The chapter provide the informations on how to draw up accounts in a small marine fishery enterprise, and how to use the results of accounting in conducting business.

Before mentioning about accounting, one has to be familiar with the situation, habits, methods of local fishermen, for there is no simple accounting method suited to all situations.

For example, there are two categories of fishermen; (i) those called skippers, who are incharge of the fishing boat. They manage the boat organise fishing, sell the catch and pay all expenses. They are regarded as chiefs of business, and also as the owners of their boats. (ii) Then, there are the fishermen (crewmen). They are casual or permanent workmen, that is, they work during certain seasons only or throughout the year. They receive wages for their work, in the form of part of the catch or part of the sales price of fish or other products caught.

But where the skippers are not the owners of their boats or of the engine or major fishing gear such as trawl-net or purse seine, they could only pay for part of it and borrowed the rest from friends or a bank or from a credit co-operative or other agency of the same nature.

For example, if the vessel costs 50000. The engine 10000 and the main fishing gear 5000, for a total of 70000 the skipper may have been able to pay 10000 out of his own pocket, to borrow 10000 from a friend, and get a loan of 50000 from a credit agency (bank, co-operatives etc.). The money will have to be paid back with interest to the friend and the credit agency. The skipper has to keep this in mind, and the accounts have to consider it as well.

To spend less money in fitting out their boats, the skippers of certain area also hire part of the equipment used by them, rather than buying it.

Example–Those who install a wireless set on board, they pay 30 a month, rather than buying a set for 1500.

In landing area or port, a tax is levied on fish sales to pay wharfage charges.

Example–If the tax is 5 per cent, this means that for every 100 of fish sold, 5 per cent is with held and paid to the port for its expenses. In other words, if the skipper sells his fish for 100, he actually gets only 95.

A tax is also paid on wages to help sick fishermen, for example, This is a tax, paid by the skipper, of 10 per cent of the wages.

Example–This means that if the wages are 100, the skipper must add 10 which is paid to the aid fund or to the health insurance or to other agency of the same nature.

Before putting to sea the skipper buys ice, fuel, oil for the engine, thread and other things needed to repair the fishing gear, food for himself and the crew. Then he puts to sea to fish.

On his return, the skipper sells his fish, receives money for the fish (less tax). Then before paying his crew, he deducts from the money received the expenses made before sailing, the purchase of fuel, oil for the engine, ice, food, and items for repairing fishing gear. The remainder is divided into two parts the part of the vessel is kept by the skipper and the part of the crew which, in turn, is divided into as many shares as there are men on board, including the skipper, who receives his share as a seagoing man like others. The part of the vessel serves to pay all expenses of the boat. The part of the crew is the wage divided among all men who work on board.

Accounting

Accounting serves to calculate all the expenses and all the receipts of a fishery enterprise to see what remains or what is lacking, that is, the profit or the loss of the enterprise or of the sea fishermen. So, it is necessary to write down all expenses and all receipts, without forgetting any item.

Example–When a fishing boat puts to sea, then in a very simple way, for calculating, on a sheet of paper, the skipper has to write what has happened. For instance, one can write down all the figures in two coloumns–one for the receipts, the other for the expenses as below:

Items	*Receipts*	*Expenses*
Purchase of fuel		150
Purchase of ice		20
Purchase of oil		50
Purchase of material		30
Purchase of food		60
Sale of fish		1000
5 per cent sales tax		50
Crew's wages		320
Wage tax		32
Total		**712**

The skipper should be left with 1000 – 712 = 288. This is infact what remains for all his expenses on the boat, plus 80 which is his share as a sea going man. But he must not mix the two figures because the 80 have to be used for his family's livelihood and the 288 for expenses on the boat. That is what is required to be done always–the wage of the skipper should never be mixed, which should be considered like all the other wages, with the boat account which is on the account of the enterprise.

The example given is a simple one calculating all receipts and expenses of the boat. For better accounting and in a very practical way it is seen, that before sailing, the skipper has two expenses that are different from earlier example, and which are not deducted from the sales price before he makes the division, for they are expenses which, is this area, are paid by the owner of the vessel. This is what happens as the skipper is followed:

1. Before sailing he buys 180 worth of fuel.
2. He also buy 15 worth of ice.
3. He then goes to pay the fee for hiring a radio 30.
4. He buys paint worth 85 for the boat.
5. And 50 worth of food.
6. He then puts to sea.
7. On return, he sales his catch for 300.
8. 5 per cent amounting to 15 is with held.
9. From the remainder 285, he deducts expenses before sailing 245 (he does not deduct expenses on radio and paint).
10. He divides the remainder 40 into two parts of 20.
11. He divides one part 20 among 4 sea going fishermen, 5 each.
12. Finally he pays 10 per cent on the wages paid, 2.

The expenses and receipts of second sailing can be calculated in the some way as in first sailing.

Items	*Receipts*	*Expenses*
First sailing total	1000	712
Purchase of fuel		180
Purchase of ice		15
Radio hire		30

Items	*Receipts*	*Expenses*
Paint		85
Food		50
Sale of fish	300	
5 per cent sales tax		15
Crew's wages		20
Wage tax		2
	1300	1109

Here again, it is seen what is left for the skipper's boat, 1300 – 1109 =191. Thus, the second sailing has cost him money because it has left him less money than the first sailing. This is why the figures all days or of all sailings have to be added up to find out after a fairly long period whether or not the boat earns money. The result of one trip is not enough.

The required for the whole year, there need another way of keeping accounts (writing down the figures) without the risk of making mistakes, because of too long columns of figures. Besides, in addition to showing the total of receipts and expenses, the accounting can also do other things.

Figures have to be arranged in another way to make fewer mistakes and to get a clear picture of receipts and expenses. The figures can be written quite simply by following what happens whenever the skipper makes his calculations.

To see how one should go about it, the first example is taken at the moment when the skipper returns from fishing.

A copy book or a register can be organised in the following way. This is what is going to be done in following the accounts of the skipper:

1. The skipper sells his fish for 1000
2. A tax of 5 per cent is with held = 50
3. To calculate the shares in this area, the skipper deducts all expenses paid jointly, that is, fuel, oil for the engine, ice, small fishing material and food.
4. One must open two columns for the division (into two parts) of the remainder, between the crew and the boat.

Here one can separate the accounts of the boat because they have to be followed separately. In the example it is seen that there are certain expenses that are paid only by the skipper out of the part of the boat. The first column can be for the receipts of the boat (the parts of the boat).

5. One can now easily draw all the columns need to enter the figures mentioned in earlier two examples and the figures of the two examples can be written one after the other.

Date	*Sale of Fish*	*Tax 5%*	*Fuel*	*Oil*	*Ice*	*Fishing Gear*	*Food*	*Wages*	*Boat*	*10% Tax on Wages*	*Radio*	*Paint*	*Tax*
9 Jan,	1000	50	150	50	20	30	60	320	320	32			32
12 Jan,	300	15	180		15		50	20	20	2	30	85	117

This is very clear and quick method of doing it, because when one have prepared all the necessary columns, one only have to write all the figures of a sailing on the same line, putting each figure in the correct column (under the figures of the same category). On the right side, in the part reserved for the boat, one can do as has been done in the table above. A column of total can be opened in which the expenses for the boat is added. Thus it can be seen that the first catch made money, as the receipts are higher than the expenses and the second catch resulted in loss. The skipper was able to pay for his purchases of the second sailing only because he had savings out of the first.

But as already been said that a fairly long period of time is necessary to find out whether a boat makes or loses money. This is why one must enter all expenses and all receipts throughout the year in the account book (copy book). If this done one will see that new advantages will appear in classifying the figures as below.

Let us consider a page of the account book where all figures of all sailings have been entered.

At the bottom of the account sheet the figures in each column is added and can be checked to see if there have been any mistake. The check is simple. If it is remembered that all figures of the columns up to the share of the boat are parts of the money from the sales of fish, it is sufficient to add up all totals of each of these columns to see, whether together they tally with the total of sales. If there is a difference, there must be a mistake somewhere and that is to be found out and correct it.

Date	*Sale of Fish*	*Tax 5%*	*Fuel*	*Oil*	*Ice*	*Fishing Gear*	*Food*	*Wages*	*Boat*	*10% Tax on Wages*	*Radio*	*Paint*	*Total*
9 Jan	1000	50	150	50	20	30	60	320	320	32			32
12 Jan	300	15	180		15		50	20	20	2	30	15	117
15 Jan	600	30	140		20	45	65	150	150	15			15
23 Jan	1200	60	200	20	30		50	420	420	42		150	192
29 Jan	200	10	150				40	0	0	0			0
8 Feb	700	35	200		30	25	50	180	180	18	30		48
15 Feb	1500	75	180	65	20		60	550	550	55		200	255
20 Feb	1300	65	135		40		60	500	500	50	30		50
25 Feb	800	40	160		20	40	60	250	250	25			25
3 March	500	25	155		30		50	120	120	12			12
7 March	400	20	170	50	20		60	40	40	4	30		34
13 March	1100	55	160		30	85	50	360	360	36			36
16 March	600	30	190		20		40	160	160	16			16
27 March	800	40	160		20		60	260	260	26			26
12 April	1300	65	170	60			45	480	480	48	30		78
16 April	300	15	120		30	80	55	0	0	0		90	90
23 April	1100	55	150		30		65	400	400	40		30	70
28 April	1600	80	160	35	35		50	620	620	62			62
5 May	300	15	150		15		60	30	30	3	30	120	153
Total+++	15600	780	3080	280	425	305	1010	4860	4860	486	150	675	1311

In the same way, it can be checked whether the addition of the columns of figures for the boat is accurate or not.

If everything is correct, the writing of the skipper figures can be continued in the following page,but it must be seen that these figures continue to follow each other and that can be added up.

It is just seen how to organise accounts for the fishing enterprises. One can prepare sufficient columns for the various figures that have to be classified together. One can of course, group several kinds in the same column. For example, all maintenance costs for the boat paid by the skipper, not only the painting costs.

If accounts are properly organised, they give accurate information on what has happened in the business as to its receipts and expenses.

The following simple operation shows the gross profit of the boat, that is, the money available at the end of the year.

For instance, at the bottom of the first page of the example, the gross profit of the boat from 9 Jan to 5 May is 4860 – 1311 = 3549

But if there is a gross profit at the end of the year, it does not mean that there is actually a profit. In fact, one has to find out whether the money left at the end of the year, that is, gross profit, is sufficient to cover the wear and tear of the boat, of the engine, and of the major fishing gear. So to find out whether the profit is sufficient at the end of the year, one has to make some other calculations.

At the beginning it is said that the boat, the engine and main fishing gear had cost 70000. To pay 70000, the skipper, who had only 10000 was able to obtain a loan of 50000 from a credit office (Bank) and a loan of 10000 from a friend. This money must, of course, be paid back and the skipper must also recover has 10000. Every year, therefore, enough money must be left for this purpose.

So, to find out whether the gross profit is sufficient, it is first of all necessary to know what the skipper has to pay back to the others or to recover for himself each year.

For instance, if the friend, is lending 10000, agreed with the skipper that he repay the 10000 in 5 years, that is 2000 a year, but that he give him an additional 1000 a year as interest on the loan, the column of repayments to the friend would be as follows:

	Repayment of Loan with Interest to Friend
1st year	3000
2nd year	3000
3rd year	3000
4th year	3000
5th year	3000

If the credit office which loaned 50000 demands repayment in 10 years, that is, 5000 a year, plus 1500 as interest on the loan, the column of repayment to the credit office, added to that of the friend would be as follows:

	Repayment of Loan with Interest	
	Friend	*Credit office*
1st year	3000	6500
2nd year	3000	6500
3rd year	3000	6500
4th year	3000	6500
5th year	3000	6500
6th year	6500	
7th year	6500	
8th year	6500	
9th year	6500	
10th year	6500	

The skipper might also decide to recover his 1000 in 10 years. In this case, 10 repayments of 1000 should be added in his column, over the same period as that of the credit office

But the skipper must perhaps recover more money or recover his 1000 faster because part of his equipment will wear out rapidly, therefore, it is necessary to recover enough money quickly enough to replace the equipment. This is called amortization of equipment.

For instance, the fishing gear, which cost 5000, has to be replaced after 4 years. To get the money needed for its replacement the skipper, therefore, must save 1250 a year.

In this case, instead of opening a column for the skipper's repayments, a column for the major fishing gear and another for the engine be opened.

The amortization of the book should be considered, when a replacement is expected to take place, before the end of the repayment of the loan obtained to buy it.

This will show the annual gross profit needed for all those charges or repayments.

Gross Profit		Repayment of Loan with Interest		Replacement of Equipments	
		Friend	Credit Office	Fishing Gear	Engine
1st year	13250	3000	6500	1250	2500
2nd year	13250	3000	6500	1250	2500
3rd year	13250	3000	6500	1250	2500
4th year	13250	3000	6500	1250	2500
5th year	13250	3000	6500	1250	2500
6th year	10250	6500	1250	2500	
7th year	10250	6500	1250	2500	
8th year	10250	6500	1250	2500	
9th year	10250	6500	1250	2500	
10th year	10250	1250	2500		
11th year	3750	1250	2500		
12th year	3750	1250	2500		

In this example, the gross profit in the first 5 years would have to be at least 13250; in the following 5 years it must be at least 10250; from the 11th year it must be at least 3750.

If the gross profit is greater than these amounts, the remainder is called net profit. In this case the boat is profitable. If the gross profit is not sufficient to repay these amounts or to make the necessary savings, the difference is called deficit. In that case, the boat is not profitable.

One should not merely calculate the gross profit (the money left at the end of the year), but must also calculate the repayments and savings to be made. The boat is really profitable only if there is enough money for the skipper's repayment and savings (the savings which, will enable to replace worn out equipment).

From the figures of the example, it can be seen that if the boat continues like this until the end of the year, it will not be profitable. In fact, from 9 Jan to 5 May it has made a gross profit of only 3549 (in 5 months); if it continues like that, it will make in 12 months

$$\frac{3549 \times 12}{5} = 8520 \text{ approximately}$$

The amount needed is 13250. Thus there is a shortage of 13250 – 8520 = 4730

One should not lose too much time before doing something about it. This is where management comes in.

What is Management? What is its purpose?

It is seen that accounting serves to clearly classify the figures of receipts and expenses, then to find out whether or not a business is profitable.

Management consists in using these figures or in thinking on the basis of the figures to improve the profitability of the business.

In going through accounts, one can consider and suggest improvements.

For instance, if one carefully look at the accounts of sailing of boat, it is seen that two sailings one on 29 Jan and another on 16 April failed to yield anything to the skipper of the boat or to the crew. While they produced no gross profit, it is known that they actually cost money, because they used equipment and supplies. If the results are bad because the sea was stormy, and it is always the same story, it is better not to put to sea under such circumstances; there will be less risk for the men, and the enterprise will make savings.

The cost of wrong economy can also be calculated. For instance, from the account book, it is seen that the skipper did not spend enough money on the upkeep of boat engine. As a result of this, there were three engine breakdowns.

1. One between 15 and 13 January which cost him 150 in repairs.
2. One between 8 and 15 February which cost him 200 in repairs.
3. One between 27 March and 12 April which cost nothing in repair but caused him to lose two sailings.

On an average, the sailings of this account book produced for the boat; 4860–1311 = 3549 in 19 sailings or 3549:19 = 186 per sailing, as well as 64 per sailing on average for each crew member. Therefore it can be said that the breakdowns cost the enterprise

Repairs 150 + 200

Sailing loss–186 × 2

Total–722

And each crew member (including the skipper) lost 128 in wages. So better maintenance can prevent one or two repetitions of break-downs during the year, and a small additional expenditure say, 200, may raise earnings by 1500, which the enterprise needs badly.

Better upkeep of the engine may even avoid other considerable losses. For instance, the engine may lasts one year less owing to the bad maintenance of the engine.

In 6 years it costs–15000 *i.e.*, 2500 a year

If its life is reduced to 5 years due to bad maintenance, then it costs 15000/5 = 3000, a year or an additional 500 a year.

By accumulating these things, a business often makes or loses money. Therefore, one has to be very careful, take a close look at the accounts and calculate things, as it is done for the engine.

So far what is seen concerns the skipper's method of work. Similarly improvements concerning other things can be thought of.

Was the skipper of this example right in borrowing money to buy his boat?

From the table of repayments and amortization, that has been worked out, if it is assumed that, instead of borrowing 10000 from a friend, he would have borrowed the money from the credit agency with 3 per cent interest a year, then the credit agency would have told him to repay 60000, not 50000 in 10 years, that is, 6000 a year plus 1800 interest.

In the first case, the total repayment would have been 15000 to the friend in 5 years and 65000 to credit agencies in 10 years totaling 65000+15000 = 80000. Whereas, in the second case, repayment of 7800 a year for 10 years to the credit agencies. Therefore, there would have been a savings of 2000 in 10 years which is not considerable. In addition, from the sixth year the repayments would have been higher than in the first case and this may be awkard because from that moment growing wear and tear of the boat calls for more maintenance and therefore more costs.

Alternately, when the skipper obtained a 10 year loan of 50000 for the boat from the credit agencies, and that for the engine and fishing gear he had borrowed money from a friend and paid with his own money.

This system has forced the skipper to repay his loans, that is to pay after the purchase, part of the engine and of the fishing gear (the 10000 he did not have). But at the same time, owing to the increasing wear and tear of the engine and fishing gear, he was obliged to make additional savings to buy them again at the right moment. Thus he had to pay twice for what he could not pay the first time.

But the skipper could have adopted another method. Instead of acting as he did, the skipper could have acted as follows:

To buy the boat–10,000 of his own money (to be recovered in 10 years) plus a loan from the credit agencies of 40000 at 3 per cent interest over 10 years.

Annual repayment would have been 4000 + 1200 = 5200

To buy the engine–A loan from the credit agencies of 15000 at 3 per cent interest over 6 years

Annual repayment would have been 2500+450 = 2950

To buy fishing gear–A loan from the credit agencies of 5000 at 3 per cent interest over 4 years

Annual repayment would have been 1250+150=1400

To be sure, a new loan on the same terms could be obtained after 6 years; likewise for fishing gear after 4 years.

In this case the annual repayment table would be, without forgetting the amount of 10000 which is skipper's own money to be recovered over 10 years, that is, 1000 a year.

	Gross Profit	*Repayment of Loan with Interest*			
		Skipper	*For Boat*	*Fishing Gear*	*Engine*
1st year	10550	1000	5200	1400	2950
2nd year	10550	1000	5200	1400	2950
3rd year	10550	1000	5200	1400	2950
4th year	10550	1000	5200	1400	2950
5th year	10550	1000	5200	1400	2950
6th year	10550	1000	5200	1400	2950
7th year	10550	1000	5200	1400	2950

Gross Profit		*Repayment of Loan with Interest*			
		Skipper	*For Boat*	*Fishing Gear*	*Engine*
8th year	10550	1000	5200	1400	2950
9th year	10550	1000	5200	1400	2950
10th year	10550	1000	5200	1400	2950
11th year	4350	1400	2950		
12th year	4350	1400	2950		

This formula is more advantageous if the skipper seeks the one that requires the smallest annual gross profit.

However, in this way the skipper cannot do better than recover his own money. He is forced to ask for a new loan whenever he has to buy a new engine or new fishing gear. In 11 th and 12 th years cost him more than in the second formula.

It is therefore, concluded from these calculations that depending on the method of borrowing the skipper will have more or less difficulty in making has business profitable, or in achieving the necessary gross profit, which amounts to the same thing.

Whenever a skipper has to build a boat, change an engine, or make major purchases, it is important to calculate all possible solutions to compare their advantages and drawbacks.

Collective Enterprises

So far the problems and solutions of individual skipper has been discussed to find improvements in his working methods, that is not putting to sea in bad weather, better upkeep of the engine, better borrowing of the money needed for the boat and major equipments.

Now it is time to think about collective solutions to be found and organised for a number of fishermen at the same time, because they can make for other improvements in the profitability of enterprises.

Consider the year end results of the boat analysed so far. If a number of boats of the same place, the same village, the same port have about the same major figures as this accounting model, or face the same problems that have been noticed (upkeep of engine for instance) a collective improvements, which apply to all fishermen can be found out. For example, a sales co-operative, with an effort to improve the quality of fish (better preparation of fish or better handling, when it is caught). For instance if by doubling the cost of preparation (processing), one can manage to sell the fish at a price that is 10 per cent higher, then,

(*i*) Expenses would rise by about 1000.

(*ii*) Receipts would rise by about 3700.

that is, there would be considerable increase in the earnings of the enterprises.

It is seen from the accounts that if a fuel co-operative are organized and makes for a savings of 5 per cent on the fuel cost; it would offer an important advantage. Remembering the bad upkeep of the engine, a well organized maintenance and repair workshop can be of considerable improvement for the profit earning capacity of fishery enterprises. In this case, the object is not necessarily a reduction of expenses, but above all the losses suffered; the money the skipper fails to earn when his boat is laid up because of a breakdown, his loss when the engine wears out faster.

If one wants to organise a collective improvement (such as, a sales co-operative, a fuel co-operative, a maintenance and repair workshop), the accounts of the boats must be carefully kept to measure (check) whether results (improvements) are really achieved (whether the skippers really profit by them).

For this reason, it is necessary to help them in keeping their accounts correctly and methodically. As a priority, therefore, an Accounting and Management centre is proposed which can offer many advantages. In this way, one can help the fishermen to a better knowledge of what they have earned and what they have spent, and explain what they should do to work better and improve their income, or the profitability of their boats, which is the same thing for them.

In keeping the accounts of a number of fishermen, one can also make comparison between several boats, or several working methods, and one can also help the fishermen in their major decisions.

For example, in order to increase his earnings, the skipper, whose accounts have been followed up to now, would like to buy a bigger boat because he has seen that another boat of his area brings in twice as much fish.

If the accounts of the other boat is known, it is easy to consider, in bringing in more fish, it is more profitable, as bigger boat is not necessarily the one that makes for higher earnings.

The same holds true if a fisherman has to choose between two different kinds of fishing gears, two different engines etc.

As it is seen, for management, that is, for taking decisions enabling to improve the profitability of the business, of the boat and the fisherman's income, accounts need be kept properly. This means one must know how to keep accounts in a clear and simple manner, that are easy to keep. But one should not hesitate to change the accounting methods if he need them to take good decisions.

For instance, the fishermen of an area sell their fish to two kinds of buyers, merchants and a co-operative. If one wants to know which kind of buyer pays a better price for the fish, then he must not hesitate to draw the necessary columns of weight and value for both type of buyers in the account book. At the end of the year it is possible to compare the average prices of fish sold to the co-operative and to the merchants by dividing the proceeds from sales by the weight of fish sold.

$$\frac{\text{Proceeds from Sales}}{\text{Weight of Fish}} = \text{Average Price of the Year}$$

The skippers are too poor to buy major fishing gear. They think that money than could be lent by the credit office is too difficult to repay. Now merchants buy them fishing gear on condition that they sell all their fish to them. But since they have given the fishing gear, they withhold 5 per cent on all fish sales, that is, they pay 5 per cent less for the fish. To calculate exactly if the fishermen should accept this method, two things are to be checked; (i) to know whether the merchant takes only 5 percent, and then to know the cost of the system to the fisherman, and whether it is to his own advantage to proceed in this way, two columns can be drawn, one for the weight, the other for the price of fish sold to calculate the average price. In this way, one can compare, at the end of the year, with the average price of fish sold by other fishermen to other merchants. The fisherman can thereby evaluate (measure) whether the merchant has not kept more than 5 percent.

One can also compare the amount kept by the merchant with the cost of fishing gear if the fisherman had borrowed money to buy it.

Some time fishermen dry their fish, store it and sell it in major or minor quantities. In this case one must, if possible, is to set the sales against the fishing periods corresponding to the fish sold. If the fisherman sell their fish about one month after catching it then one put February sales (total fish sales) against total expenses of January, that is the month when the fish sold in February was caught. Then the sales of March will correspond to the fish caught in February etc.

Chapter 9

Tasks of Fishery Administration

Promotion of Fisheries and Stabilized Supply of Fishery Products

Fishing Port and Fishing Village

1. Improvement and expansion of fishing ports.
2. Improvement of ports of distress and restoration of fishing ports after disasters.
3. Improvement of land infrastructural facilities required for a base of production and distribution.
4. Improvement of living environment in coastal fishing villages.

Stabilization of Management of Fishing Households and Fisheries Enterprises

1. Improvement and development of fishing techniques.
2. Accomodation of long–term low interest fund system.
3. Securing of labour force and technical training.
4. Diagnosis and guidence of fishery management.

Fishery Co-operative Associations

1. Improvement and reinforcement of organization in fisheries co-operative associations.
2. Intensification of economic activity in the markets of producing areas of fishery co-operative associations.
3. Improvement of fishing boat insurance and fishery mutual aid systems.

Efficient Use of Fishing Grounds and Resources in the Seas (Coastal fisheries and offshore fisheries)

Coastal Fisheries

1. Survey and development of resources.
2. Improvement and development of coastal fishing grounds (for fishing boat fisheries and culture fisheries).
3. Increment of resources by the production and stocking of artificial seeds.
4. Promotion of culture fisheries.
5. Appropriate management and adjustment for efficient use of fishing grounds.

Offshore Fisheries

1. Survey and development of resources.
2. Appropriate management and adjustment for efficient use of fishing grounds.
3. Shift to labour-saving and resource saving type fisheries.

Measures for Efficient Pelagic Fisheries in the 200-mile Age

1. Securing fishing grounds by promotion of fishery deplomacy.
2. Survey and development of new resources and new fishing grounds.
3. Shift to resource–saving type fisheries.

International Co-operation

1. Technical co-operation with developing countries.

2. Economic co-operation with developing countries.
3. Conclusion and observance of international treaties and agreements.

Efficient Use of Fishery Products through Processing

1. Efficient use of mass-catch fishes.
2. Use of unused resources by processing.
3. Reuse of fishery wastes.

For Distribution

1 Stabilization and rationalization of prices of fishery products.
2. Improvement of existing fish market systems and development of new distribution systems.
3. Improvement of various distribution facilities in each fishing base.
4. Propagation of fish consumption.

Infrastructure Establishment for Coastal Fisheries Development

For the sound development of fisheries and stabilization of fishery management, close connections among three branches, fishing, processing and distribution along with their well-balanced development are necessary. In a fishing port which will be a base for fishery, the following facilities related to fishery must be fully provided, in addition to basic facilities such as, break water, quay wall, mooring stations for fishing boats and cargo work facilities.

1. Fish market.
2. Processing factory.
3. Storage facilities for fishes (ice storage house with ice machine and cold storage).
4. Supply facilities (ice supply facilities, oil storage and supply facility).
5. Repair and storage facilities (store house for fishing gear, repair shop for fishing boats, engines and fishing nets).

6. Transporation facilities (roads, railways, trucks, carrier boats etc.).

Only after organizations and systems which will operate and control these facilities are established in addition to the above mentioned facilities a true "infra-structure" of fishery be established.

Classification of Fishing Ports by Function

1st Class–fishing port functioning at a base of production for the local coastal fisheries.

2nd Class–fishing port functioning as a base of production of the regional fisheries including neighbouring fishing villages.

3rd Class–fishing ports functioning as a base for nation wide distribution and processing of marine products.

4th Class–ports of distress for fishing boats.

The fishery budget may be outlined on the following items.

Exploitation of Fishery Resources within the 200 mile Limits and Promotion of Aquaculture

1. Survey of resources within 200 mile waters and compilation of fishing and sea condition informations.
2. Promotion of coastal fishing ground improvement projects, should be continued as nation–wide program.
3. Promotion of aquaculture, promotion of culture fishery, technical development of marine stock farms.
4. Promotion of inland water fisheries.
5. Countermeasures for fish diseases.
6. Control and management of the Fishing Zone Act.

Exploitation of Resources in Distant Seas and Development of Pelagic Fisheries

1. Survey and exploitation of new fishing grounds.
2. Promotion of fisheries diplomacy.
3. Preparation of a Fisheries Promotion Fund.
4. Expansion of overseas fisheries co-operation.
5. Enforcement of treaty and agreement.

Measures for price stabilization of fisheries products and expansion of measures for distribution and processing.

1. Measures for price stabilization of fishery products.
2. Measures for improvement of distribution and processing of fishery products.
3. Accomodation of fisheries processing management stabilizing funds.
4. Improvement of fisheries product consumption.

Reinforcement Measures for Coastal Fisheries Development

1. Promotion of structural reform projects.
2. Promotion of fishing villages.
3. Establishment of reform funds for fisheries management and techniques.
4. Expansion measures for game fishing.

Improvement of Fishing Ports and Environment of Fishing Villages

1. Improvement of fishing ports.
2. Improvement of sea shores.
3. Promotion of natural disasters relief projects.
4. Improvement of environment of fishery communities.

Enrichment of Fisheries Financing System

1. Accomodation of fisheries management stabilizing funds.
2. Accomodation of fisheries modernization funds.
3. Promotion of fisheries loan security system.
4. Accomodation of funds from Agriculture, Forestry and Fisheries Corporation.

Measures for Improvement of Fishing Ground Environment and Reinforcement of Fishery Accident Compensation Systems

1. Promotion measures for protection of fishing grounds against pollution.
2. Enforcement of fishery accident compensation systems.
3. Enforcement of fishing boat accident compensation system.

Other Measures

1. Promotion of fisheries improvement and extension projects.
2. Improvement of fisheries co-operative associations.
3. Improvement of coastal ratio for fishery.
4. Improvement of facilities in aquaculture research, and research institutes for fishery technology.
5. Promotion of measures for fishing labour improvement.

Fisheries Development and Five Year Plans

From 1951, the beginning of the First FiveYear Plan, fisheries and aquaculture in the country have undergone an impressive transformation from a highly traditional activity to one based on a well developed and diversified infrastructure with immense potential for industrialisation. The fishery sector, particularly during the recent past has played on important role in the Indian economy through employment generation, enhanced income, and through earning valuable foreign exchange.

The fishery sector provides gainful employment to about 3.84 million full time or part time fishermen with an equally impressive segment of the population engaged in ancillary activities associated with fisheries and aquaculture. The contribution of fishery sector to the net domestic product has shown an eight-fold increase from Rs. 8.06 billion in 1980-81 to Rs. 67.5 billion in 1993-94 at current prices, when compared to only the four-fold increase in agriculture during the same period. The share of fisheries in GDP from agriculture has almost doubled from 1.97 per cent in 1980-81 to 3.89 per cent in 1993-94.

India is endowed with a vast expanse for water resources a menable for capture fishery and aquaculture. While marine resources are spread in the three seas encompassing the 8129 km coastline, the inland waters in the form of rivers, canals, estuaries, lagoons, natural and man-made lakes, back waters and brackish water impoundments and mangrove wetlands constitute the bedrock of inland fishery and aquaculture. Based on the available scientific informations, exploratory surveys, experimental fishing and other data available, the potential harvestable yield of the Indian Exclusive Economic Zone (EEZ) has been estimated at 3.9 million tonnes. In the inland sector, the resource potential has been estimated at

4.5 mt, which takes into account, the production from both capture and culture fisheries.

Fish production in the country has increased from 2801 mt in 1984-85 to 4789 mt. during 1994-95. The average annual growth rate in the fish production during the period 1984-85 (base year in the Seventh Plan) to 1994-95 is 5.6 per cent. The growth rates in the marine and inland sectors have been 5.0 per cent and 6.7 per cent per annum respectively during the period. The average annual growth rate from 1990-91 to 1994-95 is 5.7 per cent. The target of fish production during 1995-96 was 4950 mt. For the year 1996-97 (terminal year of 8th Plan), fish production target has been fixed at 5140 mt, consisting of 2857 mt from marine sector and 2283 mt from inland sector. Fish seed production during 1995-96 was expected to be 15,700 million fry against the production of 14.544 million fry in the previous year and 10,332 million fry in 1990-91. The production of shrimp seed is about 1000 million per year at present. Though fish production has increased at a higher rate than food grains, milk and many other land based items, yet the consumption of fish among fish eating population is still about 8 kg per annum.

There has been a significant increase in quantum and value of export of marine products and, for the first time, the value of export of marine products crossed the one billion dollar mark in 1994-95. the export of marine products in terms of value registered a level of Rs. 3575.27 crores in 1994-95 against Rs. 2503.62 crores during 1993-94. Though shrimp accounted for 35 per cent of quantity and about 71 per cent of value of exports in recent years, there has been diversification also and the country is now exporting frozen squid, cuttle fish, fillets in large quantities.

The fact that fisheries have been recognized as a thrust area within the agriculture sector can be recogniged from the successive plan outlays. From Rs. 5.13 crores in the First Five Year Plan (1951-56), the total outlay has increased to Rs. 1172.32 crores during Eighth Plan, which comprises Rs. 400 crores for the central schemes (both central sector and centrally sponsored) and Rs. 772.32 crores for the States Schemes.

Inspite of the emphasis on the overall development of fisheries for employment and income generation and export earnings, the sector suffers from weakness in the most essential component of organised marketing, both domestic and to some extent for export. The fish farmer/fisherman, as primary producer needs to be provided remunerative prices on one hand and make the fish products available to the consumers at reasonable rates on the other. In the marine sector, utilisation of the by-catch and diversification in both harvest and post-harvest activities need priority attention. To propel Indian fisheries into the 21st century, the quality, technical skills and management of fisheries manpower must improve in consonance with the rapidly changing needs of the society, both nationally and internationally.

To maintain the pace of growth witnessed by the fisheries sector in the recent past, the efforts may have to be probably larger and faster by several times more than made earlier. For such an effort adequate funding to strengthen and streamline organisations infrastructure, manpower would be the basic requirement.

In the development of fisheries, preoccupation with bio-physical processes at the expense of social processes would be detrimental to the overall progress. Man should remain at the centre stage and the benefits of various development programmes should percolate to the poor fisher folk and fish farmers. In regard to gender issues, the fishery sector offers maximum equality from among the agriculture and allied sectors.

Programmes During Ninth Five Year Plan

Increase in productivity and production of fish would be one of the accepted programmes during the Ninth Five Year Plan. The thrust areas will be:

1. Enhancing the production of fish and the productivity of fishermen, fisher women, fish farmers and the fishing industries.
2. Generating employment and higher income in fisheries sector.
3. Improving the socio-economic conditions of traditional fisherfolk and fish farmers.
4. Augmenting the export of marine, brackish water and fresh water fin and shell fishes and other aquatic species.
5. Increasing the per capita availability and consumption of fish to about 11 kg per annum.

6. Adopting an integrated approach to marine, inland fisheries and aquaculture taking into account the needs for responsible and sustainable fisheries and aquaculture.
7. Conservation of aquatic resources and genetic diversity.

Marine Sector

In pursuance of the objectives set for fisheries development in the country and taking into account the problems and potentialities of the vast and varied marine fisheries, the following thrust areas have been considered for taking up during Ninth Five Year Plan:

1. Upgradation of fishing capabilities of existing mechanised vessels (below 20 m OAL) by providing them with navigational aids (fish finders, communication equipments, and increasing their fish hold capacity.
2. Introduction of intermediate range of fishing vessels (15-19 m OAL) with capacity to fish in depths of 70-150 m (or even upto 200 m). These vessels would be a combination of trawlers, long liners, purse seiners etc.
3. Diversification of fishing by existing vessels (wherever there is extra fishing pressure) from trawlers to long lining, purse seining, squid jigging etc.
4. Continuation of assistance for procurement of patrol boats and their operations for enforcement of MFRA with increased involvement of States.
5. Motorisation of traditional craft with OBM/IBM, including the gear component, particularly for States with weaker response to the on-going scheme.
6. Introduction of new hull materials, such as, FRP and PC for fishing vessels.
7. Provision of central exicise duty exemption on HSD oil supplied to vessels of below 20 m OAL.
8. Inclusion of turtle excluder device in trawl nets
9. Utilisation of potential fisheries zone forcasting.
10. Setting up of artificial reefs and fish aggregating devices.

Mariculture production world-wide is growing at the rate of 5-7 per cent annually, but in India, it is yet to be taken up in any appreciable scale. Mariculture includes sea farming of shrimps, oysters, mussels, clams, sea weeds, sea cucumbers, and fishes in floating cages and rafts. Technologies in these regard are available and they need to be demonstrated and popularised among coastal communities.

Inland Sector

For optimum utilisation and sustainable development of the inland fishery, the following areas have been proposed to be attended during Ninth Five Year Plan, so as to not only conserve the different inland fishery resources highly amenable for fish yield optimisation, but also to augment the fish production:

1. Development of large, medium and small reservoirs for fish yield optimisation.
2. Development of flood plains and lakes for ecosystem improvement and fish yield optimisation.
3. Impact assessment of river valley projects on fisheries.
4. Monitoring and surveillance of inland and coastal water pollution.
5. Seed rearing and production infrastructure exclusively for stocking and management of reservoirs.

Accordingly the following thrust areas have been identified for fresh water aquaculture during Ninth Plan period:

1. Development of pond aquaculture through Fish Farmer Development Agencies and strengthening of the technical wings of the FFDAS to provide assistance to fish farmers, entrepreneurs on various technology packages of aquaculture.
2. Popularisation of fresh water prawn farming, including setting up of small scale prawn hatcheries.
3. Setting up medium sized feed units (for carp and prawn farming) especially in the North-East and other remote areas.
4. Establishment of Fishery Estates for the benefits, particularly of weaker sections of the society.

For coastal aquaculture

1. Development of shrimp farming through Brackish Water Fish Farmer Development Agencies.
2. Popularisation of stabilization ponds for waste water management in shrimp farms.
3. Preparation of coastal land use maps on 1:25000 scale in collaboration with ISRO for effective utilisation of the brackish water potential.
4. Setting up of artemia production units in potential areas.
5. Setting up of composite laboratories for regular monitoring of water and soil qualities of ponds and tanks and identification treatment of common fin and shell fish diseases as disease diagnostic centres as a service to the farmers, entrepreneurs and for training.

In the field of cold water fisheries,

1. Establishment of trout hatcharies and raceways.
2. Establishment of hatcheries and farms for mahaseer species.
3. Development of selected high altitude lake and hill stream for food and game fisheries development.

Setting up of ornamental fish hatcheries and propagation of ornamental species for both domestic and export purpose is given importance during Ninth Plan considering the simple technique and moderate investment is involved.

Infrastructure development, such as, fish landing centres, much needed shore based facilities, inland fish marketing units and development of retail markets in selected cities are considered important in Ninth Plan.

Setting up of model fishermen villages, assistance to fishermen co-operative societies and introduction of insurance cover for traditional craft and gear are some of the urgent areas, attracted attention during the plan period.

Proposals of export of diversified fishing products and markets have been identified as thrust area during Ninth Plan. The trade in frozen fish, fish fillets and surimi is promising and the industry has to be geared and equipped to handle and export them in value–added packages. Fresh water species, such as, carps to have potential market, especially in West Asia. Quality control, export promotion and marketing stratigies need to be pursued more aggressively, keeping in view the contemporary requirements of the export markets.

Targets

With the growth rate of 3 per cent per annum in marine sector and 6 per cent per annum in inland fisheries sector, the fish production target of about 63.67 lakh tonnes have been kept by the end of Ninth Five Year Plan *i.e.,* 2001–2002.

During the period, the fish seed production of major carps and exotic carps would be 23000 million fry. Taking into account the three important shrimp species (*Penaeus monodon, P indicus* and *P. merguiensis*) 16,000 million post-larvae would be produced in the terminal year (2001-2002) of the Ninth Plan. Also 100 million fry of commercial fin fish species (sea bass, grey mullet, grouper, snapper, milk fish etc.) would be produced by 2001-2002.

For implementing the programmes mentioned, an allocation of about Rs. 1200 crores have been made to the Fisheries Division of Department of Agriculture as against the allocation of about Rs. 400 crores during the current plan period.

Credit for Fisheries Developmental Planning

Credit is a crucial input which plays a catalytic role in the development of any economic activity, more so incase of a fast growing sector like fisheries. The fast growth in this sector was witnessed during last five years in terms of fish production and exports, would call for careful deployment of scare resource in order to sustain the developmental momentum generated on account of economic liberisation and commercialisation of aquaculture. It has also created demand for newer technologies involving overseas collaboration. This is a task which the financial institutions have to shoulder and would call for planning of both physical and financial resources besides playing an important role in technology transfer.

The apex financial institutions like NABARD, Reserve Bank of India (RBI) and other financial institutions and banks play important role in fisheries development through credit and non-credit functions.

The RBI plays certain non-credit functions which can be classified as policy issues related to macro financial environment within which the banking sector operates in the country and thereby ensures orderly development of all sectors. RBI also looks after the health of financial institutions through statutory inspection of banks. The Commercial Banks, Developmental Financial Institutions, Regional Rural Banks and Land Development Banks perform all credit related functions. Besides this, the banks are also involved in credit planning and entrepreneurship development. NABARD, which is the apex agricultural and rural development institution performs developmental banking role through both its credit and non-credit functions.

The NABARD through its conscious effort identifies gaps and supports research to standardise new technologies and operationalise development technologies. Through its appraisal mechanisms the NABARD also renders assistance to financial sector in evolving newer technologies. Credit planning exercise of the NABARD offers scope for large scale adoption of newer technologies in resource-specific manner.

The Bank undertakes monitoring of on going projects to study problems in implementation if any. It also conducts evaluation of completed projects to study the impact of the project visa-a-vis its objectives for planning future development in the sector.

Potential linked Credit Plans (PLP)

For ensuring integrated development of rural economy, the NABARD took the initiative to prepare Potential Linked Credit Plans (PLP) for each district in the year 1987-88. The PLPs represent a viable model of decentralised credit planning at district level with realistic assessment of resource potential, availability of infrastructure, marketing support, credit requirement of different sectors, credit absorption capacity and other strengths and weaknesses of both credit institutions and developmental agencies. To co-ordinate the developmental planning efforts and their implementation the NABARD has posted District Development Managers (DDM) in majority of the districts, who play an important role in co-ordinating the efforts of various agencies at district level.

Service Area Approach

The RBI introduced "Service Area Approach" for the bank branches for the rural lending from April, 1989. The approach aims at planning from below, at village level. Each bank branch are allocated 15 to 20 villages. The branches have to conduct village wise surveys their allocated villages (Service Area) to assess the potential for lending and identify the beneficiaries for assistance. The survey also indicates infrastructure facilities and linkages available in the area and those required to be developed to support bank credit. Based on the survey annual credit plans are prepared.

References

Agbayani, R.F. Marketing system in the Aquaculture Industry. Aquaculture Department, SEAFDEC, Training and Extension, Tigbaum, Iloilo, Philippines, 1986.

Albert. J. Tacon. Aquaculture production trends analysis–FAO Fisheries Circular No 886. Rev 2–FIRI/C 886, FAO Rome, 2003.

Agarwal, S.C. Fishery management, Ashis Publishing House, New Delhi, 1990.

Anderson, A. M. Developing markets for unfamiliar species. J. Fish. Res. Board, Canada, 1973.

Beverton, R.J.H and J. J. Holt. On the dynamics of Exploited Fish Population, Fish Invest, London.Ser. 2, No 19, 1957.

Bhat, R. V. Financing of Fishing Industry : Policies of the Shipping Credit and Investment Company of India, The Second IFF Proceedings, 1990.

Bilio, M, H. Rosenthal. Realism in Aquaculture, Achievements, Constraints, and C. J. Sindermann Prospectives, European Aquaculture Society, Bredene, 1986.

Black, J. D. Farm Management, Mackmillan, New York, 1947.

Biswas, K. P. Shrimp Farming, an ecological and economic consideration, Everyman's Science, Vol. 34, No. 4, 2000.

Bakun, A. The California current, Benguela Current and South Western Atlantic Shelf Ecosystems, a comparative approach to identifying factors regulating biomass yield, AAAS Press, Washington, D.C. USA, 1993.

Bakun, A, C. Roy. Coastal upwelling and other process regulating ecosystem productivity and S. Lluch–cota and fish production in the western Indian Ocean. Blackwell Science Inc, Malden (Mass) USA.

Caddy, J. F and L. Apparent changes in the trophic composition of the World marine.

Garibaldi harvests, the perspective from the FAO Capture database, Ocean and Coastal Management 43, 2000.

Camillo, Catarci. World markets and industry of selected commercially exploited aquatic species with international conservation profile, FAO. Fisheries Circular No. 990, FAO, Rome, 2004.

Clark, C.W. Bioeconomic modelling and fisheries management, John Wiley Sons, Inc. New York, 1985.

Chaston, I. Marketing in Fisheries and Aquaculture, Fishing News Books, Oxford, 1983.

Chaston, I. Business Management in Fisheries and Aquaculture, Farnhain Survey England, Fishing News Books Ltd., 1984.

Christy, F. T and A. D. Scott.The common wealth in Ocean Fisheries, John Hopkin Press, Baltimore, 1965.

Clark, C. W. Bioeconomic modelling and fisheries management, John wilby and Sons, New York, 1985.

Dasgupta, P. The economics of common property resources, a dynamic formulation of the fisheries problem, Indian Economic Rev. 16(3), 1981.

Daniel Pauli, Villy Christensen, Sylve, Guenette, T. J. Pitcher, U. Rashid, S. Carl, J. Walters, R. Watson and D. Zeller. Towards sustainability in World Fisheries, Nature, Vol 418/8 2002.

Dalton, R. Fishing for trouble, Nature, Vol. 431, 2004.

Dwivedi, S. N. Long term variability in the food chains, biomass yield and oceanography of the Bay of Bengal ecosystem, AAAS Press Washington, DC, USA, 1993.

Griffin, W., A. Lawrence and M. Johns. Economics of penaeid culture in America, Proc. First. Inter. Conf. penaeid Prawns / Shrimps, Iloilo City, Philippines, 1984.

Garibaldi, L. and L. Limongelli. Trends in oceanic captures and clusters of large marine ecosystems, FAO Fisheries Technical Paper 435, FAO of United Nations, Rome, 2003.

Giachelli, J. W., R.E. Mississippi farm–raised catfish, cost of production estimates.

Coats and J. E. Waldrop Mississippi State Univ. Agri, Economics Res, No 143, 1982.

Gordon, H.S. An economic approach to the optimum utilization of fishery resources, J. Fish. Res. Board of Canada Vol 10, 1953.

Gordon, H.S. The economic theory of a common property resources, The Fishery, J. Political Economy, Vol. 62, No. 2, 1954.

Gulland, J. A. The concept of maximum sustainable yield and fishery management, FAO, Rome, 1968.

Giriappa, S. Role of Fisheries in rural development, Daya Publishing House, Delhi.

Govt. of India. Marketing of fish in India, 3rd. Ed. Directorate of Marketing and Inspection, 1971.

Hiroshi, K. Artificial tideland and its effect on prawn breeding, Fishery Journal No. 4.

Halwart, M., S. The role of aquaculture in rural development, FAO Fisheries.

Fungi-Smith and J. Moehl. Circulate, No. 886, Rev. 2–FIRI/C 886, FAO Rome, 2003.

Hirasawa, Y. Economics of shrimp culture in Asia, Proc. First. International conference, penaeid Prawns / Shrimps, Ililio City, Philippines, 1984.

Hirono, Y. Preliminary report on shrimp culture activities in Ecuador, J. World Mariculture Society, No. 14, 1983.

Hanson, J.S., W.L. Griffin, J.W. Richardson, and C. Nixon. Survival of shrimp farming in Texas, an investment analysis, Working paper, Texas, A & M Univ. College Stn, Texas, 1984.

Hough C and P. Bueno. Producer Associations and Farmer Societies, Support to sustainable development and management of aquaculture FAO Fisheries Circular No 886, Rev 2–FIFI / C 886, FAO Rome, 2003.

Israel, D. C. Economic Feasibility Analysis of Aquaculture Products, A Review, Training and Extension, Aquaculture Department SEAFDEC, Tigbaum, Iloilo, Philippines, 1986.

Ibrahim, P. Fisheries Development in India, Classical Publications, New Delhi, 1992.

Kondo, K. Sardine resources and life pattern, Fishery Journal No. 20, 1978.

Kaoru, T. Relation between primary production and the commercial fishery production in the fishing ground, Bull. Nansei, Reg. Fish. Res Lab (13), 1991.

Kaoru, T. Productivity of the inshore fishing grounds, Bull, Nansei, Reg. Fish. Res. Lab (13), 1991.

Konda, S. Fisheries investments in the developing countries of Asia, FAO/ICIF, Rome, 1969.

Kamat, G. S. New dimensions of co-operative management, Himalaya Publishing House, 1978.

Leeds, R. Financing Aquaculture Projects, Aquaculture Engineering, 5, 1986.

Lawson, R. M. Economics of Fisheries Development, Frances Pinter (Pub) London, 1984.

MPEDA. Hand book on shrimp farming, MPEDA House, Panampilly, Avenue, Kochi, India, 1992.

Marr, J. C. Management and Development of Fisheries in the Indian Ocean, J. Fish. Res. Board of Canada, Vol 30, No. 2, Part 2, 1973.

Ministry of Agriculture. Hand book on fisheries statistics, 1991, Govt. of India, New Delhi, 1992.

Matsuda, K and C. Matsuda. Research concerning gill net fishing gear, Bull. Jap. Soc. Sci. Fisheries, 1976.

Pantulu, V. R. Economic evaluation of fish culture enterprises, J. Inland. Fish. Soc. India, 6, 1974.

Pathak S. C. and. Fisheries development through Industrial Credit Support in India, The First Indian Fisheries Forum, Proceedings, AFS Indian Branch, Mangalore, 1988.

P. Mohanakrishnan and Pillay, T.V.R. Planning of Aquaculture Development–An Introductory guide, Fishing News Books, Oxford, 1977.

Powell, K. Eat your veg–Nature, Vol 426/27, Nov, 2003.

Quirin, S. How many more fish in the sea? Nature, Vol 419/17, October, 2002.

Reid, T. R. The great Tokyo fish Market, Tsukiji–National Geographic, Vol 188, No 5, Nov, 1995.

Schaefer, M. B. Some aspects of the dynamics of populations important to the management of commercial marine fisheries, Inter American Tropical Tuna Commission Bull No 1 (2), 1954.

Schaefer, M. B. Some considerations of population dynamics and economics in relation to the management of commercial marine fisheries, J. Fish. Res. Board of Canada, Vol 14. No 5, 1957.

Sakae, M. and Y. Ikeda. What is making set net fishery more solid? Fishery Journal No 22.

Smith, L.J. and S. Peterson. Aquaculture development in less developed countries-social, economic and political problems, West View Press, Boulder, London, 1982.

Shang, Y.C. Aquaculture Economics, Basic Concepts and Methods of Analysis West View Press, Boulder, London, 1981.

Shepherd C. J. The economic of aquaculture–a review, Oceanography and Marine Biology Annual Review, 13. 1974.

Subasinghe, R. P. An outlook for aquaculture development, major issues, opportunities and challenges–FAO Fisheries Circular No 886, FAO Rome, 2003.

Subasinghe, R.P., D. Curry, S. E. Mc. Gladdry and D. Bartley. Recent technological innovations in aquaculture–FAO Fisheries Circular No 886, FAO, Rome, 2003.

Sherman, K and Q. Tang. Large marine ecosystems of the Pacific Rim, assessment, sustainability and management, Blackwell Science Inc, Malden (Mass) USA, 1999.

Sherman, K. Sustainability, biomass yields and health of coastal ecosystems and ecological perspective, Mar. Ecol. Prog. Ser. 112, 1994.

Tsuyoshi, K. Resources of surface fish, Fishery Journal No 26.

Tamenori, N. Structure of drive–in type bottom lift net fishery in Kozushima, Fishery Journal No 24.

Vannuccini, S. Shark utilization, marketing and trade, FAO Fisheries Technical Paper No. 389, FAO, Rome, 1999.

Webber, H. H. Risks to the aquaculture enterprise, Aquaculture 2 (2), 1973.

Watson, R. and D. Pauli. Systematic distortions in world fisheries catch trend, Nature, Vol 414/29, November, 2001.

Yamaha. Fish community and human society, Fishery Journal, No 41, November, 1993.

Yamaha. Quality in white meat fish, Fishery Journal No 37, October, 1997.

Yamaha. Simpler and more effective modernized methods–Pot fishing, Fishery Journal No. 27.

Yamaha. A desired solid system of fish distribution, processing and cold storage, Fishery Journal No. 21.

Yamaha. House wives work together, Fishery Journal No 19.

Yamaha. Single person operation, Fishery Journal No 17.

Yamaha. Traditional fisheries, maturization and modernization, Fishery Journal No 16.

Yamaha. Selection in crab fishery, Fishery Journal, No 15.

Yamaha. Eels are eaten when they have grown 1000 times, Fishery Journal No 14.

Yamaha. A prime fish of coastal fishery, Fishery Journal No. 8.

Yamaha. Coastal Fisheries Development, Fishery Journal No 7.

Yamaha. Gill net fishing is playing an important role in the promotion of coastal fisheries, Fishery Journal No 6.

Yamaha. Highly economical skip jack and tuna boats for coastal and off shore waters, Fishery Journal No 5.

Yamaha. More wealth and stability to coastal fishing communities, Fishery Journal No. 4.

Yamaha. Crisp voices, decisive gesture, bidding at an auction in the fish market, Fishery Journal No. I.

Index